Synthesis of Computational Structures
for Analog Signal Processing

Cosmin Radu Popa

Synthesis of Computational Structures for Analog Signal Processing

 Springer

Cosmin Radu Popa
Faculty of Electronics, Telecommunications
and Information Technology
University Politehnica of Bucharest
Bucharest, Romania
cosmin_popa@yahoo.com

ISBN 978-1-4939-0015-2   ISBN 978-1-4614-0403-3 (eBook)
DOI 10.1007/978-1-4614-0403-3
Springer New York Dordrecht Heidelberg London

Printed on acid-free paper

Springer is part of Springer Science+Business Media (www.springer.com)

This book is dedicated to my beloved daughter Ilinca Maria.

Preface

Signal processing represents an important domain of electronics, in the last years many efforts being directed for improving the performances of these structures.

The approach of the signal processing from an analog perspective presents the advantage of allowing an important reduction of the circuits' power consumption. The compact implemented structures are compatible with ultimate low-power designs and find a lot of applications such as portable equipments, wireless nano-sensors or medical implantable devices. Even the power consumption is continuous for analog circuits comparing with digital structures that consume only in the switching intervals, the possibility of an important reduction of designs' complexities and of the number of their constitutive active devices strongly decrease the medium power consumption per unity of time for analog designs. Moreover, low-power analog signal processing circuits are often implemented using subthreshold-operated MOS transistors, having extremely low values of drain currents, this fact producing an additionally lowering of the total power requested by the analog computational structures. The original approach of designing analog signal processing circuits using multifunctional structures also contributes to the decreasing of power consumption per implemented function.

Another important advantage of analog signal processing is that the speed of circuits is usually greater than the speed of digital computational circuits, allowing a real time signal processing.

Two important classes of analog signal processing circuits can be identified. The first class corresponds to linear structures, such as differential amplifier structures, multiplier circuits or active resistor structures, being necessary to develop particular linearization techniques in order to improve their general performances. The second class of analog signal processing circuits covers the area of nonlinear structures: squaring or square-rooting circuits, exponential structures or vector summation and Euclidean distance circuits. In this case, the most important goal is to minimize the approximation error of the implemented function. In order to improve the circuits' frequency response, a part of analog signal processing circuits are implemented using exclusively MOS transistors biased in saturation region. In cases in which the low-power operation is crucial, the subthreshold operation of MOS active devices represents the single choice for the designer.

In order to obtain an important reduction of design costs and of power consumption for the designed circuits, multifunctional computational structures can be implemented. Their principle of operation is based on the possibility of a multiple use of the same functional cell that is named multifunctional circuit core. As the design effort is mostly focused on the improving of the core performances and because the most important silicon area is consumed by the multifunctional core, the reutilization of this part of the multifunctional structure for all circuit functions will strongly decrease the complexity and power consumption per implemented function. The multifunctional structures present the important advantage of a relatively simple reconfiguration, small changing of the design allowing to obtain all necessary linear or nonlinear circuit functions.

The first chapter is dedicated to the presentation of linearization techniques for improving the performances of CMOS differential structures, fundamental circuits in VLSI analog and mixed-signal designs. The mathematical fundamentals are structured in eight different elementary mathematical principles, each of them being illustrated by concrete implementations in CMOS technology of their functional relations.

As it exists a relative limited number of mathematical principles that are used for implementing the multiplier circuits, the first part of Chap. 2 is dedicated to the analysis of the mathematical relations that represent the functional core of the designed circuits. In the second part of the chapter, starting from these elementary principles, there are analyzed and designed concrete multiplier circuits, grouped according to their constitutive mathematical principles. Both current and voltage multiplier circuits are presented, their operation being extensively described in Chap. 2.

The squaring function can be relatively easily obtained considering the intrinsic squaring characteristic of the MOS transistor biased in saturation region. Referring to the input variable, the squaring circuits can be clustered in two important classes: voltage squarers and current squarers, for both of them, the output variable being, usually, a current. The first part of Chap. 3 is dedicated to the analysis of the mathematical relations that represent the functional core of the designed circuits, while, in the second part of the chapter, starting from these elementary principles, there are analyzed and designed concrete squaring circuits, clustered according to their constitutive mathematical principles.

An important class of VLSI computational structures is represented by the square-root circuits. Frequently implemented using a translinear loop, they exploit the squaring characteristic of MOS transistors biased in saturation region. The presented design techniques are based on five different elementary mathematical principles, each of them being illustrated in Chap. 4 by concrete implementations in CMOS technology.

Exponential circuits represent important building blocks with many applications in VLSI designs. In CMOS technology, the exponential law is available only for the weak inversion operation of MOS transistor, the circuits designed using subthreshold-operated MOS active devices having the disadvantage of a poor frequency response. Thus, circuits realized in CMOS technology that require a good frequency response can be designed using exclusively MOS transistors biased in saturation region. The first part of Chap. 5 is dedicated to the analysis of the mathematical relations that represent the functional core of the designed circuits. In the second part of the chapter, using these elementary principles, there are analyzed and designed concrete exponential circuits, grouped according to the mathematical principles they are based on.

Chap. 6 is dedicated to the analysis and design of Euclidean distance circuits, classified (depending on their input variable), in computational structures having current-input or voltage-input vectors.

Functionally equivalent with a classical resistor, but presenting many important advantages in comparison with them, active resistor structures are extensively analyzed in Chap. 7. The goal of designing this class of active structures is mainly related to the possibility of an important reduction of the silicon area, especially for large values of the simulated resistances. The techniques presented for designing active resistor structures are based on six different elementary mathematical principles, each of them being illustrated by concrete implementations in CMOS technology.

A multitude of fundamental linear or nonlinear analog signal processing blocks can be realized starting from the same core, the optimization techniques implemented for the core being efficient for all derived circuits. The structures that can be realized starting from an improved performance multifunctional core are: differential amplifiers, multiplier circuits, active resistors (having both positive and negative controllable equivalent resistance), squaring, square-rooting or exponential circuits. Additionally, developing proper approximation functions, multifunctional structures are able to generate any continuous mathematical function. The circuits shown in Chap. 8 are based on four different elementary mathematical principles, being also presented concrete implementations in CMOS technology of these complex computational structures.

Bucharest, Romania   Cosmin Radu Popa

Contents

Chapter 1
Differential Structures

1.1 Mathematical Analysis for Synthesis of Differential Amplifiers

Elementary mathematical principles represent the functional basis for designing differential structures [1–55], each theoretical principle corresponding to a class of differential amplifiers. Usually, the proper operation of these circuits uses a biasing in saturation of MOS active devices. The notations of variables are: V_1 and V_2 represent the input potentials, I_{OUT} signifies the output current, while, usually, V_O, V_{O1} and V_{O2} constant voltages are introduced for modeling a voltage shifting. In order to obtain a differential structure able to amplify with small distortions an input signal, a linear behavior of the circuit must be implemented.

1.1.1 First Mathematical Principle (PR 1.1)

The first mathematical principle used for implementing differential amplifiers is based on the following relations:

$$I_{OUT} = A\left(\sqrt{I_2} - \sqrt{I_1}\right)$$

$$I_1 = \frac{K}{2}(V_{GS1} - V_T)^2 \qquad \Rightarrow I_{OUT} = A\sqrt{\frac{K}{2}}(V_2 - V_1)$$

$$I_2 = \frac{K}{2}(V_{GS2} - V_T)^2$$

$$V_{GS2} - V_{GS1} = V_2 - V_1 \tag{1.1}$$

The differential amplifiers based on the previous relation compute a current proportional with the differential input voltage, $V_2 - V_1$

C.R. Popa, *Synthesis of Computational Structures for Analog Signal Processing*, DOI 10.1007/978-1-4614-0403-3_1, © Springer Science+Business Media, LLC 2011

1.1.2   Second Mathematical Principle (PR 1.2)

The mathematical relation that models this principle is:

$$\left(\frac{V_1}{2} - V_{O1}\right)^2 - \left(\frac{V_2}{2} - V_{O1}\right)^2 + \left(V_{O2} - \frac{V_2}{2}\right)^2 - \left(V_{O2} - \frac{V_1}{2}\right)^2$$

$$= \frac{V_1 - V_2}{2}\left(\frac{V_1 + V_2}{2} - 2V_{O1}\right) + \frac{V_1 - V_2}{2}\left(2V_{O2} - \frac{V_1 + V_2}{2}\right)$$

$$= (V_{O2} - V_{O1})(V_1 - V_2) = ct.\,(V_1 - V_2) \qquad (1.2)$$

The circuits that use this principle generate a current proportional with the differential input voltage, $V_1 - V_2$.

1.1.3   Third Mathematical Principle (PR 1.3)

This principle is illustrated by the following mathematical relation:

$$\left[A(V_1 - V_2)^2 + B(V_1 - V_2) + C\right] - \left[A(V_1 - V_2)^2 - B(V_1 - V_2) + C\right]$$

$$= 2B(V_1 - V_2) \qquad (1.3)$$

The output current will be also proportional with the differential input voltage, $V_1 - V_2$

1.1.4   Fourth Mathematical Principle (PR 1.4)

The mathematical relation that models this principle is:

$$\left(V_C - V_O - V_T + \frac{V_1}{2}\right)^2 + \left(V_O - V_T + \frac{V_1}{2}\right)^2 - \left(V_C - V_O - V_T - \frac{V_1}{2}\right)^2$$

$$- \left(V_O - V_T - \frac{V_1}{2}\right)^2 = V_1(2V_C - 2V_O - 2V_T)$$

$$+ V_1(2V_O - 2V_T) = 2V_1(V_C - 2V_T) \qquad (1.4)$$

The output current of the differential amplifier is linearly dependent on the input voltage, V_1.

1.1.5 Fifth Mathematical Principle (PR 1.5)

The fifth mathematical principle can be written as follows:

$$I_{OUT} = \frac{V_1 - V_2}{2} \sqrt{4K\left[I_O + \frac{K}{4}(V_1 - V_2)^2\right] - K^2(V_1 - V_2)^2}$$
$$= \sqrt{KI_O}(V_1 - V_2) \tag{1.5}$$

1.1.6 Sixth Mathematical Principle (PR 1.6)

The method modeled by this mathematical principle, used for linearizing the transfer characteristic of differential structures, uses an anti-parallel connection of two differential amplifiers, the controlled asymmetries between their biasing currents, also between the aspect ratios of their transistors fulfilling this desiderate.

1.1.7 Seventh Mathematical Principle (PR 1.7)

This principle is useful for obtaining a rail-to-rail operation of a differential structure, based on a parallel connection of two complementary differential amplifiers. The mathematical relations of this principle are shown in the following lines:

$$I_{OUT} = \sqrt{2K}\left(\sqrt{I_{Op}} + \sqrt{I_{On}}\right)(V_1 - V_2) \tag{1.6}$$

$$\sqrt{I_{On}} + \sqrt{I_{Op}} = 2\sqrt{I_O} \tag{1.7}$$

resulting:

$$I_{OUT} = \sqrt{8KI_O}(V_1 - V_2) \tag{1.8}$$

1.1.8 Different Mathematical Principle for Differential Amplifiers (PR 1.D)

There are some circuits based on different mathematical principles that are useful for linearizing the behavior of differential amplifiers.

1.2   Analysis and Design of Differential Structures

The classical MOS differential presents a strong nonlinear behavior, as a result of the quadratic characteristic of their constitutive transistors biased in saturation region. In order to improve the linearity of the structure, it is necessary to develop efficient linearization techniques, functional mathematical principles being elaborated for fulfilling this desiderate.

Based on the previous presented mathematical analysis, it is possible to design different types of differential structures, included in eight classes, corresponding to the previous presented eight mathematical principles (PR 1.1 – PR 1.7 and PR 1.D).

The MOS differential amplifier represents a fundamental block in analog design, having a large area of applications. The analysis of the large signal operation for the classical MOS differential structure (Fig. 1.1) [1, 2] can quantitatively evaluate the circuit's nonlinearity, being possible to determine the weight of each superior-order distortion introduced by the structure nonlinearity.

The $V_I = V_1 - V_2$ differential input voltage can be expressed as follows:

$$V_I = V_{GS1} - V_{GS2} = \left(V_T + \sqrt{\frac{2I_1}{K}}\right) - \left(V_T + \sqrt{\frac{2I_2}{K}}\right) = \sqrt{\frac{2}{K}}(\sqrt{I_1} - \sqrt{I_2}) \quad (1.9)$$

Squaring and replacing the sum $I_1 + I_2$ with I_O, it results:

$$2\sqrt{I_1(I_O - I_1)} = I_O - \frac{KV_I^2}{2} \quad (1.10)$$

The resultant second-order equation will be:

$$I_1^2 - I_O I_1 + \frac{1}{4}\left(I_O - \frac{KV_I^2}{2}\right)^2 = 0 \quad (1.11)$$

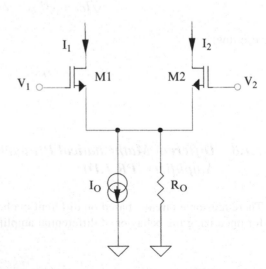

Fig. 1.1 Classical MOS differential structure

having the following solutions:

$$(I_1)_{1,2} = \frac{I_O}{2} \pm \frac{I_O}{2} \sqrt{\frac{KV_I^2}{I_O} - \frac{K^2 V_I^4}{4I_O^2}} \qquad (1.12)$$

so:

$$I_1 = \frac{I_O}{2} + \frac{I_O}{2} \sqrt{\frac{KV_I^2}{I_O} - \frac{K^2 V_I^4}{4I_O^2}}; \quad I_2 = \frac{I_O}{2} - \frac{I_O}{2} \sqrt{\frac{KV_I^2}{I_O} - \frac{K^2 V_I^4}{4I_O^2}} \qquad (1.13)$$

The output differential current will be:

$$I_2 - I_1 = -I_O \sqrt{\frac{KV_I^2}{I_O} - \frac{K^2 V_I^4}{4I_O^2}} = -\frac{V_I}{2} \sqrt{4KI_O - K^2 V_I^2} \qquad (1.14)$$

The $(I_2 - I_1)(V_I)$ function is strongly nonlinear, the quantitative evaluation of its nonlinearity being possible using a Taylor series expansion. So, it is necessary to compute the superior-order derivates of the following function:

$$f(V_I) = \sqrt{4KI_O - K^2 V_I^2} \qquad (1.15)$$

and their values for $V_I = 0$ The first-order derivate is:

$$f'(V_I) = -K^2 V_I (4KI_O - K^2 V_I^2)^{-1/2} \qquad (1.16)$$

while the second-order one has the following expression:

$$f''(V_I) = -4K^3 I_O (4KI_O - K^2 V_I^2)^{-3/2} \qquad (1.17)$$

resulting:

$$f'(V_I)|_{V_I=0} = 0 \qquad (1.18)$$

$$f''(V_I)|_{V_I=0} = -\frac{1}{2} K^{3/2} I_O^{-1/2} \qquad (1.19)$$

The Taylor series expansion of the function (1.14) gives:

$$(I_2 - I_1)(V_I) = -K^{1/2} I_O^{1/2} V_I + \frac{K^{3/2}}{8 I_O^{1/2}} V_I^3 + \frac{K^{5/2}}{128 I_O^{3/2}} V_I^5 + \cdots \qquad (1.20)$$

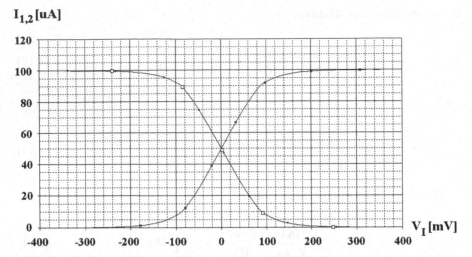

Fig. 1.2 The $I_1(V_I)$ and $I_2(V_I)$ dependencies for the classical differential amplifier

or:

$$(I_2 - I_1)\,(V_I) = a_1 V_I + a_3 V_I^3 + a_5 V_I^5 + \cdots \tag{1.21}$$

The first term is linearly dependent on the input voltage, while the last two terms model the third-order and fifth-order nonlinearities of the differential structure.

The dependencies of the drain currents I_1 and I_2 on the differential input voltage V_I for the differential amplifier from Fig. 1.1 are presented in Fig. 1.2.

Considering a load resistance $R_L = 10\,\text{k}\,\Omega$, the simulation of the transfer characteristic $V_O(V_I) = R_L(I_2 - I_1)\,(V_I)$ for the differential amplifier presented in Fig. 1.1 is shown in Fig. 1.3.

The simulation of the transfer characteristic $V_O(V_I)$ for a maximal input range between $-0.4\,\text{V}$ and $0.4\,\text{V}$ and a biasing current I_O having the values 0.1 mA, 0.2 mA and 0.3 mA is shown in Fig. 1.4. It could be remarked an increasing of the differential-mode voltage gain for an increasing of the biasing current I_O (using relation (1.20)), the doubling of the biasing current generating an increasing of $\sqrt{2}$ of the voltage gain.

The voltage gain is -10.95 (using relation (1.20)), while the simulated value is -10.43.

Fig. 1.5 presents the simulation of the transfer characteristic of the differential amplifier for different values of the common-mode input voltage $V_C = (V_1 + V_2)/2$ (between 1 V and 1.3 V), showing a minimal value of V_C of about 1.2 V. The simulation was made considering a passive load attached to the differential amplifier from Fig. 1.1, having $R_1 = R_2 = 10\,\text{k}\,\Omega$ and a supply voltage $V_{DD} = 9$ V.

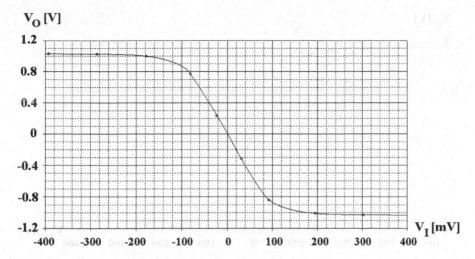

Fig. 1.3 The $V_O(V_I)$ dependence for the classical differential amplifier

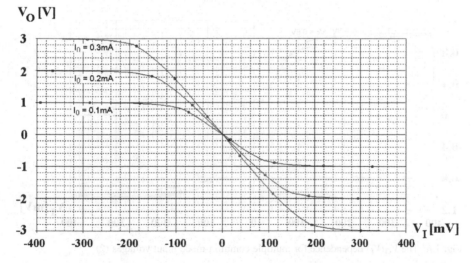

Fig. 1.4 Parametric $V_O(V_I)$ dependence for the classical differential amplifier

Figure 1.6 represents the simulation of the transfer characteristic of the differential amplifier for different values of the common-mode input voltage $V_C = (V_1 + V_2)/2$ (between 8.9 V and 9.1 V), showing a maximal value for V_C of about 9 V. The simulation was made considering a particular implementation of the current source I_O from Fig. 1.1 using a classical current mirror.

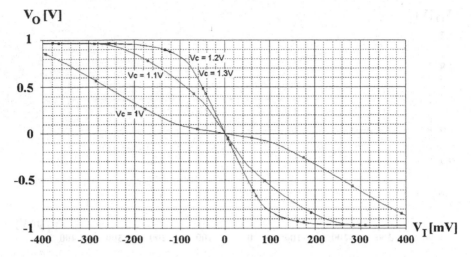

Fig. 1.5 The $V_O(V_I)$ dependence for multiple common-mode input voltages (1)

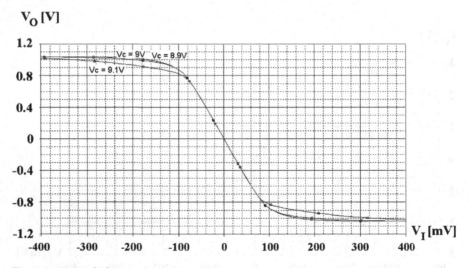

Fig. 1.6 The $V_O(V_I)$ dependence for multiple common-mode input voltages (2)

1.2.1 Differential Structures Based on the First Mathematical Principle (PR 1.1)

The method for obtaining a linear transfer characteristic of the differential amplifier based on the first mathematical principle (PR 1.1) uses the compensation of the squaring characteristic of the MOS transistor biased in saturation using complementary square-root circuits.

Fig. 1.7 Differential
structures (1) based on PR 1.1

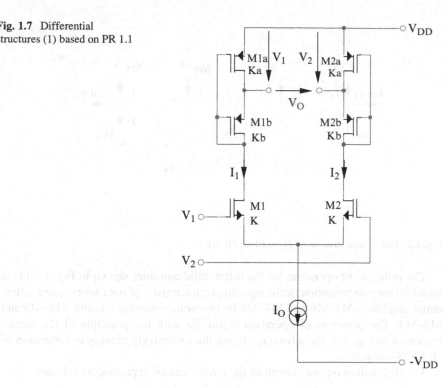

The first circuit using this method is shown in Fig. 1.7 [3] and it uses two square-root circuits for improving the linearity of the differential amplifier.

Using relation that describes the operation of M1a–M1b and M2a–M2b square-root circuits, the output voltage of the differential amplifier from Fig. 1.7 will be:

$$V_O = V_2 - V_1 = \sqrt{2I_2}\left(\sqrt{\frac{1}{K_a}+\frac{1}{K_b}} - \sqrt{\frac{1}{K_b}}\right) - \sqrt{2I_1}\left(\sqrt{\frac{1}{K_a}+\frac{1}{K_b}} - \sqrt{\frac{1}{K_b}}\right) \quad (1.22)$$

Because:

$$\sqrt{I_2} - \sqrt{I_1} = \sqrt{\frac{K}{2}}(V_{GS2} - V_{GS1}) = \sqrt{\frac{K}{2}}(V_2 - V_1) \quad (1.23)$$

it results:

$$V_O = \sqrt{K}\left(\sqrt{\frac{1}{K_a}+\frac{1}{K_b}} - \sqrt{\frac{1}{K_b}}\right)(V_2 - V_1) \quad (1.24)$$

equivalent with a linear dependence of the output voltage on the differential input voltage.

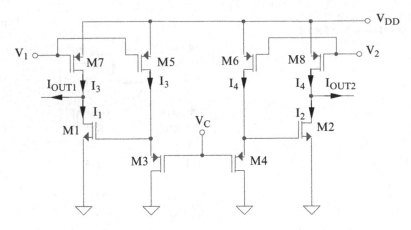

Fig. 1.8 Differential structures (2) based on PR 1.1

The principle of operation for the differential amplifier shown in Fig. 1.8 [4] is based on the compensation of the squaring characteristic of parallel-coupled differential amplifiers M5–M6 and M7–M8 by two square-rooting circuits, M1–M3 and M2–M4. The principle of operation is similar with the principle of the circuit presented in Fig. 1.7, the advantage being the exclusively biasing in saturation of all MOS transistors.

The differential output current of the circuit can be expressed as follows:

$$I_{OUT1} - I_{OUT2} = (I_3 - I_1) - (I_4 - I_2) \tag{1.25}$$

The constant potential V_C is equal with the difference between two gate-source voltages. Supposing a biasing in saturation of all identical MOS transistors, it results:

$$V_C = V_{GS1} - V_{SG3} = \sqrt{\frac{2I_1}{K}} - \sqrt{\frac{2I_3}{K}} \tag{1.26}$$

So:

$$\sqrt{I_1} = \sqrt{I_3} + \sqrt{\frac{K}{2}} V_C \tag{1.27}$$

resulting:

$$I_1 = I_3 + \frac{K}{2} V_C^2 + \sqrt{2KI_3} V_C \tag{1.28}$$

and, similarly:

$$I_2 = I_4 + \frac{K}{2} V_C^2 + \sqrt{2KI_4} V_C \tag{1.29}$$

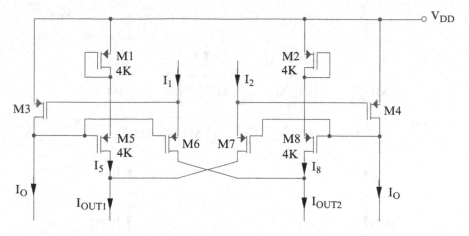

Fig. 1.9 Current-mode square-root circuit (1)

From (1.25), (1.28) and (1.29), it results a square-root dependence of the differential output current on the input currents:

$$I_{OUT1} - I_{OUT2} = \sqrt{2K}\, V_C \left(\sqrt{I_4} - \sqrt{I_3} \right) \tag{1.30}$$

The differential expression $\sqrt{I_4} - \sqrt{I_3}$ is a function of the input potentials V_1 and V_2, as follows:

$$\sqrt{I_4} - \sqrt{I_3} = \sqrt{\frac{K}{2}}(V_{SG6} - V_T) - \sqrt{\frac{K}{2}}(V_{SG5} - V_T)$$

$$= \sqrt{\frac{K}{2}}(V_{SG6} - V_{SG5}) = \sqrt{\frac{K}{2}}(V_1 - V_2) \tag{1.31}$$

From (1.30) and (1.31), it results a linear dependence of the output current on the differential input voltage:

$$I_{OUT1} - I_{OUT2} = KV_C(V_1 - V_2) \tag{1.32}$$

Alternate implementations of square-root circuits used for linearizing the transfer characteristic of the classical differential amplifier are shown in Figs. 1.9–1.13 [4].

For the square-root circuit shown in Fig. 1.9, the translinear loop realized using M1, M3, M5 and M6 transistors has the following characteristic equation:

$$V_{SG3} + V_{SG6} = V_{SG1} + V_{SG5} \tag{1.33}$$

So:

$$\sqrt{I_O} + \sqrt{I_1} = \sqrt{I_5} \tag{1.34}$$

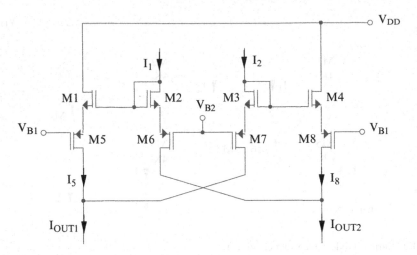

Fig. 1.10 Current-mode square-root circuit (2)

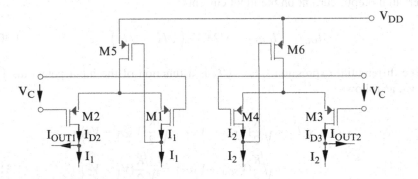

Fig. 1.11 Current-mode square-root circuit (3)

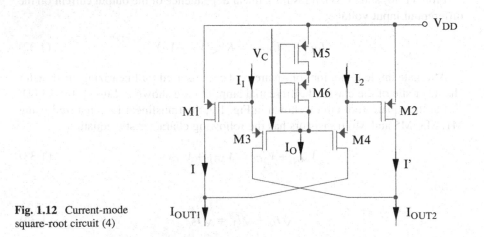

Fig. 1.12 Current-mode square-root circuit (4)

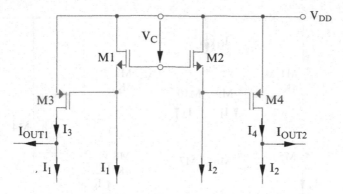

Fig. 1.13 Current-mode square-root circuit (5)

Squaring the previous relation, it is possible to write that:

$$I_5 = I_O + I_1 + 2\sqrt{I_O I_1} \qquad (1.35)$$

In a similar way, it results, from the right part of the circuit:

$$I_8 = I_O + I_2 + 2\sqrt{I_O I_2} \qquad (1.36)$$

The differential output current of the squaring structure can be expressed as follows:

$$I_{OUT1} - I_{OUT2} = (I_5 + I_2) - (I_8 + I_1) = 2\sqrt{I_O}\left(\sqrt{I_1} - \sqrt{I_2}\right) \qquad (1.37)$$

In order to obtain a linear differential amplifier, the square-root circuit shown in Fig. 1.9 must have as input currents the drain currents I_1 and I_2 of a classical differential amplifier M9–M10, resulting the circuit presented in Fig. 1.14. [4]

The differential output current of this circuit can be obtained replacing the expressions of I_1 and I_2 drain currents by their squaring dependencies on the gate-source voltages:

$$I_{OUT1} - I_{OUT2} = 2\sqrt{I_O}\sqrt{\frac{K}{2}}(V_{SG9} - V_{SG10}) = \sqrt{2KI_O}(V_2 - V_1) \qquad (1.38)$$

For the square-root circuit presented in Fig. 1.10, the translinear loop achieved using M1, M2, M5 and M6 transistors has the following characteristic equation:

$$V_{B2} - V_{B1} = (V_{GS1} + V_{SG5}) - (V_{GS2} + V_{SG6}) \qquad (1.39)$$

resulting:

$$(V_{B2} - V_{B1})\sqrt{\frac{K}{2}} = 2\sqrt{I_5} - 2\sqrt{I_1} \qquad (1.40)$$

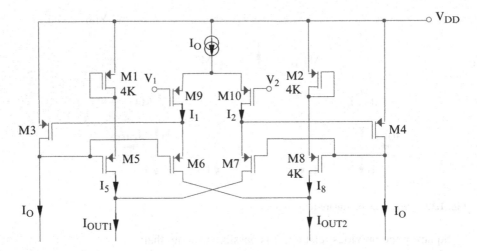

Fig. 1.14 Differential structures (3) based on PR 1.1

or:

$$\sqrt{I_5} = \sqrt{I_1} + \frac{V_{B2} - V_{B1}}{2}\sqrt{\frac{K}{2}} \tag{1.41}$$

Squaring the previous relation, it results:

$$I_5 = I_1 + \sqrt{\frac{KI_1}{2}}(V_{B2} - V_{B1}) + K\frac{(V_{B2} - V_{B1})^2}{8} \tag{1.42}$$

and, similarly:

$$I_8 = I_2 + \sqrt{\frac{KI_2}{2}}(V_{B2} - V_{B1}) + K\frac{(V_{B2} - V_{B1})^2}{8} \tag{1.43}$$

The differential output current can be expressed as follows:

$$I_{OUT1} - I_{OUT2} = (I_5 + I_2) - (I_8 + I_1) \tag{1.44}$$

From (1.42)–(1.44), the expression of the differential output current becomes:

$$I_{OUT1} - I_{OUT2} = (V_{B2} - V_{B1})\sqrt{\frac{K}{2}}(\sqrt{I_1} - \sqrt{I_2}) \tag{1.45}$$

The implementation of a differential amplifier using this square-root circuit is shown in Fig. 1.15 [4].

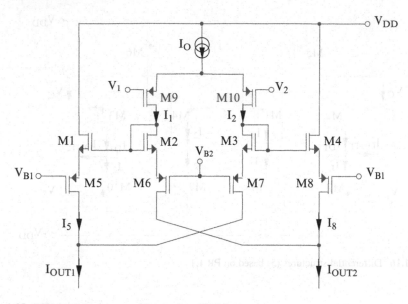

Fig. 1.15 Differential structures (4) based on PR 1.1

The differential output current of the linear differential amplifier from Fig. 1.15 can be obtained replacing I_1 and I_2 currents by their squaring dependencies on the gate-source voltages:

$$I_{OUT1} - I_{OUT2} = \frac{K}{2}(V_{B2} - V_{B1})(V_{SG9} - V_{SG10}) = \frac{K}{2}(V_{B2} - V_{B1})(V_2 - V_1) \quad (1.46)$$

equivalent with:

$$G_m = \frac{K}{2}(V_{B2} - V_{B1}) \quad (1.47)$$

The square-root circuit presented in Fig. 1.11 is composed from two identical cores (M1–M2 and M3–M4), each of them computing the square-root function of an input current. For M1–M2 pair, V_C input voltage can be expressed as follows:

$$V_C = V_{SG2} - V_{SG1} = \sqrt{\frac{2}{K}}\left(\sqrt{I_{D2}} - \sqrt{I_1}\right) \quad (1.48)$$

So:

$$\sqrt{I_{D2}} = \sqrt{\frac{K}{2}}V_C + \sqrt{I_1} \quad (1.49)$$

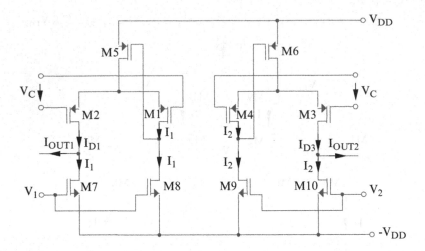

Fig. 1.16 Differential structures (5) based on PR 1.1

or:

$$I_{D2} = I_1 + \frac{K}{2}V_C^2 + \sqrt{2K}\, V_C \sqrt{I_1} \tag{1.50}$$

The expression of the first output current will be:

$$I_{OUT1} = I_{D2} - I_1 = \frac{K}{2}V_C^2 + \sqrt{2K}\, V_C \sqrt{I_1} \tag{1.51}$$

Similarly:

$$I_{OUT2} = \frac{K}{2}V_C^2 + \sqrt{2K}\, V_C \sqrt{I_2} \tag{1.52}$$

resulting the following expression of the differential output current:

$$I_{OUT1} - I_{OUT2} = \sqrt{2K}\, V_C \left(\sqrt{I_1} - \sqrt{I_2} \right) \tag{1.53}$$

Attaching two parallel-connected classical differential amplifiers to the double square-root circuit presented in Fig. 1.11, it results a differential amplifier with a linear transfer characteristic (Fig. 1.16) [4].

For this circuit, the differential input voltage can be expressed as follows:

$$V_1 - V_2 = V_{GS7} - V_{GS10} = \sqrt{\frac{2}{K}} \left(\sqrt{I_1} - \sqrt{I_2} \right) \tag{1.54}$$

From (1.53) and (1.54), it is possible to obtain the expression of the differential output current as a function on the differential input voltage:

$$I_{OUT1} - I_{OUT2} = KV_C(V_1 - V_2) \tag{1.55}$$

so an equivalent transconductance of the circuit expressed by:

$$G_m = KV_C \tag{1.56}$$

For the square-root circuit shown in Fig. 1.12, the translinear loop realized using M1, M3, M5 and M6 transistors has the following characteristic equation:

$$V_{SG1} + V_{SG3} = V_{SG5} + V_{SG6} \tag{1.57}$$

Noting $V_{SG5} + V_{SG6}$ with V_C, it results:

$$\sqrt{I} = \sqrt{\frac{K}{2}(V_C - 2V_T)} - \sqrt{I_1} \tag{1.58}$$

So:

$$I = I_1 + \frac{K}{2}(V_C - 2V_T)^2 - \sqrt{2KI_1}(V_C - 2V_T) \tag{1.59}$$

The expression of the first output current will be:

$$I_{OUT1} = I + I_2 = I_1 + I_2 + \frac{K}{2}(V_C - 2V_T)^2 - \sqrt{2KI_1}(V_C - 2V_T) \tag{1.60}$$

and, similarly:

$$I_{OUT2} = I_1 + I_2 + \frac{K}{2}(V_C - 2V_T)^2 - \sqrt{2KI_2}(V_C - 2V_T) \tag{1.61}$$

The differential output current can be expressed as follows:

$$I_{OUT1} - I_{OUT2} = \sqrt{2K}(V_C - 2V_T)(\sqrt{I_2} - \sqrt{I_1}) \tag{1.62}$$

The biasing voltage is equal with:

$$V_C = V_{SG5} + V_{SG6} = 2V_T + 2\sqrt{\frac{2I_O}{K}} \tag{1.63}$$

Replacing (1.63) in (1.62), it can be obtained:

$$I_{OUT1} - I_{OUT2} = 4\sqrt{I_O}(\sqrt{I_2} - \sqrt{I_1}) \tag{1.64}$$

Fig. 1.17 Differential
structures (6) based on PR 1.1

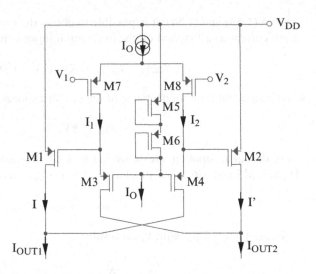

The implementation of a differential amplifier using this square-root circuit is presented in Fig. 1.17 [4, 5].

The differential output current of the linear differential amplifier from Fig. 1.17 can be obtained replacing the currents I_1 and I_2 by their squaring dependencies on the gate-source voltages:

$$I_{OUT1} - I_{OUT2} = 4\sqrt{I_O}\sqrt{\frac{K}{2}}(V_{SG8} - V_{SG7}) = \sqrt{8KI_O}(V_1 - V_2) \qquad (1.65)$$

The equivalent transconductance of the circuit can be expressed as follows:

$$G_m = \sqrt{8KI_O} \qquad (1.66)$$

Another possible implementation of a square-root circuit based on a similar principle is shown in Fig. 1.13.

The V_C potential can be expressed as follows:

$$V_C = V_{SG3} - V_{GS1} = \sqrt{\frac{2}{K}}\left(\sqrt{I_3} - \sqrt{I_1}\right) \qquad (1.67)$$

resulting:

$$\sqrt{I_3} = \sqrt{I_1} + \sqrt{\frac{K}{2}}V_C \qquad (1.68)$$

Squaring the previous relation, the expression of I_3 current will be:

$$I_3 = I_1 + \frac{K}{2}V_C^2 + \sqrt{2KI_1}\,V_C \qquad (1.69)$$

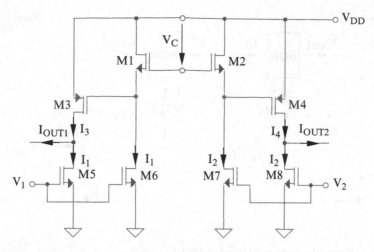

Fig. 1.18 Differential structures (7) based on PR 1.1

and, similarly:

$$I_4 = I_2 + \frac{K}{2}V_C^2 + \sqrt{2KI_2}\,V_C \tag{1.70}$$

The output currents have the following expressions:

$$I_{OUT1} = I_3 - I_1 = \frac{K}{2}V_C^2 + \sqrt{2KI_1}\,V_C \tag{1.71}$$

and:

$$I_{OUT2} = I_4 - I_2 = \frac{K}{2}V_C^2 + \sqrt{2KI_2}\,V_C \tag{1.72}$$

So, the differential output current will be:

$$I_{OUT1} - I_{OUT2} = \sqrt{2K}\,V_C\left(\sqrt{I_1} - \sqrt{I_2}\right) \tag{1.73}$$

The implementation of a differential amplifier using this square-root circuit is presented in Fig. 1.18 [4]. The I_1 and I_2 currents are generated by two parallel-connected differential amplifiers, M5–M8 and M6–M7, respectively.

The differential output current of the linear differential amplifier from Fig. 1.18 can be obtained replacing the I_1 and I_2 currents by their squaring dependencies on the gate-source voltages of M5–M8 transistors:

$$I_{OUT1} - I_{OUT2} = \sqrt{2K}\,V_C\sqrt{\frac{K}{2}}(V_{GS5} - V_{GS8}) = KV_C(V_1 - V_2) \tag{1.74}$$

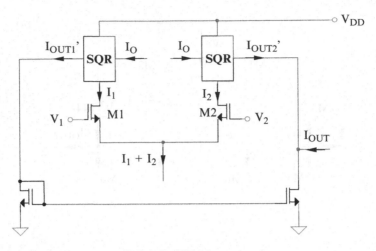

Fig. 1.19 Differential structures (8) based on PR 1.1

resulting:

$$G_m = KV_C \tag{1.75}$$

The same method for improving the linearity of the classical differential amplifier is used for designing the following differential amplifier. The principle is presented in Fig. 1.19, while the implementation of square-root circuits from Fig. 1.19 [4, 6] is shown in Fig. 1.20a [7].

1.2.1.1 The FGMOS Transistor

The multiple-input floating-gate transistor (M from Fig. 1.20a) is an ordinary MOS device whose gate is floating. The basic structure of a n-channel floating-gate MOS transistor is shown in Fig. 1.20b. The first silicon layer forms the floating-gate over the channel, while the second polysilicon layer forms the multiple input gates, which is located over the floating-gate. This floating-gate is capacitive coupled to the multiple input gates. The symbolical representation of such devices with n inputs is shown in Fig. 1.20c.

The drain current of a FGMOS transistor with n-input gates working in the saturation region is given by the following equation:

$$I_D = \frac{K}{2} \left[\sum_{i=1}^{n} k_i(V_i - V_S) - V_T \right]^2 \tag{1.76}$$

where $K = \mu_n C_{ox}(W/L)$ is the transconductance parameter of the transistor, μ_n is the electron mobility, C_{ox} is the gate oxide capacitance, W/L is the transistor aspect

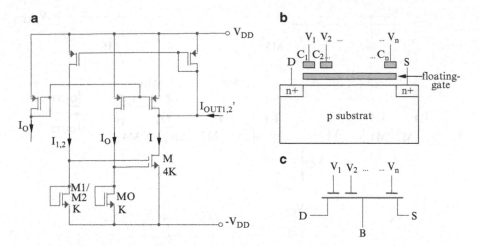

Fig. 1.20 Current-mode square-root circuit (6)

ratio, k_i, $i = 1, ..., n$ are the capacitive coupling ratios, V_i is the ith input voltage, V_S is the source voltage and V_T is the threshold voltage of the transistor. The capacitive coupling ratio is defined as:

$$k_i = \frac{C_i}{\sum\limits_{i=1}^{n} C_i + C_{GS}} \qquad (1.77)$$

C_i represent the input capacitances between the floating-gate and each of the i-th input and C_{GS} is the gate-source capacitance. Equation (1.76) shows that the FGMOS transistor drain current in saturation is proportional with the square of the weighted sum of the input signals, where the weight of each input signal is determined by the capacitive coupling ratio of the input.

The drain current of M transistors from Fig. 1.20a can be expressed as follows:

$$I = \frac{4K}{2} \left(\frac{V_{GSO} + V_{GS1,2}}{2} - V_T \right)^2 \qquad (1.78)$$

while V_{GSO} and $V_{GS1,2}$ expressions can be obtained from the squaring dependencies of the drain currents of MO and M1/M2 transistors on their gate-source voltages $V_{GSO} = V_T + \sqrt{2I_O/K}$ and $V_{GS1,2} = V_T + \sqrt{2I_{1,2}/K}$. So:

$$I = \left(\sqrt{I_O} + \sqrt{I_{1,2}} \right)^2 = I_O + I_{1,2} + 2\sqrt{I_O I_{1,2}} \qquad (1.79)$$

resulting:

$$I_{OUT1,2}' = I - I_O - I_{1,2} = 2\sqrt{I_{1,2} I_O} \qquad (1.80)$$

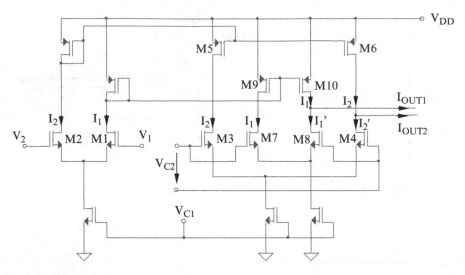

Fig. 1.21 Differential structures (9) based on PR 1.1

The circuit shown in Fig. 1.19 will have a linear transfer characteristic:

$$I_{OUT} = I_{OUT1}' - I_{OUT2}' = 2\sqrt{I_O}\left(\sqrt{I_1} - \sqrt{I_2}\right)$$

$$= 2\sqrt{I_O}\sqrt{\frac{K}{2}}(V_{GS1} - V_{GS2}) = \sqrt{2KI_O}(V_1 - V_2) = G_m(V_1 - V_2) \qquad (1.81)$$

$G_m = \sqrt{2KI_O}$ being the circuit transconductance. So, in a first-order analysis, the dependence of the output current of the differential circuit on its differential input voltage is perfectly linear.

A similar method used for linearizing the transfer characteristic of the differential amplifier is presented in Fig. 1.21 [8].

The elementary differential amplifier is composed by M1 and M2 transistors. The linearization of its transfer characteristic is realized using two square-root circuits (M3–M6 and M7–M10), their operation being characterized by the following relations:

$$V_{C2} = V_{GS7} - V_{GS8} = \sqrt{\frac{2}{K}}\left(\sqrt{I_1} - \sqrt{I_1'}\right) \qquad (1.82)$$

resulting:

$$I_1' = I_1 + \frac{K}{2}V_{C2}^2 - \sqrt{2K}\,V_{C2}\sqrt{I_1} \qquad (1.83)$$

The output current of the first square-root circuit is:

$$I_{OUT1} = I_1 - I_1' = -\frac{K}{2}V_{C2}^2 + \sqrt{2K}\,V_{C2}\sqrt{I_1} \qquad (1.84)$$

Similarly, for the second square-root circuit, the output current can be expressed as follows:

$$I_{OUT2} = I_2 - I_2' = -\frac{K}{2}V_{C2}^2 + \sqrt{2K}\,V_{C2}\sqrt{I_2} \qquad (1.85)$$

Because:

$$I_{OUT} = I_{OUT2} - I_{OUT1} \qquad (1.86)$$

it can be obtained:

$$I_{OUT} = \sqrt{2K}\,V_{C2}\left(\sqrt{I_1} - \sqrt{I_2}\right) \qquad (1.87)$$

As all MOS transistors are biased in saturation region, the differential input voltage of the M1–M2 differential amplifier depends on the difference of the square-roots of the drain currents, I_1 and I_2:

$$V_1 - V_2 = V_{GS1} - V_{GS2} = \sqrt{\frac{2}{K}}\left(\sqrt{I_1} - \sqrt{I_2}\right) \qquad (1.88)$$

From the previous relations it results:

$$I_{OUT} = KV_{C2}(V_1 - V_2) \qquad (1.89)$$

equivalent with a constant transconductance of the entire differential amplifier presented in Fig. 1.21, $G_m = KV_{C2}$.

Comparing with the previous similar circuit, the linear differential amplifier from Fig. 1.21 presents the important advantage of implementing also the multiplier function (V_{C2} and $V_1 - V_2$ can be considered as input voltages).

1.2.2 Differential Structures Based on the Second Mathematical Principle (PR 1.2)

The symmetrical circuit presented in Fig. 1.22 [9] represents a differential amplifier having the transfer characteristic linearized using the second mathematical principle (PR 1.2).

The gate-source voltages of M1a and M4a transistors are equal because they are identical and are biased at the same drain current I_{1a}. As $V_1 = V_{GS1a} + V_{GS4a}$,

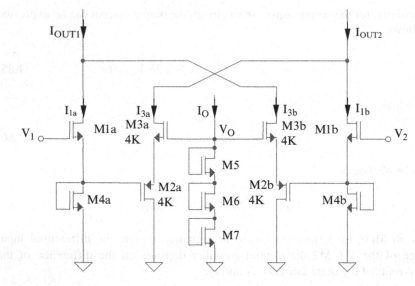

Fig. 1.22 Differential structures (1) based on PR 1.2

it results that the gate potential of M2a transistor is $V_1/2$. Similarly, the gate-source voltages of M2a and M3a are equal, resulting:

$$V_O - \frac{V_1}{2} = 2V_{GS3a} = 2\left(V_T + \sqrt{\frac{2I_{3a}}{4K}}\right) \tag{1.90}$$

equivalent with:

$$I_{3a} = \frac{K}{2}\left(V_O - \frac{V_1}{2} - 2V_T\right)^2 \tag{1.91}$$

and:

$$I_{1a} = \frac{K}{2}\left(\frac{V_1}{2} - V_T\right)^2 \tag{1.92}$$

The difference between the previous currents can be expressed as follows:

$$I_{1a} - I_{3a} = \frac{K}{2}(V_O - 3V_T)(V_1 - V_O + V_T) \tag{1.93}$$

Similarly, the $I_{1b} - I_{3b}$ differential current will have the following expression:

$$I_{1b} - I_{3b} = \frac{K}{2}(V_O - 3V_T)(V_2 - V_O + V_T) \tag{1.94}$$

Fig. 1.23 Differential
structures (2) based on PR 1.2

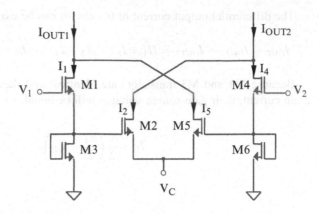

Thus, the total differential output current of the circuit, $I_{OUT1} - I_{OUT2}$, will be:

$$I_{OUT1} - I_{OUT2} = (I_{1a} + I_{3b}) - (I_{1b} + I_{3a}) = (I_{1a} - I_{3a})$$
$$- (I_{1b} - I_{3b}) = \frac{K}{2}(V_O - 3V_T)(V_1 - V_2) \tag{1.95}$$

The V_O biasing voltage is implemented as a current-controlled voltage source, this realization having the advantage of removing the dependence of the circuit performances on the variation of the threshold voltage with temperature and technological parameters:

$$V_O = 3V_{GS5} = 3\left(V_T + \sqrt{\frac{2I_O}{K}}\right) \tag{1.96}$$

For this particular realization of the voltage source V_O, the total differential output current becomes:

$$I_{OUT1} - I_{OUT2} = 3\sqrt{\frac{KI_O}{2}}(V_1 - V_2) \tag{1.97}$$

It was obtained a linear dependence of the output current on the differential input voltage, resulting an equivalent transconductance of the entire structure that can be controlled by the I_O biasing current:

$$G_m = \frac{I_{OUT1} - I_{OUT2}}{V_1 - V_2} = 3\sqrt{\frac{KI_O}{2}} \tag{1.98}$$

An alternate implementation of a linear differential amplifier using the same linearization principle is presented in Fig. 1.23 [9].

The differential output current of the circuit can be expressed as follows:

$$I_{OUT} = I_{OUT1} - I_{OUT2} = (I_1 + I_5) - (I_2 + I_4) = (I_1 - I_2) - (I_4 - I_5) \qquad (1.99)$$

Because M1 and M3 transistors are identical and they are biased at the same drain current, their gate-source voltages will be equal, so:

$$I_1 = \frac{K}{2}\left(\frac{V_1}{2} - V_T\right)^2 \qquad (1.100)$$

and:

$$I_2 = \frac{K}{2}\left(\frac{V_1}{2} - V_C - V_T\right)^2 \qquad (1.101)$$

Similarly, for the right part of the circuit, the expressions of the drain currents are:

$$I_4 = \frac{K}{2}\left(\frac{V_2}{2} - V_T\right)^2 \qquad (1.102)$$

and:

$$I_5 = \frac{K}{2}\left(\frac{V_2}{2} - V_C - V_T\right)^2 \qquad (1.103)$$

resulting:

$$I_{OUT} = \frac{K}{2}V_C(V_1 - V_C - 2V_T) - \frac{K}{2}V_C(V_2 - V_C - 2V_T) = \frac{K}{2}V_C(V_1 - V_2) \qquad (1.104)$$

the equivalent transconductance being expressed by:

$$G_m = \frac{K}{2}V_C \qquad (1.105)$$

The circuit presented in Fig. 1.23 presents the disadvantage of requiring a current from the V_C voltage source. In order to avoid a current consumption from the external V_C voltage source, the circuit presented in Fig. 1.23 can be modified, as it is shown in Fig. 1.24.

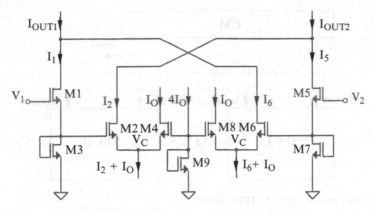

Fig. 1.24 Differential structures (2) based on PR 1.2 with implementation of V_C source

The V_C potential is obtained as the difference between the gate-source of M9 transistor (biased at $4I_O$ current) and the gate sources of M4 or M8 transistors (working at I_O current):

$$V_C = V_{GS9} - V_{GS4} = V_{GS9} - V_{GS8}$$

$$= \left(V_T + \sqrt{\frac{8I_O}{K}}\right) - \left(V_T + \sqrt{\frac{2I_O}{K}}\right) = \sqrt{\frac{2I_O}{K}} \tag{1.106}$$

Replacing (1.106) in (1.105), it results:

$$I_{OUT} = \sqrt{\frac{KI_O}{2}}(V_1 - V_2) \tag{1.107}$$

In this case, the equivalent transconductance can be controlled by the reference current I_O:

$$G_m = \sqrt{\frac{KI_O}{2}} \tag{1.108}$$

A differential difference amplifier (DDA) can be designed using two previous presented differential amplifiers, the block diagram of the DDA being shown in Fig. 1.25. These circuits present a multitude of applications such as amplifying or comparing of differential input voltages and designing complex active filters.

The output current of the DDA can be expressed as follows:

$$I_{OUT} = (I_{OUT1-I} + I_{OUT2-II}) - (I_{OUT2-I} + I_{OUT1-II})$$

$$= (I_{OUT1-I} - I_{OUT2-I}) - (I_{OUT1-II} - I_{OUT2-II}) \tag{1.109}$$

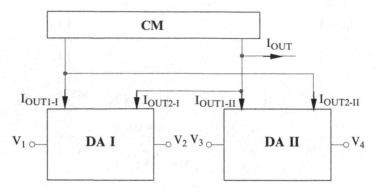

Fig. 1.25 The block diagram of a DDA circuit

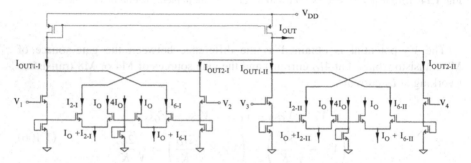

Fig. 1.26 DDA circuit

Considering that DA I and DA II differential amplifiers from Fig. 1.25 are identical, the expression of the output current becomes:

$$I_{OUT} = G_m[(V_1 - V_2) - (V_3 - V_4)] \qquad (1.110)$$

with G_m expressed by (1.108). The complete implementation of the DDA structure is presented in Fig. 1.26.

The differential amplifier shown in Fig. 1.27 [10] presents a linear transfer characteristic, obtained using the same mathematical principle, its equivalent transconductance being controlled by a reference voltage.

The expression of I_3 current is:

$$I_3 = \frac{K}{2}(V_{GS5} - V_T)^2 = \frac{K}{2}(V_{SG3} - V_T)^2 \qquad (1.111)$$

Because M1 and M3 are identical and they are biased at the same drain current, their source-gate voltages will be equal, so:

$$I_3 = \frac{K}{2}(V_{SG1} - V_T)^2 = \frac{K}{2}(V_{DD} - V_2 - V_T) \qquad (1.112)$$

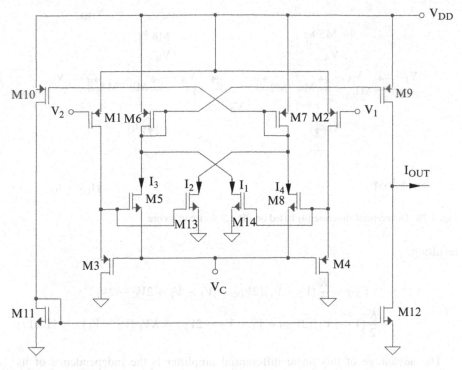

Fig. 1.27 Differential structure (3) based on PR 1.2

Similarly:

$$I_4 = \frac{K}{2}(V_{DD} - V_1 - V_T) \tag{1.113}$$

The I_1 current can be expressed as follows:

$$I_1 = \frac{K}{2}(V_{GS14} - V_T)^2 = \frac{K}{2}(V_{SG4} + V_C - V_T)^2$$
$$= \frac{K}{2}(V_{DD} - V_1 + V_C - V_T)^2 \tag{1.114}$$

and, similarly:

$$I_2 = \frac{K}{2}(V_{DD} - V_2 + V_C - V_T)^2 \tag{1.115}$$

The expression of the differential output current is:

$$I_{OUT} = (I_1 + I_3) - (I_2 + I_4) \tag{1.116}$$

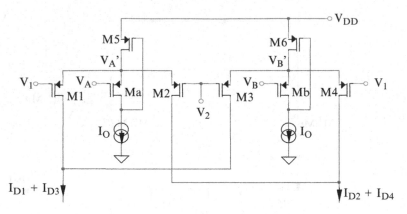

Fig. 1.28 Differential structure (4) based on PR 1.2 – circuit's core

resulting:

$$I_{OUT} = \frac{K}{2}(V_2 - V_1)(2V_{DD} - V_1 - V_2 + 2V_C - 2V_T)$$

$$-\frac{K}{2}(V_2 - V_1)(2V_{DD} - V_1 - V_2 - 2V_T) = KV_C(V_2 - V_1) \qquad (1.117)$$

The advantage of this linear differential amplifier is the independence of its equivalent transconductance on the threshold voltage, the result being an important increasing of the circuit accuracy.

The linearization technique for the differential amplifier presented in Fig. 1.28 [11, 12] exploits the second mathematical principle (PR 1.2). Its output current is made to be linearly dependent on the drain currents of M1–M4 transistors using a current mirror (not shown in Fig. 1.28), $I_{OUT} = (I_{D1} + I_{D3}) - (I_{D2} + I_{D4})$:

$$I_{OUT} = \frac{K}{2}(V_A' - V_1 - V_T)^2 + \frac{K}{2}(V_B' - V_2 - V_T)^2$$

$$-\frac{K}{2}(V_A' - V_2 - V_T)^2 - \frac{K}{2}(V_B' - V_1 - V_T)^2 \qquad (1.118)$$

resulting:

$$I_{OUT} = \frac{K}{2}(V_2 - V_1)(2V_A' - V_1 - V_2 - 2V_T)$$

$$+\frac{K}{2}(V_1 - V_2)(2V_B' - V_1 - V_2 - 2V_T) \qquad (1.119)$$

and:

$$I_{OUT} = K(V_B' - V_A')(V_1 - V_2) \qquad (1.120)$$

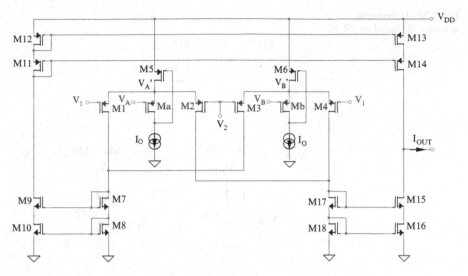

Fig. 1.29 Differential structure (4) based on PR 1.2 – complete implementation

Because:

$$V_A' = V_A + V_{SGa} = V_A + V_T + \sqrt{\frac{2I_O}{K}} \qquad (1.121)$$

and:

$$V_B' = V_B + V_{SGb} = V_B + V_T + \sqrt{\frac{2I_O}{K}} \qquad (1.122)$$

it results:

$$I_{OUT} = K(V_B - V_A)(V_1 - V_2) \qquad (1.123)$$

The equivalent transconductance of the entire structure presented in Fig. 1.28 can be expressed as follows:

$$G_m = \frac{I_{OUT}}{V_1 - V_2} = K(V_B - V_A) \qquad (1.124)$$

The complete implementation of previous differential amplifier (Fig. 1.29) [11] contains two cascode current mirrors, M7–M10 and M15–M18, in order to increase the circuit accuracy.

The efficiency of the second mathematical principle can be illustrated by the differential amplifier presented in Fig. 1.30 [13], having a very simple implementation comparing with the previous circuits. The V_{C1} and V_{C2} voltages represent

Fig. 1.30 Differential
structure (5) based on PR 1.2

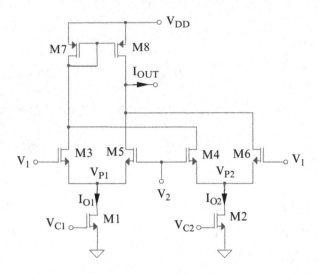

constant external potentials that impose the drain currents of M1 and M2 transistors to be equal with I_{O1} and I_{O2}, respectively. The $V_{P2} - V_{P1}$ voltage is considered to be imposed by an external circuit.

The differential amplifier is composed from two parallel-connected differential stages (M3–M5 and M4–M6), having different biasing currents (I_{O1} and I_{O2}, respectively). The output current of the entire structure can be expressed as follows:

$$I_{OUT} = I_{D3} + I_{D4} - I_{D5} - I_{D6} \tag{1.125}$$

equivalent with:

$$I_{OUT} = \frac{K}{2}(V_1 - V_{P1} - V_T)^2 + \frac{K}{2}(V_2 - V_{P2} - V_T)^2$$
$$- \frac{K}{2}(V_2 - V_{P1} - V_T)^2 - \frac{K}{2}(V_1 - V_{P2} - V_T)^2 \tag{1.126}$$

or:

$$I_{OUT} = K(V_1 - V_2)(V_{P2} - V_{P1}) \tag{1.127}$$

As V_{P1} and V_{P2} potentials are fixed, the behavior of the circuit shown in Fig. 1.30 is linear, having an equivalent transconductance expressed as:

$$G_m = K(V_{P2} - V_{P1}) \tag{1.128}$$

A method for obtaining a linear differential amplifier is presented in Fig. 1.31.

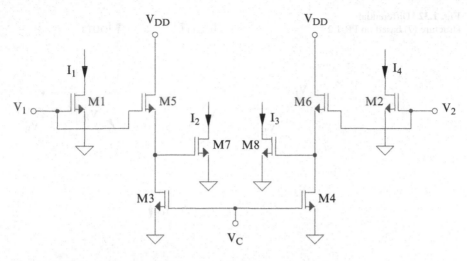

Fig. 1.31 Differential structure (6) based on PR 1.2

Because M5 and M3 transistors are biased at the same drain current, $V_{GS5} = V_{GS3} = V_C$, for the left part of the circuit, the differential output current will have the following expression:

$$I_1 - I_2 = \frac{K}{2}(V_1 - V_T)^2 - \frac{K}{2}(V_1 - V_C - V_T)^2 = \frac{K}{2}V_C(2V_1 - V_C - 2V_T) \quad (1.129)$$

Similarly, for the right part of the circuit, it results:

$$I_4 - I_3 = \frac{K}{2}V_C(2V_2 - V_C - 2V_T) \quad (1.130)$$

Using a current mirror (not shown in Fig. 1.31), the differential output current of the differential amplifier presented in Fig. 1.31 is designed to be:

$$I_{OUT} = (I_1 - I_2) - (I_4 - I_3) = KV_C(V_1 - V_2) \quad (1.131)$$

The equivalent transconductance of the structure is independent on the threshold voltage, with the result of reducing the circuit errors.

Another implementation of a linear differential amplifier based on the same mathematical principle is presented in Fig. 1.32 [14].

The expressions of I_{OUT1} and I_{OUT2} currents are:

$$I_{OUT1} = \frac{K}{2}(V_L - V_T)^2 + \frac{K}{2}(V_R - V_B - V_T)^2 \quad (1.132)$$

Fig. 1.32 Differential
structure (7) based on PR 1.2

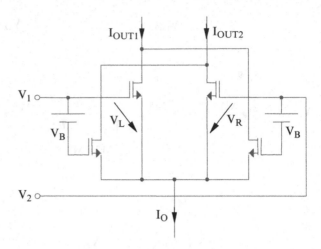

and:

$$I_{OUT2} = \frac{K}{2}(V_R - V_T)^2 + \frac{K}{2}(V_L - V_B - V_T)^2 \qquad (1.133)$$

Implementing an output current I_{OUT} as the difference between I_{OUT1} and I_{OUT2}, it results:

$$I_{OUT} = I_{OUT1} - I_{OUT2} = \frac{K}{2}(V_L - V_R)(V_L + V_R - 2V_T)$$
$$+ \frac{K}{2}(V_R - V_L)(V_L + V_R - 2V_B - 2V_T) \qquad (1.134)$$

So:

$$I_{OUT} = KV_B(V_L - V_R) = KV_B(V_1 - V_2) \qquad (1.135)$$

The equivalent transconductance of the circuit can be controlled by the biasing voltage V_B.

1.2.3 *Differential Structures Based on the Third Mathematical Principle (PR 1.3)*

The circuit shown in Fig. 1.33 [15] represents a differential amplifier having the transfer characteristic linearized using the third mathematical principle (PR 1.3). The advantage of the following circuits is represented by the possibility of implementing also the squaring function, by considering the sum of their output currents.

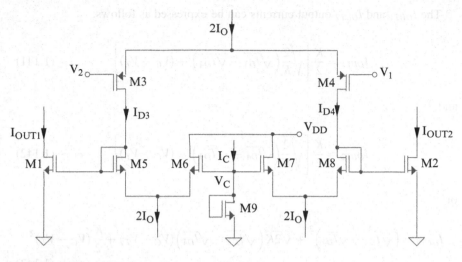

Fig. 1.33 Differential structure (1) based on PR 1.3

The current sources and the circuit's connections impose the following relation between the currents:

$$I_{D3} + I_{D6} = I_{D4} + I_{D7} = I_{D3} + I_{D4} = 2I_O \qquad (1.136)$$

resulting $I_{D6} = I_{D4}$ and $I_{D7} = I_{D3}$. The translinear loops containing M1, M5, M6 and M2, M7, M8 transistors have the following characteristic equations:

$$V_{GS1} - V_C = V_{GS5} - V_{GS6} \qquad (1.137)$$

and:

$$V_{GS2} - V_C = V_{GS8} - V_{GS7} \qquad (1.138)$$

resulting:

$$V_T + \sqrt{\frac{2I_{OUT1}}{K}} - V_C = \sqrt{\frac{2}{K}}\left(\sqrt{I_{D3}} - \sqrt{I_{D4}}\right) \qquad (1.139)$$

and:

$$V_T + \sqrt{\frac{2I_{OUT2}}{K}} - V_C = \sqrt{\frac{2}{K}}\left(\sqrt{I_{D4}} - \sqrt{I_{D3}}\right) \qquad (1.140)$$

The I_{OUT1} and I_{OUT2} output currents can be expressed as follows:

$$I_{OUT1} = \frac{K}{2}\left[\sqrt{\frac{2}{K}}\left(\sqrt{I_{D3}} - \sqrt{I_{D4}}\right) + (V_C - V_T)\right]^2 \qquad (1.141)$$

and:

$$I_{OUT2} = \frac{K}{2}\left[\sqrt{\frac{2}{K}}\left(\sqrt{I_{D4}} - \sqrt{I_{D3}}\right) + (V_C - V_T)\right]^2 \qquad (1.142)$$

or:

$$I_{OUT1} = \left(\sqrt{I_{D3}} - \sqrt{I_{D4}}\right)^2 + \sqrt{2K}\left(\sqrt{I_{D3}} - \sqrt{I_{D4}}\right)(V_C - V_T) + \frac{K}{2}(V_C - V_T)^2 \qquad (1.143)$$

and:

$$I_{OUT2} = \left(\sqrt{I_{D4}} - \sqrt{I_{D3}}\right)^2 + \sqrt{2K}\left(\sqrt{I_{D4}} - \sqrt{I_{D3}}\right)(V_C - V_T) + \frac{K}{2}(V_C - V_T)^2 \qquad (1.144)$$

The circuit's differential input voltage is equal with the difference between two source-gate voltages:

$$V_1 - V_2 = V_{SG3} - V_{SG4} = \sqrt{\frac{2}{K}}\left(\sqrt{I_{D3}} - \sqrt{I_{D4}}\right) \qquad (1.145)$$

From (1.143), (1.144) and (1.145), the expressions of the output currents become:

$$I_{OUT1} = \frac{K}{2}(V_1 - V_2)^2 + K(V_1 - V_2)(V_C - V_T) + \frac{K}{2}(V_C - V_T)^2 \qquad (1.146)$$

and:

$$I_{OUT2} = \frac{K}{2}(V_1 - V_2)^2 - K(V_1 - V_2)(V_C - V_T) + \frac{K}{2}(V_C - V_T)^2 \qquad (1.147)$$

As V_C voltage is equal with the gate-source of M9 transistor, that is biased at the I_C constant current, the previous relations become:

$$I_{OUT1} = \frac{K}{2}(V_1 - V_2)^2 + \sqrt{2KI_C}(V_1 - V_2) + I_C \qquad (1.148)$$

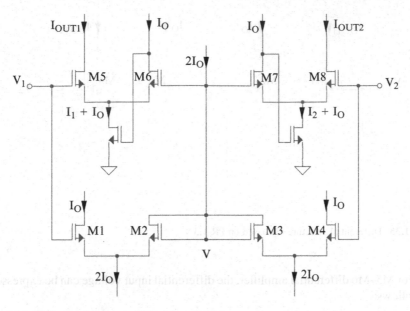

Fig. 1.34 Differential structure (2) based on PR 1.3

and:

$$I_{OUT2} = \frac{K}{2}(V_1 - V_2)^2 - \sqrt{2KI_C}(V_1 - V_2) + I_C \qquad (1.149)$$

The differential output current will have the following expression:

$$I_{OUT1} - I_{OUT2} = \sqrt{8KI_C}(V_1 - V_2) \qquad (1.150)$$

so, the circuit implements a linear dependence of the differential output current on the differential input voltage, the equivalent transconductance being:

$$G_m = \sqrt{8KI_C} \qquad (1.151)$$

A realization of a linear differential amplifier using the computation of the arithmetical mean for input potentials is presented in Fig. 1.34.

As M1–M4 transistors implement an arithmetical mean circuit, the expression of V potential will be:

$$V = \frac{V_1 + V_2}{2} \qquad (1.152)$$

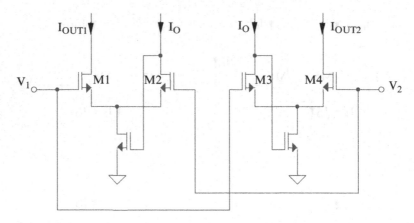

Fig. 1.35 Differential structure (3) based on PR 1.3

For M5–M6 differential amplifier, the differential input voltage can be expressed as follows:

$$V_1 - V = V_{GS5} - V_{GS6} = \sqrt{\frac{2}{K}}\left(\sqrt{I_{OUT1}} - \sqrt{I_O}\right) \qquad (1.153)$$

Replacing (1.152) in (1.153), it results:

$$I_{OUT1} = I_O + \sqrt{\frac{KI_O}{2}}(V_1 - V_2) + \frac{K}{8}(V_1 - V_2)^2 \qquad (1.154)$$

Similarly, for M7–M8 differential amplifier, it can be obtained:

$$I_{OUT2} = I_O - \sqrt{\frac{KI_O}{2}}(V_1 - V_2) + \frac{K}{8}(V_1 - V_2)^2 \qquad (1.155)$$

The output current of the differential amplifier circuit shown in Fig. 1.34 will be:

$$I_{OUT} = I_{OUT1} - I_{OUT2} = \sqrt{2KI_O}(V_1 - V_2) \qquad (1.156)$$

A differential amplifier with linear transfer characteristic can be implemented using two classical differential amplifiers (Fig. 1.35) [16].

For M1–M2 differential amplifier, the differential input voltage can be expressed as follows:

$$V_1 - V_2 = \sqrt{\frac{2}{K}}\left(\sqrt{I_{OUT1}} - \sqrt{I_O}\right) \qquad (1.157)$$

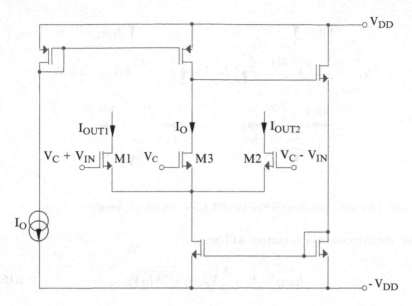

Fig. 1.36 Differential structure (4) based on PR 1.3

resulting:

$$I_{OUT1} = I_O + \sqrt{2KI_O}(V_1 - V_2) + \frac{K}{2}(V_1 - V_2)^2 \qquad (1.158)$$

Similarly, for M3–M4 differential amplifier, the expression of I_2 current will be:

$$I_{OUT2} = I_O - \sqrt{2KI_O}(V_1 - V_2) + \frac{K}{2}(V_1 - V_2)^2 \qquad (1.159)$$

The output current of the differential amplifier will be linearly dependent on the differential input voltage:

$$I_{OUT} = I_{OUT1} - I_{OUT2} = \sqrt{8KI_O}(V_1 - V_2) \qquad (1.160)$$

The circuit presented in Fig. 1.36 [17] is used for linearizing the transfer characteristic of a classical differential amplifier.

The difference between gate-source voltages of M1 and M3 transistors can be expressed as follows:

$$V_{GS1} - V_{GS3} = (V_C + V_{IN}) - V_C \qquad (1.161)$$

For a biasing in saturation of all MOS transistors from Fig. 1.36, it results:

$$V_{IN} = \sqrt{\frac{2}{K}}\left(\sqrt{I_{OUT1}} - \sqrt{I_O}\right) \qquad (1.162)$$

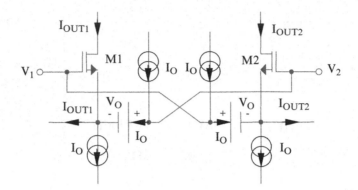

Fig. 1.37 Differential structure (5) based on PR 1.3 – principle of operation

So, the expression of I_1 current will be:

$$I_{OUT1} = I_O + \frac{K}{2}V_{IN}^2 + \sqrt{2KI_O}\, V_{IN} \tag{1.163}$$

Similarly, computing the difference between the gate-source voltages of M2–M3 transistors, it results:

$$I_{OUT2} = I_O + \frac{K}{2}V_{IN}^2 - \sqrt{2KI_O}\, V_{IN} \tag{1.164}$$

The differential output current for the circuit presented in Fig. 1.36 will be:

$$I_{OUT} = I_{OUT1} - I_{OUT2} = \sqrt{8KI_O}\, V_{IN} \tag{1.165}$$

The following presented principle for linearizing the transfer characteristic of a differential structure is based on the constant sum of gate-source voltages, the circuit's core being represented by a particular implementation of a differential amplifier (Fig. 1.37) [18] that provides a linear behavior of the circuit.

For a biasing in saturation of MOS transistors from Fig. 1.37, the $V_1 - V_2$ differential input voltage can be expressed as follows:

$$V_1 - V_2 = V_{GS1} - V_O \tag{1.166}$$

and:

$$V_1 - V_2 = V_O - V_{GS2} \tag{1.167}$$

resulting the expressions of the sum and difference between gate-source voltages:

$$V_{GS1} + V_{GS2} = 2V_O \tag{1.168}$$

and:

$$V_{GS1} - V_{GS2} = 2(V_1 - V_2) \tag{1.169}$$

The differential output current I_{OUT} is:

$$I_{OUT} = I_{OUT1} - I_{OUT2} = \frac{K}{2}(V_{GS1} - V_T)^2 - \frac{K}{2}(V_{GS2} - V_T)^2 \tag{1.170}$$

equivalent with:

$$I_{OUT} = \frac{K}{2}(V_{GS1} - V_{GS2})(V_{GS1} + V_{GS2} - 2V_T) \tag{1.171}$$

Replacing (1.168) and (1.169) in (1.171), it results a linear transfer characteristic of the differential amplifier presented in Fig. 1.37:

$$I_{OUT} = 2K(V_O - V_T)(V_1 - V_2) \tag{1.172}$$

Usually, the V_O voltage sources are implemented as current-controlled voltage sources. The simplest way to realize these sources, having the advantages of simplicity, also of minimizing the errors introduced by the bulk effect is to use the gate-source voltage of a MOS transistor in saturation, biased at a constant current, I_O:

$$V_O = V_{GSO} = V_T + \sqrt{\frac{2I_O}{K}} \tag{1.173}$$

Replacing these particular expressions of V_O voltage sources in the general expression (1.172) of the output current of the differential amplifier, it can be obtained:

$$I_{OUT} = \sqrt{8KI_O}(V_1 - V_2) \tag{1.174}$$

so, an equivalent transconductance of the differential amplifier $G_m = \sqrt{8KI_O}$ that can be very easily controlled by the biasing current, I_O.

A similar linearization technique is used for the circuit shown in Fig. 1.38 [19]. The V_O voltage sources from Fig. 1.37 are implemented in Fig. 1.38 using the gate-source voltages of M3 and M4 transistors, biased at constant current, I_O.

Considering a biasing in saturation of MOS transistors, the output differential current I_{OUT} can be expressed using (1.172) and (1.174):

$$I_{OUT} = 2K(V_1 - V_2)(V_O - V_T) = \sqrt{8KI_O}(V_1 - V_2) \tag{1.175}$$

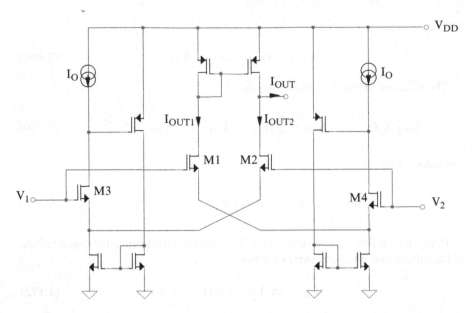

Fig. 1.38 Differential structure (5) based on PR 1.3 – first implementation

The equivalent transconductance of the structure is:

$$G_m = \frac{I_{OUT}}{V_1 - V_2} = \sqrt{8KI_O} \tag{1.176}$$

In Fig. 1.39 [19], the V_O voltage sources from Fig. 1.37 are realized as current-controlled voltage sources, the gate-source voltages of M3 and M5 transistors being dependent on the I_O biasing current. Similarly with the previous circuits, the I_{OUT} output differential current can be expressed as:

$$I_{OUT} = I_{OUT1} - I_{OUT2} = 2K(V_1 - V_2)(V_O - V_T) = \sqrt{8KI_O}(V_1 - V_2) \tag{1.177}$$

The equivalent transconductance of the structure is:

$$G_m = \frac{I_{OUT}}{V_1 - V_2} = \sqrt{8KI_O} \tag{1.178}$$

A differential amplifier with linear transfer characteristic based on the same mathematical principle can be designed (Fig. 1.40) [20] using two translinear loops implemented using M1, M3, M8, M7 and M5, M6, M4, M2 transistors, respectively.

Using the notation $V_{GS}(I)$ for the absolute value of the gate-source voltage of a MOS transistor biased at a drain current equal with I, it is possible to write:

$$V_1 - V_2 = 2V_{GS}(I_O) - 2V_{GS}(I_{OUT1}) = 2\sqrt{\frac{2}{K}}\left(\sqrt{I_O} - \sqrt{I_{OUT1}}\right) \tag{1.179}$$

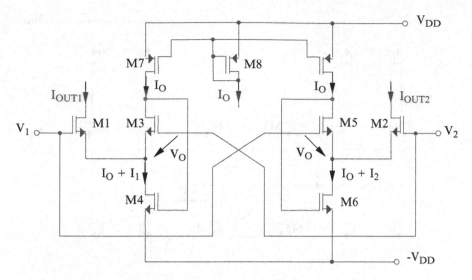

Fig. 1.39 Differential structure (5) based on PR 1.3 – second implementation

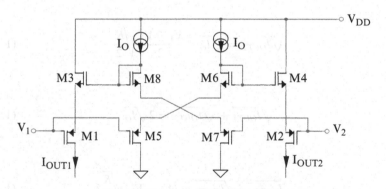

Fig. 1.40 Differential structure (5) based on PR 1.3 – third implementation

and:

$$V_1 - V_2 = 2V_{GS}(I_{OUT2}) - 2V_{GS}(I_O) = 2\sqrt{\frac{2}{K}}\left(\sqrt{I_{OUT2}} - \sqrt{I_O}\right) \qquad (1.180)$$

It results:

$$\sqrt{I_{OUT1}} = \sqrt{I_O} - \frac{V_1 - V_2}{2}\sqrt{\frac{K}{2}} \qquad (1.181)$$

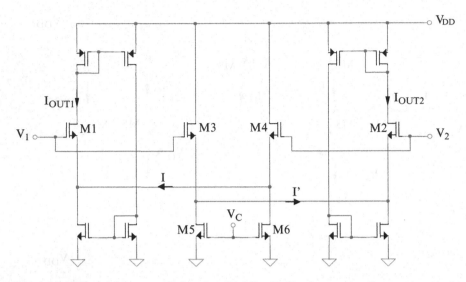

Fig. 1.41 Differential structure (5) based on PR 1.3 – third implementation (improved version)

and:

$$\sqrt{I_{OUT2}} = \sqrt{I_O} + \frac{V_1 - V_2}{2}\sqrt{\frac{K}{2}} \qquad (1.182)$$

So:

$$\sqrt{I_{OUT1}} + \sqrt{I_{OUT2}} = 2\sqrt{I_O} \qquad (1.183)$$

and:

$$\sqrt{I_{OUT2}} - \sqrt{I_{OUT1}} = (V_1 - V_2)\sqrt{\frac{K}{2}} \qquad (1.184)$$

The differential output current will have the following expression:

$$I_{OUT2} - I_{OUT1} = \left(\sqrt{I_{OUT2}} - \sqrt{I_{OUT1}}\right)\left(\sqrt{I_{OUT2}} + \sqrt{I_{OUT1}}\right)$$
$$= \sqrt{2KI_O}(V_1 - V_2) = G_m(V_1 - V_2) \qquad (1.185)$$

The equivalent transconductance of the differential amplifier is:

$$G_m = \sqrt{2KI_O} \qquad (1.186)$$

The principle of operation of the differential amplifier presented in Fig. 1.41 [21] is similar with the general principle described for the previous circuit. The V_O voltage

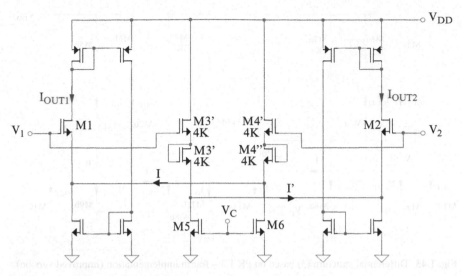

Fig. 1.42 Differential structure (5) based on PR 1.3 – fourth implementation

sources from Fig. 1.37 are implemented in Fig. 1.41 using the gate-source voltages of M3 and M4 transistors. Because, from the current equations, $I = I' = 0$, the M3–M6 transistors will be biased at the same drain currents, imposed by the V_C potential. Since they are identical, their gate-source voltages will be equal, so $V_O = V_{GS3} = V_{GS4} = V_C$. Using the general relation (1.172), the differential output current can be expressed as follows:

$$I_{OUT} = I_{OUT1} - I_{OUT2} = 2K(V_C - V_T)(V_1 - V_2) \qquad (1.187)$$

the equivalent transconductance being:

$$G_m = 2K(V_C - V_T) \qquad (1.188)$$

In order to avoid the dependence of the equivalent transconductance on threshold voltage, the M3 and M4 transistors from Fig. 1.41 have been replaced in Fig. 1.42 [21] with two series-connected MOS transistors, M3'–M3" and M4'–M4". The voltage sources V_O will have the following expression:

$$V_O = V_{GS3}' + V_{GS3}'' = 2\left(V_T + \sqrt{\frac{2I_{D5}}{4K}}\right) = 2V_T + \sqrt{\frac{2}{K}\frac{K}{2}(V_C - V_T)^2} = V_C + V_T$$

$$(1.189)$$

Replacing (1.189) in (1.172), it results:

$$I_{OUT} = 2KV_C(V_1 - V_2) \qquad (1.190)$$

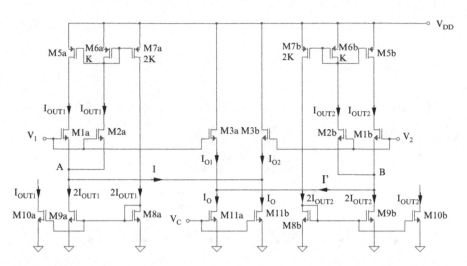

Fig. 1.43 Differential structure (5) based on PR 1.3 – fourth implementation (improved version)

So, the equivalent transconductance, $G_m = 2KV_C$ will be not dependent on the threshold voltage and V_C voltage can control it.

Another realization of a differential amplifier with linear behavior based on the third mathematical principle is presented in Fig. 1.43 [22]. The M1 and M2 transistors from Fig. 1.37 have been replaced in Fig. 1.43 by two series-connected transistors, M1a–M2a and M1b–M2b, while the voltage sources V_O from Fig. 1.37 have been practically implemented in Fig. 1.43 by the gate-source voltages of M3a and M3b transistors. The current equations impose $I = I' = 0$, so $I_{O1} = I_{O2} = I_O$. Because M3a, M3b, M11a and M11b transistors are identical and biased at the same current, their gate-source voltages will be identical, so $V_O = V_{GS3a} = V_{GS3b} = V_C$. Using (1.172), it can be obtained:

$$I_{OUT} = I_{OUT1} - I_{OUT2} = 2K(V_C - V_T)(V_1 - V_2) \tag{1.191}$$

The equivalent transconductance is:

$$G_m = 2K(V_C - V_T) \tag{1.192}$$

The replacing of M3a and M3b transistors with two series-connected MOS transistors, M3a'–M3a" and M3b'–M3b" (Fig. 1.44) [21, 22], allows to remove the dependence of the circuit equivalent transconductance on the threshold voltage:

$$G_m = 2KV_C \tag{1.193}$$

The advantage of the circuits presented in Figs. 1.43 and 1.44 with respect to the circuits shown in Figs. 1.41 and 1.42 consists in the availability of the output currents I_{OUT1} and I_{OUT2} as external currents, being possible to process them

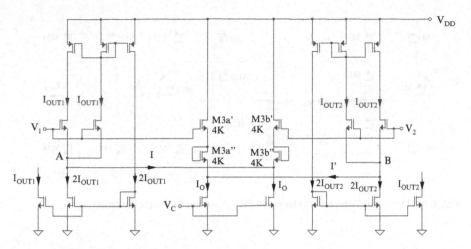

Fig. 1.44 Differential structure (6) based on PR 1.3

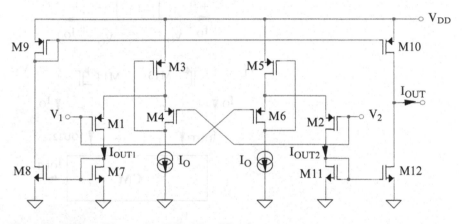

Fig. 1.45 Differential structure (7) based on PR 1.3

in order to implement other classes of circuits (active resistors, multipliers or multifunctional structures).

A differential amplifier with linear behavior obtained using the same mathematical principle is presented in Fig. 1.45, the NMOS transistors being replaced by complementary PMOS active devices. The V_O voltage sources from Fig. 1.37 are practically implemented in Fig. 1.45 [23] using the source-gate voltages of M4 and M6 transistors, biased at constant current I_O.

$$V_O = V_{SG4} = V_{SG6} = V_T + \sqrt{\frac{2I_O}{K}} \tag{1.194}$$

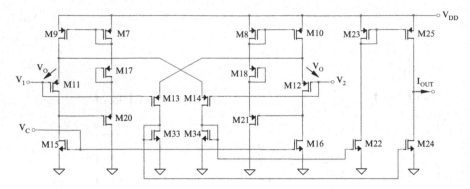

Fig. 1.46 Differential structure (8) based on PR 1.3 – complete implementation

Fig. 1.47 Differential
structure (8) based on PR 1.3
– principle of operation

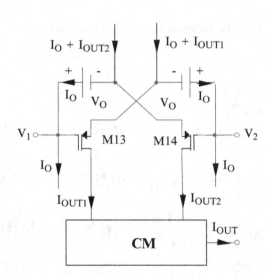

Using the general relation (1.172), the differential output current can be
expressed as follows:

$$I_{OUT2} - I_{OUT1} = 2K(V_O - V_T)(V_1 - V_2) = 2\sqrt{2KI_O}(V_1 - V_2) \qquad (1.195)$$

resulting an equivalent transconductance of the entire structure, given by:

$$G_m = 2\sqrt{2KI_O} \qquad (1.196)$$

A similar method based on the constant sum of gate-source voltages is used in
Fig. 1.46 [23].

The principle diagram, equivalent with the previous circuit is presented in Fig. 1.47.

The core of the differential amplifier is represented by the M13 and M14 transistors, while the V_O current-controlled voltage sources are implemented using M11 and M12 transistors, biased at a current imposed by M15 and M16 transistors (having the gate-source voltages determined by the V_C control voltage). The current mirror from Fig. 1.47 is implemented in Fig. 1.46 using M34–M22, M33–M24 and M23–M25 pairs, with the goal of realizing the difference of the drain currents of M13 and M14 transistors. For the circuit presented in Fig. 1.47, the differential input voltage can be expressed as follows:

$$V_1 - V_2 = V_{SG14} - V_O \tag{1.197}$$

and:

$$V_1 - V_2 = V_O - V_{SG13} \tag{1.198}$$

resulting:

$$V_{SG14} - V_{SG13} = 2(V_1 - V_2) \tag{1.199}$$

and:

$$V_{SG14} + V_{SG13} = 2V_O \tag{1.200}$$

The differential output current for the circuit presented in Fig. 1.47 will be:

$$
\begin{aligned}
I_{OUT2} - I_{OUT1} &= \frac{K}{2}(V_{SG14} - V_T)^2 - \frac{K}{2}(V_{SG13} - V_T)^2 \\
&= \frac{K}{2}(V_{SG14} - V_{SG13})(V_{SG13} + V_{SG14} - 2V_T)
\end{aligned} \tag{1.201}
$$

Replacing (1.199) and (1.200) in (1.201), it results:

$$I_{OUT2} - I_{OUT1} = 2K(V_1 - V_2)(V_O - V_T) \tag{1.202}$$

For the complete circuit presented in Fig. 1.46, the V_O voltage sources will have the following relations:

$$V_O = V_{SG11} = V_{SG12} = V_{GS15} = V_{GS16} = V_C \tag{1.203}$$

Because M11, M12, M15 and M16 transistors are identical and they work at the same drain current, it results:

$$I_{OUT2} - I_{OUT1} = 2K(V_1 - V_2)(V_C - V_T) \tag{1.204}$$

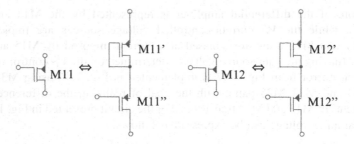

Fig. 1.48 Improving method for the differential structure (8) based on PR 1.3

so, an equivalent transconductance of the circuit expressed by:

$$G_m = 2K(V_C - V_T) \qquad (1.205)$$

The disadvantage of the circuit is represented by the dependence of its equivalent transconductance on the threshold voltage, concretized in the existance of errors introduced by the bulk effect. In order to avoid this aspect, M11 and M12 transistors from Fig. 1.46 must be replaced (Fig. 1.48) with two series connections of two transistors, M11'–M11' and M12'–M12", respectively, each of them having $K' = K'' = 4K$.

For this new configuration, the expression of V_O becomes:

$$V_O = V_{SG11}' + V_{SG11}'' = 2V_{SG11}' \qquad (1.206)$$

$$V_O = 2\left(V_T + \sqrt{\frac{2I_{D11}'}{4K}}\right) = 2V_T + \sqrt{\frac{2}{K}\frac{K}{2}(V_C - V_T)^2} \qquad (1.207)$$

resulting:

$$V_O = V_C + V_T \qquad (1.208)$$

In this case, the circuit equivalent transconductance becomes independent on the threshold voltage:

$$G_m = 2KV_C \qquad (1.209)$$

A differential amplifier with linear behavior, having a similar principle of operation is presented in Fig. 1.49 (M5, M6, M12 and M13 transistors have the parameter K fourth time greater than the other circuit's transistors). The M1 and M2 transistors form the differential input stage, excited by the differential input voltage $V_1 - V_2$, while M1 and M2 transistors, together with the current mirrors M17–M18 and M19–M20, mirror the output currents from the differential amplifier (I_{OUT1} and I_{OUT2}) for obtaining the differential output current I_{OUT}. The V_O voltage sources

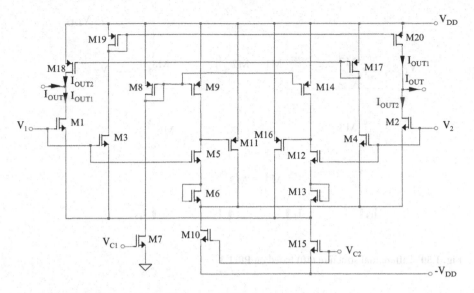

Fig. 1.49 Differential structure (9) based on PR 1.3

from Fig. 1.37 are implemented in Fig. 1.49 using M5–M6 and M12–M13 pairs, biased at a constant current. This current is imposed using M8–M9–M14 current mirror by M7 transistor, having the gate-source voltage determined by V_{C1} control potential.

The voltage V_O can be expressed as follows:

$$V_O = V_{GS5} + V_{GS6} = 2\left(V_T + \sqrt{\frac{2I_{D5}}{4K}}\right)$$

$$= 2V_T + \sqrt{\frac{2}{K}\frac{K}{2}(V_{C1} - V_T)^2} = V_{C1} + V_T \qquad (1.210)$$

So, using relation (1.172), it results:

$$G_m = 2KV_{C1}. \qquad (1.211)$$

The advantage of the circuit consists in the possibility of controlling G_m transconductance by V_{C1} potential, as well as the independence of the equivalent transconductance on the bulk effect.

A differential amplifier with linear behavior based on a translinear loop is presented in Fig. 1.50 [24].

The translinear loop containing M2, M3, M4 and M7 transistors has the following characteristic equation:

$$V_1 - V_2 = 2V_{GS}(I_O) - 2V_{GS}(I_{OUT1}) = 2\sqrt{\frac{2}{K}}\left(\sqrt{I_O} - \sqrt{I_{OUT1}}\right) \qquad (1.212)$$

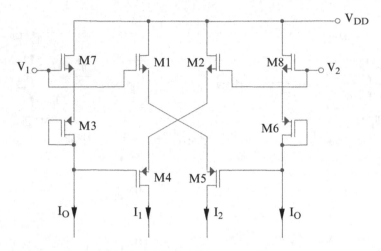

Fig. 1.50 Differential structure (10) based on PR 1.3

resulting:

$$I_{OUT1} = I_O + \frac{K}{8}(V_1 - V_2)^2 - \sqrt{\frac{KI_O}{2}}(V_1 - V_2) \qquad (1.213)$$

Similarly, analyzing the translinear loop containing M1, M5, M6 and M8 transistors, it results:

$$I_{OUT2} = I_O + \frac{K}{8}(V_1 - V_2)^2 + \sqrt{\frac{KI_O}{2}}(V_1 - V_2) \qquad (1.214)$$

The differential output current of the circuit presented in Fig. 1.50 will have the following expression:

$$I_{OUT2} - I_{OUT1} = \sqrt{2KI_O}(V_1 - V_2) \qquad (1.215)$$

A linearization technique [25] using the third mathematical principle (PR 1.3), based on the utilization of translation blocks for realizing DC shifting of input potentials has the block diagram presented in Fig. 1.51 [25].

The "DA" block represents a classical active-load differential amplifier, having the common-sources point biased at a potential V fixed by the circuit "M". This circuit computes the arithmetical mean of input potentials, providing a very good linearity of the entire structure, with the contribution of "T" blocks (which are used for introducing a translation of input potentials).

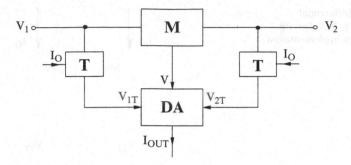

Fig. 1.51 Differential structure (11) based on PR 1.3 – block diagram

Fig. 1.52 Differential
structure (11) based on PR 1.3
– DA block implementation

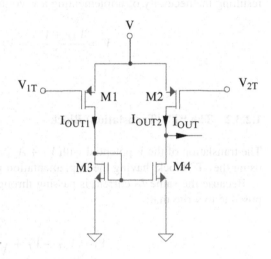

1.2.3.1 The "DA" (Differential Amplifier) Block

The "DA" block is implemented as a classical active-load differential amplifier, having the concrete realization presented in Fig. 1.52 [25–27].

Considering a biasing in saturation of all MOS devices from Fig. 1.52, the output current of the differential amplifier can be expressed as:

$$I_{OUT} = \frac{K}{2}(V_{1T} - V_{2T})(2V - V_{1T} - V_{2T} - 2V_T) \tag{1.216}$$

In order to obtain a linear transfer characteristic $I_{OUT}(V_{1T} - V_{2T})$, it is necessary that the second parenthesis from (1.216) to be constant with respect to the differential input voltage, $V_{1T} - V_{2T}$:

$$2V - V_{1T} - V_{2T} - 2V_T = A = ct. \tag{1.217}$$

Fig. 1.53 Differential
structure (11) based on PR
1.3 – T block implementation

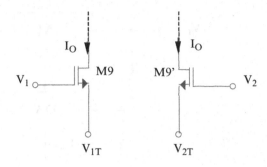

resulting the necessity of implementing a V voltage equal with:

$$V = \frac{V_{1T} + V_{2T}}{2} + V_T + \frac{A}{2} \qquad (1.218)$$

1.2.3.2 The "T" (Translation) Block

The translation of the V potential with $V_T + A/2$ (relation (1.218)) can be obtained using the "T" block, having the implementation presented in Fig. 1.53.

Because the same I_O current is passing through transistors from Fig. 1.53, it is possible to write that:

$$V_1 = V_{1T} + V_T + \sqrt{\frac{2I_O}{K}} \qquad (1.219)$$

and:

$$V_2 = V_{2T} + V_T + \sqrt{\frac{2I_O}{K}} \qquad (1.220)$$

So, both input potentials V_1 and V_2 are DC shifted with the same amount, $V_T + \sqrt{2I_O/K}$.

1.2.3.3 The "M" (Arithmetic Mean) Block

In order to obtain the arithmetic mean of input potentials expressed by (1.221) relation, the circuit from Fig. 1.54 [25] can be used, having the advantages of using only MOS transistors biased in saturation region and of avoiding any current consumption from the input voltage sources, V_1 and V_2.

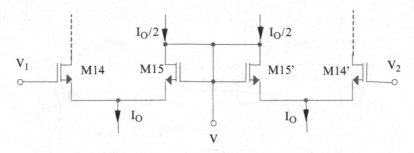

Fig. 1.54 Differential structure (11) based on PR 1.3 – M block implementation

The M14–M15 and M14'–M15' differential amplifiers are biased at the same current, I_O and, additionally, the sum of drain currents of M15 and M15' transistors are, also, equal with I_O. As a result, gate-source voltages of M14 and M15' transistors are equal and, similarly, gate-source voltages of M14' and M15 transistors are equal. In order to obtain the expression of V voltage, it can write that $V_1 - V = V_{GS14} - V_{GS15}$ and $V - V_2 = V_{GS15}' - V_{GS14}'$. Subtracting these two relations and using the previous observations, V potential will represent the arithmetical mean of V_1 and V_2 input potentials:

$$V = \frac{V_1 + V_2}{2} \tag{1.221}$$

So:

$$V = \frac{V_{1T} + V_{2T}}{2} + V_T + \sqrt{\frac{2I_O}{K}} \tag{1.222}$$

Comparing (1.218) and (1.212) relations, it results that $A = 2\sqrt{2I_O/K}$, so:

$$I_{OUT} = \frac{K}{2}(V_{1T} - V_{2T})2\sqrt{\frac{2I_O}{K}} = \sqrt{2KI_O}(V_{1T} - V_{2T}) \tag{1.223}$$

equivalent (using (1.219) and (1.220)) with:

$$I_{OUT} = \sqrt{2KI_O}(V_1 - V_2) = G_m(V_1 - V_2) \tag{1.224}$$

$G_m = \sqrt{2KI_O}$ being the equivalent transconductance of the differential amplifier.

In conclusion, a linear transfer characteristic of the differential structure having the block diagram presented in Fig. 1.51 is obtained by using a proper voltage biasing of the classical differential amplifier from Fig. 1.52. The full implementation of the linearized differential structure is presented in Fig. 1.55. An equivalent structure, obtained by replacing the "DA" block from Fig. 1.51 with a complementary circuit is presented in Fig. 1.56 [25].

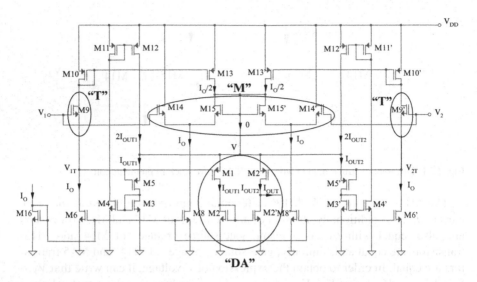

Fig. 1.55 Differential structure (11) based on PR 1.3 – complete implementation

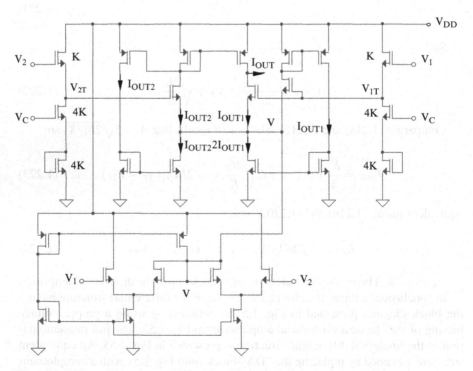

Fig. 1.56 Differential structure (11) based on PR 1.3 – alternate complete implementation

Fig. 1.57 Differential
structure (12) based on PR 1.3
– block diagram

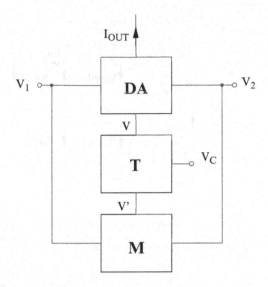

A similar behavior can be obtained based on the block diagram presented in Fig. 1.57, using another structure for replacing the block diagram from Fig. 1.51. The full implementation of the differential amplifier using this architecture is shown in Fig. 1.58.

A differential amplifier with linear transfer characteristic using FGMOS transistors is presented in Fig. 1.59 [28].

Considering that FGMOS transistors from Fig. 1.59 have different inputs, the expressions of their drain currents will be:

$$I_{OUT1} = \frac{K}{2} \left(\frac{V_1 + V_{POL1}}{2} - V - V_T \right)^2 \tag{1.225}$$

$$I_{OUT2} = \frac{K}{2} \left(\frac{V_2 + V_{POL2}}{2} - V - V_T \right)^2 \tag{1.226}$$

and:

$$I_O = \frac{K}{2} \left(\frac{V_1 + V_2 + 2V_{POL3}}{4} - V - V_T \right)^2 \tag{1.227}$$

Replacing in (1.225) and (1.226) the expression of V potential, expressed from (1.227), it results:

$$I_{OUT1} = \frac{K}{2} \left(\frac{V_1 - V_2 + 2V_{POL1} - 2V_{POL3}}{4} + \sqrt{\frac{2I_O}{K}} \right)^2 \tag{1.228}$$

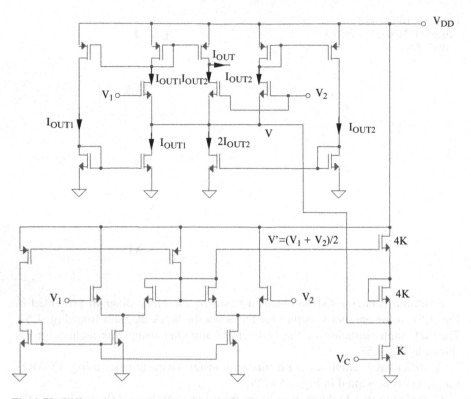

Fig. 1.58 Differential structure (12) based on PR 1.3 – complete implementation

and:

$$I_{OUT2} = \frac{K}{2}\left(\frac{V_2 - V_1 + 2V_{POL2} - 2V_{POL3}}{4} + \sqrt{\frac{2I_0}{K}}\right)^2 \tag{1.229}$$

The differential output current of the circuit presented in Fig. 1.59 will have the following expression:

$$I_{OUT1} - I_{OUT2} = \frac{K}{2}\left(\frac{V_1 - V_2}{2} + \frac{V_{POL1} - V_{POL2}}{2}\right)$$

$$\times \left(\frac{V_{POL1} + V_{POL2}}{2} - V_{POL3} + 2\sqrt{\frac{2I_0}{K}}\right) \tag{1.230}$$

In order to obtain a linear differential amplifier, the structure must be symmetrical, so $V_{POL1} = V_{POL2}$, resulting:

$$I_{OUT1} - I_{OUT2} = \frac{K}{4}(V_1 - V_2)\left(V_{POL1} - V_{POL3} + 2\sqrt{\frac{2I_0}{K}}\right) \tag{1.231}$$

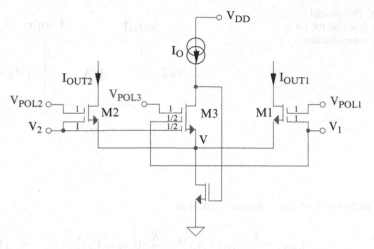

Fig. 1.59 Differential structure (13) based on PR 1.3

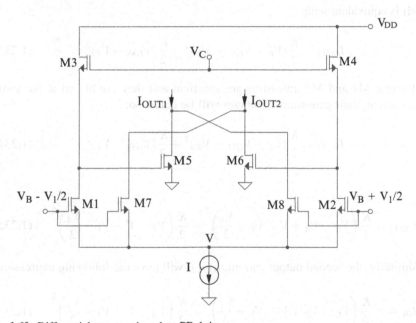

Fig. 1.60 Differential structure based on PR 1.4

1.2.4 Differential Structures Based on the Fourth Mathematical Principle (PR 1.4)

A possible illustration of the utilization of fourth mathematical principle (PR 1.4) for linearizing the transfer characteristic of a differential structure is presented in Fig. 1.60 [29]. The symmetrical structure is responsible for an important improvement of the circuit accuracy.

Fig. 1.61 Differential
structure based on PR 1.4 –
symbolic representation

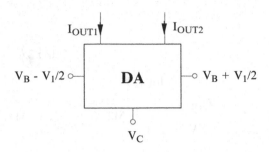

Fig. 1.61 Differential
structure based on PR 1.4 –
symbolic representation

The expression of I_{OUT1} output current is:

$$I_{OUT1} = \frac{K}{2}(V_{GS5} - V_T)^2 + \frac{K}{2}(V_{GS8} - V_T)^2 \tag{1.232}$$

which is equivalent with:

$$I_{OUT1} = \frac{K}{2}(V_C - V_{GS3} - V_T)^2 + \frac{K}{2}(V_{GS8} - V_T)^2 \tag{1.233}$$

Because M1 and M3 transistors are identical and they are biased at the same drain current, their gate-source voltages will be equal, so:

$$I_{OUT1} = \frac{K}{2}(V_C - V_{GS1} - V_T)^2 + \frac{K}{2}(V_{GS8} - V_T)^2 \tag{1.234}$$

or:

$$I_{OUT1} = \frac{K}{2}\left(V_C - V_B + V - V_T + \frac{V_1}{2}\right)^2 + \frac{K}{2}\left(V_B - V - V_T + \frac{V_1}{2}\right)^2 \tag{1.235}$$

Similarly, the second output current, I_{OUT2}, will have the following expression:

$$I_{OUT2} = \frac{K}{2}\left(V_C - V_B + V - V_T - \frac{V_1}{2}\right)^2 + \frac{K}{2}\left(V_B - V - V_T - \frac{V_1}{2}\right)^2 \tag{1.236}$$

The differential output current can be expressed as follows:

$$I_{OUT1} - I_{OUT2} = KV_1(V_C - V_B + V - V_T)$$
$$+ KV_1(V_B - V - V_T) = K(V_C - 2V_T)V_1 \tag{1.237}$$

The symbolic representation of the circuit is shown in Fig. 1.61.

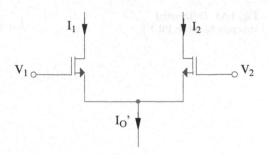

Fig. 1.62 Classical MOS differential structure

1.2.5 Differential Structures Based on the Fifth Mathematical Principle (PR 1.5)

The method for obtaining a linear behavior of the differential amplifier using the fifth mathematical principle (PR 1.5) is based on a proper biasing of the structure at a current that is dependent on the differential input voltage.

The analysis of the classical differential amplifier (Fig. 1.62) using MOS transistors biased in saturation region illustrates a strongly nonlinear behavior, that can be quantitatively evaluated by the (1.14) dependence of I_{OUT} differential output current on the differential input voltage, $V_I = V_1 - V_2$.

$$I_{OUT} = I_1 - I_2 = I_O' \sqrt{\frac{K(V_1 - V_2)^2}{I_O'} - \frac{K^2(V_1 - V_2)^4}{4I_O'^2}} \qquad (1.238)$$

equivalent with:

$$I_{OUT} = \frac{V_1 - V_2}{2} \sqrt{4KI_O' - K^2(V_1 - V_2)^2} \qquad (1.239)$$

I_O' being the biasing current of the differential structure. So, superior-order distortions will characterize the behavior of the classical differential structure, imposing the design of a linearization technique for removing the superior-order terms from its transfer characteristic.

The method [30] for obtaining a linear transfer characteristic of the differential amplifier, despite the quadratic law of its composing devices (Fig. 1.63) [30], is to obtain the I_O' biasing current of the entire differential structure as a sum of a main constant term, I_O and an additional term, proportional with the square of the differential input voltage, $I = K(V_1 - V_2)^2/4$:

$$I_O' = I_O + I = I_O + \frac{K}{4}(V_1 - V_2)^2 \qquad (1.240)$$

Fig. 1.63 Differential
structure based on PR 1.5

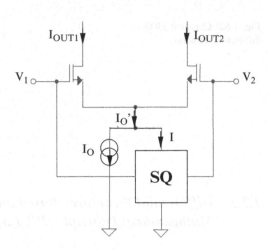

resulting, in this case, a perfect linear behavior of the differential amplifier:

$$I_{OUT} = \sqrt{KI_O}(V_1 - V_2) = G_m(V_1 - V_2) \tag{1.241}$$

$G_m = \sqrt{KI_O}$ being the equivalent transconductance of the structure, that could be very easily controlled by the biasing current, I_O. The circuit implementation is relatively simple, allowing to implement, by minor changes of the design, a multitude of circuit functions (multiplying, amplifying and simulating a positive or negative equivalent resistance).

In order to implement (1.240) relation, it is necessary to design a CMOS structure able to compute the square of the $V_1 - V_2$ differential input voltage. A possible implementation of this circuit uses the arithmetical mean of the input potentials (Fig. 1.64) [6].

$$I = I_1 + I_2 - I_3 = \frac{K}{2}(V_1 - V_T)^2 + \frac{K}{2}(V_2 - V_T)^2$$
$$- K\left(\frac{V_1 + V_2}{2} - V_T\right)^2 = \frac{K}{4}(V_1 - V_2)^2 \tag{1.242}$$

1.2.6   Differential Structures Based on the Sixth
Mathematical Principle (PR 1.6)

The general linearization technique based on the sixth mathematical principle (PR 1.6) achieves the minimization of superior-order distortions using an anti-parallel connection of two or more differential amplifiers having controlled asymmetries and different controlled biasing currents. Using (1.20), the expression

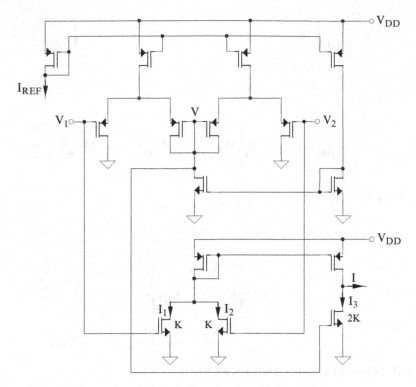

Fig. 1.64 Squaring circuit for the differential structure based on PR 1.5

of the differential output current for the classical differential amplifier from Fig. 1.62 [31] will be:

$$I_{OUT}(V_I) = -K^{1/2}I_O^{1/2}V_I + \frac{K^{3/2}}{8I_O^{1/2}}V_I^3 + \frac{K^{5/2}}{128I_O^{3/2}}V_I^5 + \cdots \qquad (1.243)$$

As a consequence of the circuit symmetry, the even order terms from the previous expansion have been cancel out:

$$I_{OUT}(V_I) = a_1V_I + a_3V_I^3 + a_5V_I^5 + \cdots \qquad (1.244)$$

a_k being constant coefficients of the expansion. For simplicity, $I_O{}'$ biasing current from Fig. 1.62 has been renamed I_O. The total harmonic distortions of the classical differential amplifier are mainly given by the third-order term from the previous expansion:

$$THD = \frac{a_3V_I^3}{a_1V_I} = \frac{K}{I_O}V_I^2 \qquad (1.245)$$

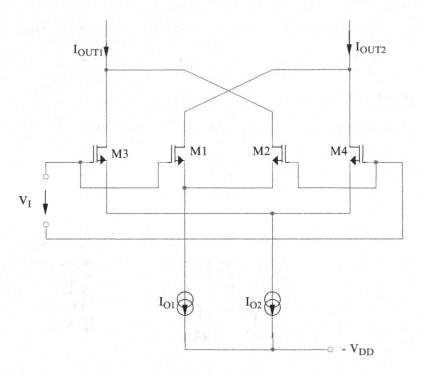

Fig. 1.65 Differential structure (1) based on PR 1.6

The first linear term is proportional with the differential-mode voltage gain of the differential amplifier, while the following two terms model the third-order and the fifth-order nonlinearities of the transfer characteristic. The most important cause of the distortions introduced by the circuit nonlinearity is the third-order term from the previous expansion.

The method for improving the circuit linearity uses an anti-parallel connection of two differential amplifiers with controlled asymmetries and different controlled biasing currents (M1–M2 and M3–M4 pairs from Fig. 1.65) [13, 31]. Because M1–M2 differential amplifier is biased at I_{O1} current and M3–M4 structure is biased at I_{O2} current, the differential output currents for these circuits are:

$$(I_{D2} - I_{D1})(V_I) = -K_{1,2}^{1/2}I_{O1}^{1/2}V_I + \frac{K_{1,2}^{3/2}}{8I_{O1}^{1/2}}V_I^3 + \frac{K_{1,2}^{5/2}}{128I_{O1}^{3/2}}V_I^5 + \cdots \qquad (1.246)$$

and:

$$(I_{D4} - I_{D3})(V_I) = -K_{3,4}^{1/2}I_{O2}^{1/2}V_I + \frac{K_{3,4}^{3/2}}{8I_{O2}^{1/2}}V_I^3 + \frac{K_{3,4}^{5/2}}{128I_{O2}^{3/2}}V_I^5 + \cdots \qquad (1.247)$$

The differential output current for the entire anti-parallel structure from Fig. 1.65 will have the following expression:

$$I_{OUT2} - I_{OUT1} = (I_{D2} + I_{D3}) - (I_{D1} + I_{D4})$$
$$= (I_{D2} - I_{D1}) - (I_{D4} - I_{D3}) \qquad (1.248)$$

Replacing (1.257) and (1.258) in (1.259), it results:

$$I_{OUT2} - I_{OUT1} = \left(K_{3,4}^{1/2} I_{O2}^{1/2} - K_{1,2}^{1/2} I_{O1}^{1/2} \right) V_I$$
$$+ \left(\frac{K_{1,2}^{3/2}}{8 I_{O1}^{1/2}} - \frac{K_{3,4}^{3/2}}{8 I_{O2}^{1/2}} \right) V_I^3 + \left(\frac{K_{1,2}^{5/2}}{128 I_{O1}^{3/2}} - \frac{K_{3,4}^{5/2}}{128 I_{O2}^{3/2}} \right) V_I^5 + \cdots \qquad (1.249)$$

The linearization technique is considered to be efficient if it is able to cancel the third-order distortion, so the condition that must be fulfilled by the design is:

$$\frac{K_{1,2}^{3/2}}{8 I_{O1}^{1/2}} = \frac{K_{3,4}^{3/2}}{8 I_{O2}^{1/2}} \qquad (1.250)$$

equivalent with:

$$\frac{I_{O1}}{I_{O2}} = \left(\frac{K_{1,2}}{K_{3,4}} \right)^3 \qquad (1.251)$$

Imposing this condition for obtaining the minimization of the circuit nonlinearity, the main distortions will be caused by the fifth-order term from the circuit transfer characteristic. Using (1.251), relation (1.249) becomes:

$$I_{OUT2} - I_{OUT1} = -K_{1,2}^{1/2} I_{O1}^{1/2} \left[1 - \left(\frac{I_{O2}}{I_{O1}} \right)^{2/3} \right] V_I$$
$$- \frac{K_{1,2}^{5/2}}{128 I_{O1}^{5/2}} \left[1 - \left(\frac{I_{O1}}{I_{O2}} \right)^{2/3} \right] V_I^5 + \cdots \qquad (1.252)$$

The total harmonic distortions of the improved linearity differential amplifier from Fig. 1.65 are mainly given by the fifth-order term from the previous expansion:

$$THD' = \frac{V_I^4}{128} \left(\frac{K_{1,2}}{I_{O1}} \right)^2 \left(\frac{I_{O1}}{I_{O2}} \right)^{2/3} \qquad (1.253)$$

resulting an important increasing of the circuit linearity (THD' is much smaller than THD).

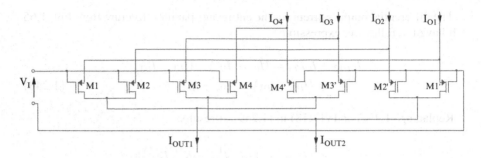

Fig. 1.66 Differential structure (2) based on PR 1.6

In order to further increase the differential amplifier linearity, the previous technique can be extended for removing both third-order and fifth-order terms from the Taylor series expansion (1.249). This is possible using a parallel connection of four differential amplifiers, having controlled asymmetries and controlled different biasing (Fig. 1.66).

Similarly with the circuit presented in Fig. 1.65, the conditions for canceling the third-order and the fifth-order distortions are:

$$\frac{K_2}{K_3} = \left(\frac{I_{O2}}{I_{O3}}\right)^{\frac{1}{3}}; \quad \frac{K_1}{K_4} = \left(\frac{I_{O1}}{I_{O4}}\right)^{\frac{1}{3}} \tag{1.254}$$

$$\frac{K_1}{K_2} = \left(\frac{I_{O1}}{I_{O2}}\right)^{\frac{3}{5}}; \quad \frac{K_3}{K_4} = \left(\frac{I_{O3}}{I_{O4}}\right)^{\frac{3}{5}} \tag{1.255}$$

K_1–K_4 being model parameters for M1–M1', M2–M2', M3–M3' and M4–M4' pairs, respectively. The total harmonic distortions of the circuit presented in Fig. 1.66 are mainly given by the fifth-order term from the Taylor series expansion of the transfer characteristic:

$$THD'' \cong \frac{1}{2^{15}} \left(\frac{K_2}{I_{O2}}\right)^3 V_{ID}^6 \frac{\left[1 - \left(\frac{K_3}{K_2}\right)^{7/2} \left(\frac{I_{O3}}{I_{O2}}\right)^{-5/2}\right] \left[1 - \left(\frac{K_4}{K_3}\right)^{7/2} \left(\frac{I_{O4}}{I_{O3}}\right)^{-5/2}\right]}{\left[1 - \left(\frac{K_3}{K_2}\right)^{1/2} \left(\frac{I_{O3}}{I_{O2}}\right)^{1/2}\right] \left[1 - \left(\frac{K_4}{K_3}\right)^{1/2} \left(\frac{I_{O4}}{I_{O3}}\right)^{1/2}\right]} \tag{1.256}$$

resulting an important improvement of the circuit linearity by applying this technique.

An alternate implementation [32] of the previous linearization technique can be obtained using differential amplifiers based on MOS transistors working in weak inversion region (Fig. 1.67) [32]. This implementation of the differential amplifier circuit allows to obtain an important reduction of the current consumption, the structure being useful for low-power designs. The drive of these transistors is also realized on their gates and bulks. The utilization of the bulk as active terminal

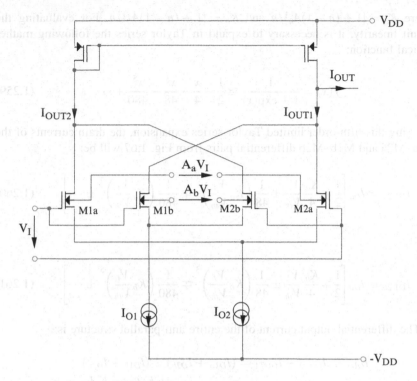

Fig. 1.67 Differential structure (3) based on PR 1.6

avoids the utilization of large aspect ratios MOS transistors. A_a and A_b represent the voltage gains used for computing the voltage drives of the bulks.

The drain currents of MOS transistors composing the M1a–M2a differential amplifier from Fig. 1.67 have the following expressions:

$$I_{D1,2a} = \frac{I_{Oa}}{1 + \exp\left(\pm \dfrac{V_{GS1a} - V_{GS2a}}{nV_{th}}\right) \exp\left(\pm \dfrac{n-1}{n} \dfrac{V_{BS1a} - V_{BS2a}}{V_{th}}\right)}$$

$$= \frac{I_{Oa}}{1 + \exp\left(\pm K_a \dfrac{V_I}{V_{th}}\right)} \tag{1.257}$$

Similarly, for the second differential amplifier, M1b–M2b, the drain currents can be expressed as follows:

$$I_{D1,2b} = \frac{I_{Ob}}{1 + \exp\left(\pm \dfrac{V_{GS1b} - V_{GS2b}}{nV_{th}}\right) \exp\left(\pm \dfrac{n-1}{n} \dfrac{V_{BS1b} - V_{BS2b}}{V_{th}}\right)}$$

$$= \frac{I_{Ob}}{1 + \exp\left(\pm K_b \dfrac{V_I}{V_{th}}\right)} \tag{1.258}$$

where $K_a = [1 + (n-1)A_a]/n$ and $K_b = [1 + (n-1)A_b]/n$. For evaluating the circuit linearity, it is necessary to expand in Taylor series the following mathematical function:

$$f(x) = \frac{1}{1 + \exp(x)} \cong \frac{1}{2} - \frac{x}{4} + \frac{x^3}{48} - \frac{x^5}{480} + \cdots \tag{1.259}$$

Using this fifth-order limited Taylor series expansion, the drain currents of the M1a–M2a and M1b–M2b differential pairs from Fig. 1.67 will be:

$$I_{D1,2a} = I_{Oa} \left[\frac{1}{2} \mp \frac{K_a}{4} \frac{V_I}{V_{th}} \pm \frac{1}{48} \left(K_a \frac{V_I}{V_{th}} \right)^3 \mp \frac{1}{480} \left(K_a \frac{V_I}{V_{th}} \right)^5 \pm \cdots \right] \tag{1.260}$$

and:

$$I_{D1,2b} = I_{Ob} \left[\frac{1}{2} \mp \frac{K_b}{4} \frac{V_I}{V_{th}} \pm \frac{1}{48} \left(K_b \frac{V_I}{V_{th}} \right)^3 \mp \frac{1}{480} \left(K_b \frac{V_I}{V_{th}} \right)^5 \pm \cdots \right] \tag{1.261}$$

The differential output current of the entire anti-parallel structure is:

$$\begin{aligned} I_{OUT} &= I_{OUT2} - I_{OUT1} = (I_{D1a} + I_{D2b}) - (I_{D1b} + I_{D2a}) \\ &= (I_{D1a} - I_{D2a}) - (I_{D1b} - I_{D2b}) = \frac{K_b I_{Ob} - K_a I_{Oa}}{2V_{th}} V_I \\ &\quad + \frac{K_a^3 I_{Oa} - K_b^3 I_{Ob}}{24 V_{th}^3} V_I^3 + \frac{K_b^5 I_{Ob} - K_a^5 I_{Oa}}{240 V_{th}^5} V_I^5 + \cdots \end{aligned} \tag{1.262}$$

The condition for cancellation of the third-order distortions introduced by the circuit is:

$$\frac{K_a}{K_b} = \sqrt[3]{\frac{I_{Ob}}{I_{Oa}}} = \frac{1 + (n-1)A_a}{1 + (n-1)A_b} \tag{1.263}$$

resulting the following expression of the I_{OUT} output current:

$$I_{OUT} \cong a_1 V_I + a_5 V_I^5 + \cdots \tag{1.264}$$

where a_1 and a_5 constants are given by:

$$a_1 = \frac{K_a}{2V_{th}} I_{Oa} \left[\left(\frac{I_{Ob}}{I_{Oa}} \right)^{2/3} - 1 \right] \tag{1.265}$$

$$a_5 = \frac{K_a^5}{240V_{th}^5} I_{Oa} \left[\left(\frac{I_{Ob}}{I_{Oa}} \right)^{-2/3} - 1 \right]$$ (1.266)

The total harmonic distortion coefficient for the linearized circuit will have the following expression:

$$THD = \frac{1}{120} \left(\frac{K_a V_I}{n V_{th}} \right)^4 \left[\frac{1 + (n-1)A_b}{1 + (n-1)A_a} \right]^2$$ (1.267)

1.2.7 Differential Structures Based on the Seventh Mathematical Principle (PR 1.7)

This principle is useful for obtaining a rail-to-rail operation of a differential structure. The general method for extending the maximal range of the common-mode input voltage is to use a parallel connection of two complementary (NMOS and PMOS) differential amplifiers. The simple parallel connection of these stages has the disadvantage of a variable value of the equivalent transconductance for the resulted structure, while the utilization of a translinear loop or of a "maximum" circuit permits to obtain an approximately constant value of the equivalent transconductance, only slightly dependent on the value of the common-mode input voltage ($R_1 = R_2 = R_3 = R_4$).

A possible method for obtaining a rail-to-rail operation of a differential structure, using two parallel-connected complementary differential amplifiers is presented in Fig. 1.68. The NMOS differential amplifier is realized using M1–M4 transistors, while the PMOS differential circuit is implemented using M5–M8 active devices. Because these differential amplifiers present complementary common-mode input ranges, their parallel connection will extend the equivalent domain of the common-mode input voltage for the entire structure. The *DIFF*1 and *DIFF*2 blocks computes the differential output voltages of the complementary differential amplifiers, while the *SUM* block realizes the summation of these two output voltages. For medium values of the common-mode input voltages, both NMOS and PMOS differential amplifiers are active, so the equivalent transconductance of the parallel-connected structure will be equal with the sum of the individual transconductances, $g_m^T = g_{mn} + g_{mp}$.

The maximal range of the common-mode input voltage for the NMOS differential amplifier, M1–M4 is included between the following limits:

$$V_{IC\,max}^{NMOS} = V_{DD} - \frac{I_O R_1}{2} - V_{DS1sat} + V_{GS1} = V_{DD} - \frac{I_O R_1}{2} + V_T$$ (1.268)

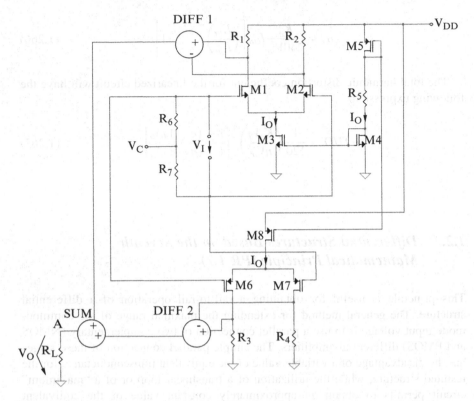

Fig. 1.68 Differential structure (1) based on PR 1.7

and:

$$V_{IC\min}^{NMOS} = V_{GS1} + V_{DS3sat} = V_{GS1} + V_{GS3} - V_T = V_T + \left(\sqrt{2}+1\right)\sqrt{\frac{I_O}{K}} \quad (1.269)$$

while the maximal range of the common-mode input voltage for the PMOS differential amplifier M5–M8 is included between the following limits:

$$V_{IC\max}^{PMOS} = V_{DD} - V_{SG6} - V_{SD8sat} = V_{DD} - V_{SG6} - V_{SG8} + V_T$$

$$= V_{DD} - V_T - \left(\sqrt{2}+1\right)\sqrt{\frac{I_O}{K}} \quad (1.270)$$

and:

$$V_{IC\min}^{PMOS} = \frac{I_O R_3}{2} + V_{SD6sat} - V_{SG6} = \frac{I_O R_3}{2} - V_T \quad (1.271)$$

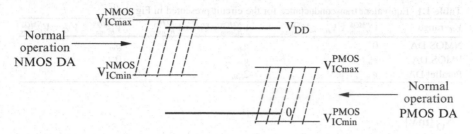

Fig. 1.69 Diagram for the range of V_{IC}

The maximal range of the common-mode input voltage for the parallel-connection from Fig. 1.68 is determined by superposing the individual common-mode input ranges for the complementary NMOS and PMOS differential pairs (relations (1.268) – (1.271)) – Fig. 1.69.

The maximal range of the common-mode input voltage for the parallel connection from Fig. 1.68 must include the supply voltage range, $[0, V_{DD}]$. From Fig. 1.69, the conditions for obtaining this goal are:

$$V_{IC\,max}^{NMOS} > V_{DD} \tag{1.272}$$

$$V_{IC\,max}^{PMOS} > V_{IC\,min}^{NMOS} \tag{1.273}$$

$$V_{IC\,min}^{PMOS} < 0 \tag{1.274}$$

equivalent with:

$$I_O R_1 < 2V_T \tag{1.275}$$

and:

$$V_{DD} > 2\left[V_T + \left(\sqrt{2}+1\right)\sqrt{\frac{I_O}{K}}\right] \tag{1.276}$$

The common-mode input ranges for the two parallel-connected differential stages are complementary, existing, however, a range of the common-mode input voltage (corresponding to medium values of this voltages), where the range of NMOS differential structure overlaps the range of the PMOS differential stage. Supposing $g_{mn} = g_{mp} = g_m^T$, the dependence of the circuit's total transconductance, g_m^T, on the common-mode input voltage is synthesized in Table 1.1.

The disadvantage of this parallel connection of two complementary differential amplifiers is a value of the equivalent transconductance g_m^T that depends on the common-mode input voltage.

Table 1.1 Equivalent transconductance for the circuit presented in Fig. 1.68

V_{IC} range	$V_{IC\,min}^{PMOS} < V_{IC} < V_{IC\,min}^{NMOS}$	$V_{IC\,min}^{NMOS} < V_{IC} < V_{IC\,max}^{PMOS}$	$V_{IC\,max}^{PMOS} < V_{IC} < V_{IC\,max}^{NMOS}$
NMOS DA	0	g_m	g_m
PMOS DA	g_m	g_m	0
Parallel DA	g_m	$2g_m$	g_m

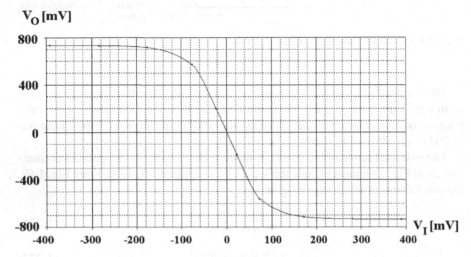

Fig. 1.70 Transfer characteristic of the M1–M4 NMOS differential amplifier

The operation of the previous circuit is verified for the following particular values of their components and model parameters:

$R_1 = R_2 = R_3 = R_4 = R_5 = R_6 = R_7 = 10\,\text{k}\,\Omega, R_L = 1\,\text{M}\,\Omega, V_T = 1\,\text{V}/-1\,\text{V}$, $\lambda = 3 \times 10^{-3}\,\text{V}^{-1}, K' = 8 \times 10^{-3}\,\text{A/V}^2, W = 30\,\mu\text{m}, L = 20\,\mu\text{m}$. V_{DD}, V_I and V_C are continuous input voltages having the following values: 3 V, 1 mV and 1.5 V, respectively.

For determination of the value of the biasing current I_O, it is possible to write the following relation:

$$V_{DD} = 2V_{GS} + \frac{KR_5}{2}(V_{GS} - V_T)^2 \qquad (1.277)$$

equivalent with:

$$60V_{GS}^2 - 118V_{GS} + 57 = 0 \qquad (1.278)$$

resulting $V_{GS} = 1.1135\,\text{V}$ and $I_O = 77.3\,\mu\text{A}$, very closely to the simulated obtained value, $I_O = 73.3\,\mu\text{A}$.

The simulation of the transfer characteristic for the NMOS differential amplifier M1–M4 is shown in Fig. 1.70. A similar characteristic can be obtained for the

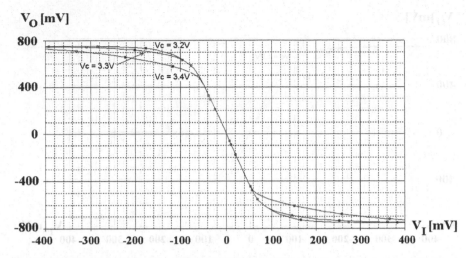

Fig. 1.71 Transfer characteristic of the M1–M4 NMOS differential amplifier (parametric) (1)

PMOS differential stage M5–M8. The differential-mode voltage gain is $A_{DD} = -g_m R_1 = -\sqrt{KI_O}R_1 = -9.63$, while the simulated value is -9.71.

Using (1.268) and (1.690), the maximal range of the common-mode input voltage for the M1–M4 NMOS differential amplifier will be between $1.1934V$ and $3.6135V$. The simulation of the transfer characteristic (Fig. 1.71) of the NMOS differential amplifier for three values of the common-mode input voltage, $V_C = 3.2V$, $V_C = 3.3V$ and $V_C = 3.4V$ highlights a maximal common-mode input voltage of about $3.3V$.

Similarly, the simulation of the transfer characteristic (Fig. 1.72) of the NMOS differential amplifier for two values of the common-mode input voltage, $V_C = 1.1V$ and $V_C = 1.2V$ points out a minimal common-mode input voltage of about $1.2V$.

Using (1.270) and (1.271), the maximal range of the common-mode input voltage for the PMOS differential amplifier M5–M8 is between $-0.6135V$ and $1.807V$. The simulation of the transfer characteristic (Fig. 1.73) of the PMOS differential amplifier for three values of the common-mode input voltage, $V_C = -0.5V$, $V_C = -0.4V$ and $V_C = -0.3V$ highlights a minimal common-mode input voltage of about $-0.4V$.

Similarly, the simulation of the transfer characteristic (Fig. 1.74) of the PMOS differential amplifier for two values of the common-mode input voltage, $V_C = 1.8V$ and $V_C = 1.9V$ points out a maximal common-mode input voltage of about $1.8V$.

In conclusion, the maximal ranges of the common-mode input voltage are $-0.4V < V_{IC} < 1.8V$ for the PMOS differential amplifier, and $1.2V < V_{IC} < 3.3V$ for the NMOS differential amplifier. The parallel connection of the complementary circuits extends the maximal range of the common-mode input voltage to $-0.4V < V_{IC} < 3.3V$, with the disadvantage of obtaining a variable value of the equivalent transconductance for the parallel-connected structure (a double transconductance for the range of the common-mode input voltage, in which both

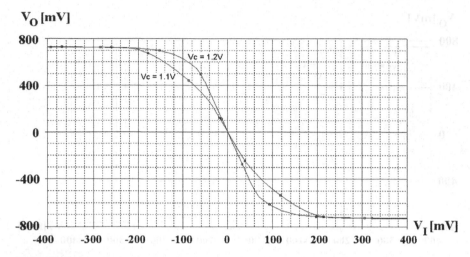

Fig. 1.72 Transfer characteristic of the M1–M4 NMOS differential amplifier (parametric) (2)

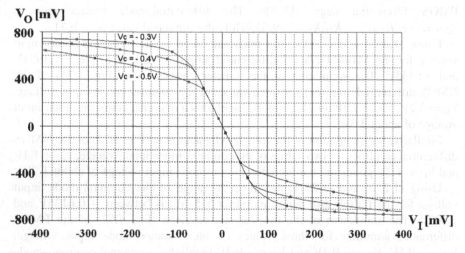

Fig. 1.73 Transfer characteristic of the M5–M8 PMOS differential amplifier (parametric) (1)

NMOS and PMOS differential amplifiers are active, $1.2V < V_{IC} < 1.8V$). So, the differential mode voltage gain will be not constant, having a value, A_{DD}, for extreme values of the common-mode input voltages ($-0.4V < V_{IC} < 1.2V$ and $1.8V < V_{IC} < 3.3V$) and a double value, $2A_{DD}$, for the medium values of the common-mode input voltages, $1.2V < V_{IC} < 1.8V$. This behavior is illustrated in Fig. 1.75, the simulation of the transfer characteristic of the parallel structure being done for five values of the common-mode input voltages:

- Two values in the area in which only one differential amplifier (NMOS or PMOS) is active ($V_C = 0.5V$ and $V_C = 2.5V$), the characteristics being approximately identical;

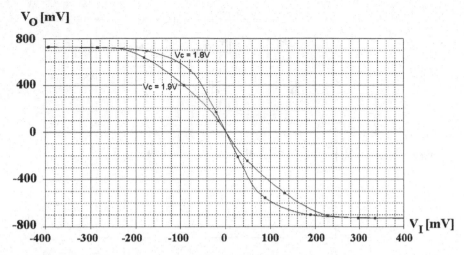

Fig. 1.74 Transfer characteristic of the M5–M8 PMOS differential amplifier (parametric) (2)

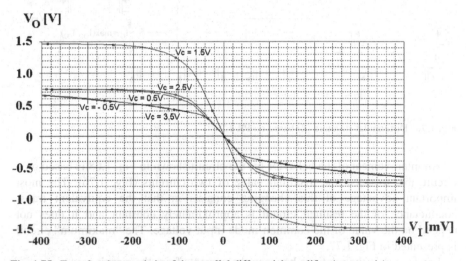

Fig. 1.75 Transfer characteristic of the parallel differential amplifier (parametric)

- Two values near the extended interval, but outside it ($V_C = -0.5V$ and $V_C = 3.5V$), remarking a small decreasing of the differential mode voltage gain with respect to the previous case, as a result of a fault operation of some transistors from the circuit
- A value placed in the center of the interval ($V_C = 1.5V$), the differential-mode voltage gain having a double value comparing with the first case, because both PMOS and NMOS differential amplifiers are active, the equivalent transconductance being the sum of each individual transconductances.

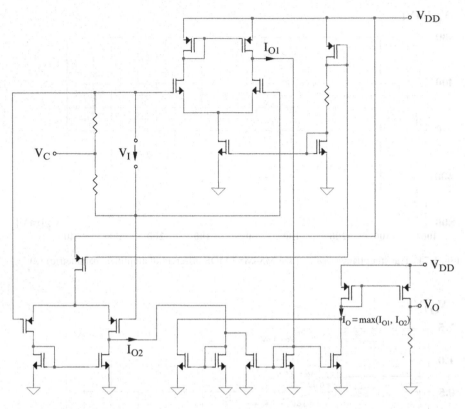

Fig. 1.76 Differential structure (2) based on PR 1.7

An improvement of the behavior for the circuit shown in Fig. 1.68 is based on a circuit able to select the maximal value of two currents (Fig. 1.76). The most important advantage of this changing consists in that the equivalent trans-conductance of the resulted circuit is approximately constant and it does not depend on the value of the common-mode input voltage. The "maximum" circuit is presented in Fig. 1.77.

For $I_{O1} > I_{O2}$, the relations between the currents from the circuit are:

$$I_{D3} = I_{D4} = I_{D5} = I_{O2} \tag{1.279}$$

$$I_{D1} = I_{D2} = I_{O1} - I_{D3} = I_{O1} - I_{O2} \tag{1.280}$$

$$I_O = I_{D1} + I_{D5} = (I_{O1} - I_{O2}) + I_{O2} = I_{O1} \tag{1.281}$$

while for $I_{O1} < I_{O2}$:

$$I_{D4} = I_{D5} = I_{O2} \tag{1.282}$$

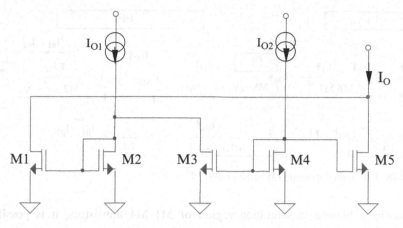

Fig. 1.77 The implementation of the "maximum" circuit

Table 1.2 Equivalent transconductance of the circuit presented in Fig. 1.76

V_{IC} range	$V^{PMOS}_{IC\,min} < V_{IC} < V^{NMOS}_{IC\,min}$	$V^{NMOS}_{IC\,min} < V_{IC} < V^{PMOS}_{IC\,max}$	$V^{PMOS}_{IC\,max} < V_{IC} < V^{NMOS}_{IC\,max}$
NMOS DA	0	g_m	g_m
PMOS DA	g_m	g_m	0
Parallel modified DA	g_m	g_m	g_m

$$I_{D3} = I_{O1}; \quad I_{D1} = I_{D2} = 0 \tag{1.283}$$

$$I_O = I_{D1} + I_{D5} = I_{O2} \tag{1.284}$$

For medium values of the common-mode input voltages, $V^{NMOS}_{IC\,min} < V_{IC} < V^{PMOS}_{IC\,max}$, the "maximum" circuit will select the maximal transconductance from g_{mn} and g_{mp}, so the equivalent transconductance of the modified parallel circuit from Fig. 1.76 will be approximately constant and independent on the common-mode input voltage the results are centralized in Table 1.2.

The circuit presented in Fig. 1.78 [33] is designed with the main goal of obtaining a rail-to-rail operation using a proper biasing of a parallel connection of two complementary classical differential amplifiers, M1–M2 and M3–M4, biased at I_{On} and I_{Op} currents, respectively. The utilization of a translinear loop, that forces a proper relation between the biasing currents I_{On} and I_{Op}, has the advantage of obtaining an approximately constant equivalent transconductance of the entire structure.

The output current of the circuit can be expressed as follows:

$$I_{OUT} = (I_{n1} - I_{n2}) + (I_{P1} - I_{P2}) = (g_{mn} + g_{mp})(V_1 - V_2) \tag{1.285}$$

where g_{mn} represents the transconductance of M1–M2 NMOS differential amplifier, while g_{mp} is the transconductance of M3–M4 PMOS differential amplifier.

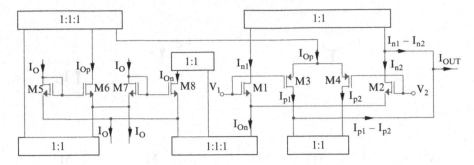

Fig. 1.78 Differential structure (3) based on PR 1.7

Supposing a biasing in saturation region of M1–M4 transistors, it is possible to write that:

$$g_{mn} = \sqrt{K_n I_{On}} \tag{1.286}$$

and:

$$g_{mp} = \sqrt{K_p I_{Op}} \tag{1.287}$$

Considering that M1–M4 transistors are identical, it results:

$$I_{OUT} = \sqrt{K}\left(\sqrt{I_{Op}} + \sqrt{I_{On}}\right)(V_1 - V_2) \tag{1.288}$$

The relation between I_{On} and I_{Op} currents is imposed by the translinear loop implemented using M5–M8 transistors:

$$V_{GS5} + V_{GS7} = V_{GS6} + V_{GS8} \tag{1.289}$$

Using the square-root dependence of the drain current on the gate-source voltage for a MOS transistor biased in saturation, it results:

$$\sqrt{I_{On}} + \sqrt{I_{Op}} = 2\sqrt{I_O} \tag{1.290}$$

From (1.288) and (1.290), the expression of the output current will be:

$$I_{OUT} = 2\sqrt{K I_O}(V_1 - V_2) \tag{1.291}$$

The advantage of this circuit consists in the rail-to-rail operation that can be obtained because of the parallel connection of two complementary differential amplifiers (M1–M2 and M3–M4). The utilization of the M5–M8 translinear loop eliminates the main disadvantage of this method (a variable equivalent transconductance of the parallel connection, depending on the common-mode input voltage).

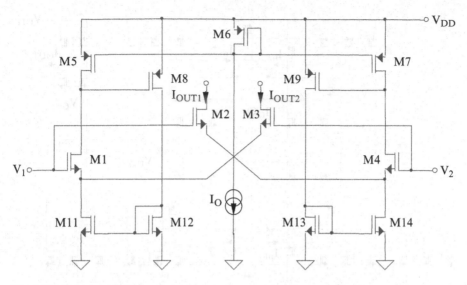

Fig. 1.79 The core of DDA

Using the same principle it is possible to design a double differential amplifier (DDA), having an extended range of the common-mode input voltage and an approximately constant equivalent transconductance.

The core of the DDA circuit is represented by a differential stage, having the circuit presented in Fig. 1.79 [34]. The differential output current of this circuit can be expressed as follows:

$$I_{OUT1} - I_{OUT2} = \sqrt{8KI_O}(V_1 - V_2) = g_{mn}(V_1 - V_2) \tag{1.292}$$

In order to extend the maximal range of the common-mode input voltage, a parallel connection of two complementary differential amplifiers are used, the selection of the active one being done using two "maximum" circuits. The differential amplifier with extended common-mode input range and constant equivalent transconductance is presented in Fig. 1.80 [34].

Relation (1.305) can be rewritten for the two complementary differential amplifiers:

$$I_{n1} - I_{n2} = \sqrt{8K_nI_O}(V_1 - V_2) = g_{mn}(V_1 - V_2) \tag{1.293}$$

and:

$$I_{p_1} - I_{p_2} = \sqrt{8K_pI_O}(V_1 - V_2) = g_{mp}(V_1 - V_2) \tag{1.294}$$

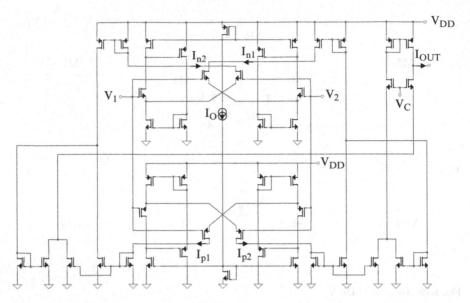

Fig. 1.80 Differential amplifier with extended common-mode input range

The output current has the following expression:

$$I_{OUT} = \max(I_{n1}, I_{p2}) - \max(I_{n2}, I_{P1}) = \sqrt{8K_nI_O}(V_1 - V_2)$$
$$= \sqrt{8K_pI_O}(V_1 - V_2) = g_m^T(V_1 - V_2) \qquad (1.295)$$

where $g_{mn} = g_{mp} = g_m^T$.

The implementation (Fig. 1.81) of the double differential amplifier is based on the previous presented differential stage (Fig. 1.79). The output current of DDA circuit has the following expression:

$$I_{OUT} = \max(I_{n_2} + I'_{n_1}, I_{p_1} + I'_{p_2}) - \max(I_{n_1} + I'_{n_2}, I_{p_2} + I'_{p_1})$$
$$= g_m^T\left[(V_{pp} - V_{pn}) - (V_{np} - V_{nn})\right] \qquad (1.296)$$

Because of the utilization of the differential amplifier with extended common-mode input range, the equivalent transconductance of the DDA, g_m^T, will be approximately constant.

1.2.8   Differential Structures Based on Different Mathematical Principle (PR 1.D)

Another possible realization of a DDA is presented in Fig. 1.81 [35], having a relatively simple implementation comparing with the previous designs of similar circuits.

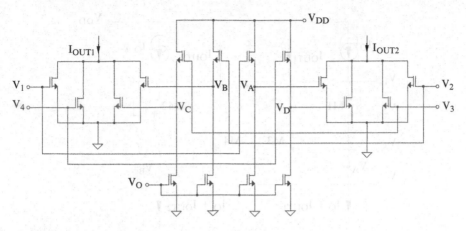

Fig. 1.81 DDA based on PR 1.D

The expressions of I_A and I_B currents are:

$$I_{OUT1} = \frac{K}{2}(V_1 - V_T)^2 + \frac{K}{2}(V_4 - V_T)^2 + \frac{K}{2}(V_C - V_T)^2 + \frac{K}{2}(V_B - V_T)^2 \quad (1.297)$$

and:

$$I_{OUT2} = \frac{K}{2}(V_2 - V_T)^2 + \frac{K}{2}(V_3 - V_T)^2 + \frac{K}{2}(V_A - V_T)^2 + \frac{K}{2}(V_D - V_T)^2 \quad (1.298)$$

The differential output current of the DDA circuit will have the following expressions:

$$I_{OUT1} - I_{OUT2} = \frac{K}{2}(V_1 - V_A)(V_1 + V_A - 2V_T)$$

$$+ \frac{K}{2}(V_4 - V_D)(V_4 + V_D - 2V_T)$$

$$+ \frac{K}{2}(V_B - V_2)(V_B + V_2 - 2V_T) + \frac{K}{2}(V_C - V_3)(V_C + V_3 - 2V_T) \quad (1.299)$$

Because:

$$V_O = V_1 - V_A = V_2 - V_B = V_3 - V_C = V_4 - V_D \quad (1.300)$$

it can be obtained:

$$I_{OUT1} - I_{OUT2} = KV_O[(V_4 - V_3) - (V_2 - V_1)] \quad (1.301)$$

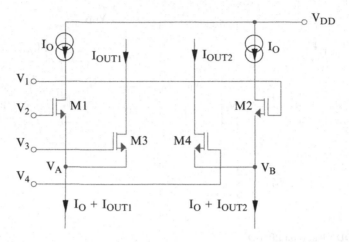

Fig. 1.82 Differential structure (1) based on PR 1.D

A linear transfer characteristic for a differential amplifier can be obtained using the circuit presented in Fig. 1.82 [36], the simplicity recommending it for a multitude of analog signal processing applications.

As M1 and M2 transistors from Fig. 1.82 are biased at I_O drain currents, V_A and V_B potentials can be expressed as follows:

$$V_A = V_2 - V_T - \sqrt{\frac{2I_O}{K}} \tag{1.302}$$

and:

$$V_B = V_1 - V_T - \sqrt{\frac{2I_O}{K}} \tag{1.303}$$

The expressions of I_{OUT1} and I_{OUT2} currents will be:

$$I_{OUT1} = \frac{K}{2}(V_3 - V_A - V_T)^2 \tag{1.304}$$

and:

$$I_{OUT2} = \frac{K}{2}(V_4 - V_B - V_T)^2 \tag{1.305}$$

Replacing (1.302) and (1.303) in (1.304) and (1.305), it results:

$$I_{OUT1} = \frac{K}{2}\left(V_3 - V_2 + \sqrt{\frac{2I_O}{K}}\right)^2 \tag{1.306}$$

Fig. 1.83 Differential
structure (2) based on PR 1.D

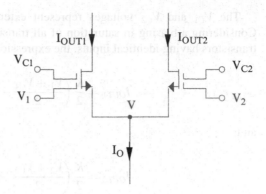

and:

$$I_{OUT2} = \frac{K}{2}\left(V_4 - V_1 + \sqrt{\frac{2I_O}{K}}\right)^2 \qquad (1.307)$$

The input potentials are chosen to have both common-mode and differential-mode components:

$$V_1 = V_C - V_{i1} \qquad (1.308)$$

$$V_2 = V_C + V_{i1} \qquad (1.309)$$

$$V_3 = V_C + V_{i2} \qquad (1.310)$$

and:

$$V_4 = V_C - V_{i2} \qquad (1.311)$$

resulting that the differential output current will have the following expression:

$$I_{OUT1} - I_{OUT2} = \frac{K}{2}\left(V_{i2} - V_{i1} + \sqrt{\frac{2I_O}{K}}\right)^2$$

$$-\frac{K}{2}\left(V_{i1} - V_{i2} + \sqrt{\frac{2I_O}{K}}\right)^2 = \sqrt{8KI_O}(V_{i2} - V_{i1}) \qquad (1.312)$$

A differential amplifier using FGMOS transistors for cancellation of the errors introduced by the input offset voltage is presented in Fig. 1.83 [37].

The V_{C1} and V_{C2} voltages represent external applied continuous voltages. Considering a biasing in saturation of all transistors from Fig. 1.83 and FGMOS transistors having identical inputs, the expressions of I_{OUT1} and I_{OUT2} currents are:

$$I_{OUT1} = \frac{K}{2}\left(\frac{V_1 + V_{C1}}{2} - V - V_T\right)^2 \tag{1.313}$$

and:

$$I_{OUT2} = \frac{K}{2}\left(\frac{V_2 + V_{C2}}{2} - V - V_T\right)^2 \tag{1.314}$$

resulting:

$$\sqrt{\frac{2}{K}}\left(\sqrt{I_{OUT1}} - \sqrt{I_{OUT2}}\right) = \frac{V_1 - V_2}{2} + \frac{V_{C1} - V_{C2}}{2} \tag{1.315}$$

Using the notations $V_{ID} = V_1 - V_2$ and $\Delta V_C = V_{C1} - V_{C2}$ and replacing the $I_{OUT1} + I_{OUT2}$ sum with I_O, the previous relation becomes:

$$\frac{8}{K}\left[I_O - 2\sqrt{I_{OUT1}(I_O - I_{OUT1})}\right] = (V_{ID} + \Delta V_C)^2 \tag{1.316}$$

After some computations, it results the following expressions of I_{OUT1} and I_{OUT2} currents:

$$I_{OUT1} = \frac{I_O}{2} + \frac{I_O}{2}(V_{ID} + \Delta V_C)\sqrt{\frac{K}{4I_O} - \frac{K^2}{64I_O^2}(V_{ID} + \Delta V_C)^2} \tag{1.317}$$

and:

$$I_{OUT2} = \frac{I_O}{2} - \frac{I_O}{2}(V_{ID} + \Delta V_C)\sqrt{\frac{K}{4I_O} - \frac{K^2}{64I_O^2}(V_{ID} + \Delta V_C)^2} \tag{1.318}$$

So, the differential output current of the circuit presented in Fig. 1.83 will have the following expression:

$$I_{OUT1} - I_{OUT2} = I_O(V_{ID} + \Delta V_C)\sqrt{\frac{K}{4I_O} - \frac{K^2}{64I_O^2}(V_{ID} + \Delta V_C)^2} \tag{1.319}$$

The expression is similar with the result obtained for the differential amplifier based on MOS transistors biased in saturation. The advantage of using FGMOS transistors for replacing classical MOS active devices is given by the possibility of compensating the input offset voltage of the differential amplifier by choosing a proper biasing voltage ΔV_C, complementary with the intrinsic offset voltage of the stage.

1.3 Conclusions

Chapter extensively presents a multitude of design techniques for improving the performances of CMOS differential structures. The main goals of these design methods are to improve the linearity of differential structures and to obtain a rail-to-rail operation, for a more efficient utilization of the available supply voltage. The functional mathematical principles of operation that represent the basis for designing the presented linearization techniques have been used in order to minimize the linearity error of the implemented CMOS differential structures.

References

1. Popa C (2006) An improved performances FGMOS voltage comparator for data acquisition systems. In: International conference on microelectronics, pp 420–423, Nis, Serbia and Montenegro
2. Popa C (2009) CMOS nanostructures with improved temperature behavior using double differential structures. In: International conference on sensor technologies and applications, pp 86–89, Athens, Greece
3. Filanovsky IM, Baltes H (1992) CMOS two-quadrant multiplier using transistor triode regime. IEEE J Solid-State Circuits 27:831–833
4. Ngamkham W, Kiatwarin N et al (2008) A linearized source-couple pair transconductor using a low-voltage square root circuit. In: International conference on electrical engineering/electronics, computer, telecommunications and information technology, pp 701–704, Krabi, Thailand
5. Popa C (2010) Improved linearity CMOS differential amplifiers with applications in VLSI designs. In: International symposium on electronics and telecommunications, pp 29–32, Timisoara, Romania
6. Manolescu AM, Popa C (2009) Low-voltage low-power improved linearity CMOS active resistor circuits. Springer J Analog Integr Circuits Signal Process 62:373–387
7. Popa C (2007) Improved performances linearization technique for CMOS differential structure. In: Instrumentation and measurement technology conference, pp 1–4, Warsaw, Poland
8. Popa C (2002) CMOS transconductor with extended linearity range. In: IEEE international conference on automation, quality and testing, robotics, pp 349–354, Cluj, Romania
9. Huang SC, Ismail M (1993) Linear tunable COMFET transconductor. Electron Lett 29:459–461
10. Aronhime P, Maundy BJ, Finvers IG (2000) Cross coupled transconductance cell with improved linearity range. IEEE international symposium on circuits and systems, pp 157–160, Geneva, Switzerland

11. Ramirez-Angulo J, Carvajal RG, Martinez-Heredia J (2000) 1.4 V supply, wide swing, high frequency CMOS analogue multiplier with high current efficiency. In: IEEE international symposium on circuits and systems, pp 533–536, Geneva, Switzerland

12. Farshidi E (2009) A low-voltage class-AB linear transconductance based on floating-gate MOS technology. In: European conference on circuit theory and design, pp 437–440, Antalya, Turkey

13. Mitrea O, Popa C, Manolescu AM, Glesner M (2003) A linearization technique for radio frequency CMOS Gilbert-type mixers. In: IEEE international conference on electronics, circuits and systems, pp 1086–1089, Dubrovnik, Croatia

14. Wang Z (1991) A CMOS four-quadrant analog multiplier with single-ended voltage output and improved temperature performance. IEEE J Solid-State Circuits 26:1293–1301

15. Klumperink E, van der Zwan E, Seevinck E (1989) CMOS variable transconductance circuit with constant bandwidth. Electron Lett 25:675–676

16. Kumar JV, Rao KR (2002) A low-voltage low power square-root domain filter. In: Asia-Pacific conference on circuits and systems, pp 375–378, Singapore

17. Zarabadi SR, Ismail M, Chung-Chih H (1998) High performance analog VLSI computational circuits. IEEE J Solid-State Circuits 33:644–649

18. Popa C (2009) High accuracy CMOS multifunctional structure for analog signal processing. In: International semiconductor conference, pp 427–430, Sinaia, Romania

19. De La Cruz Blas CA, Feely O (2008) Limit cycle behavior in a class-AB second-order square root domain filter. In: IEEE international conference on electronics, circuits and systems, pp 117–120, St. Julians, Malta

20. Zele RH, Allstot DJ, Fiez TS (1991) Fully-differential CMOS current-mode circuits and applications. IEEE international symposium on circuits and systems, pp 1817–1820, Raffles City, Singapore

21. Popa C (2002) A 0.35um CMOS linear differential amplifier independent of threshold voltage. In: International conference on advanced semiconductor devices and microsystems, pp 227–230, Slovakia

22. Sakurai S, Ismail M (1992) A CMOS square-law programmable floating resistor independent of the threshold voltage. IEEE Trans Circuits and Systems II, Analog Digit Signal Process 39:565–574

23. Demosthenous A, Panovic M (2005) Low-voltage MOS linear transconductor/squarer and four-quadrant multiplier for analog VLSI. IEEE Trans Circuits Syst I, Reg Pap 52:1721–1731

24. Lee BW, Sheu BJ (1990) A high slew-rate CMOS amplifier for analog signal processing. IEEE J Solid-State Circuits 25:885–889

25. Popa C, Manolescu AM (2007) CMOS differential structure with improved linearity and increased frequency response. In: International semiconductor conference, pp 517–520, Sinaia, Romania

26. Popa C (2004) 0.35um CMOS voltage references using threshold voltage extractors and offset voltage followers. In: International conference on optimization of electric and electronic equipment, pp 25–28, Brasov, Romania

27. Popa C (2007) CMOS nanostructure with auto-programmable thermal loop and superior-order curvature corrected technique. In: Instrumentation and measurement technology conference, pp 1–4, Warsaw, Poland

28. El Mourabit A, Lu GN, Pittet P (2005) Wide-linear-range subthreshold OTA for low-power, low-voltage, and low-frequency applications. IEEE Trans Circuits and Syst I, Reg Pap 52:1481–1488

29. Szczepanski S, Koziel S (2002) A 3.3 V linear fully balanced CMOS operational transconductance amplifier for high-frequency applications. In: IEEE international conference on circuits and systems for communications, pp 38–41, St. Petersburg, Russia

30. Popa C (2008) Programmable CMOS active resistor using computational circuits. In: International semiconductor conference, pp 389–392, Sinaia, Romania

31. Manolescu AM, Popa C (2011) A 2.5 GHz CMOS mixer with improved linearity. J Circuits 20:233–242
32. Popa C, Coada D (2003) A new linearization technique for a CMOS differential amplifier using bulk-driven weak-inversion MOS transistors. In: International symposium on circuits and systems, pp 589–592, Iasi, Romania
33. Botma JH, Wassenaar RF, Wiegerink RJ (1993) A low-voltage CMOS op amp with a rail-to-rail constant-g_m input stage and a class AB rail-to-rail output stage. In: IEEE international symposium on circuits and systems, pp 1314–1317, Chicago, USA
34. Chung-Chih H, Ismail M, Halonen K, Porra V (1997) Low-voltage rail-to-rail CMOS differential difference amplifier. IEEE international symposium on circuits and systems, pp 145–148, Hong Kong
35. Mahmoud SA, Soliman AM (1998) The differential difference operational floating amplifier: a new block for analog signal processing in MOS technology. IEEE Trans Circuits Syst II, Analog Digit Signal Process 45:148–158
36. Kimura K (1994) Analysis of "An MOS four-quadrant analog multiplier using simple two-input squaring circuits with source followers". IEEE Trans Circuits Syst I, Fundam Theory Appl 41:72–75
37. Babu VS, Rose KAA, Baiju MR (2008) Adaptive neuron activation function with FGMOS based operational transconductance amplifier. In: IEEE computer society annual symposium on VLSI, pp 353–356, Montpellier, France
38. Vlassis S, Siskos S (2000) Current-mode non-linear building blocks based on floating-gate transistors. IEEE In: International symposium on circuits and systems, pp 521–524, Geneva, Switzerland
39. Abbasi M, Kjellberg T, et al (2010) A broadband differential cascode power amplifier in 45 nm CMOS for high-speed 60 GHz system-on-chip. In: IEEE radio frequency integrated circuits symposium, pp 533–536, Anaheim, USA
40. Yonghui J, Ming L, et al (2010) A low power single ended input differential output low noise amplifier for L1/L2 band. In: IEEE international symposium on circuits and systems, pp 213–216, Paris, France
41. Ong GT, Chan PK (2010) A micropower gate-bulk driven differential difference amplifier with folded telescopic cascode topology for sensor applications. In: IEEE international midwest symposium on circuits and systems, pp 193–196, Seattle, USA
42. Vaithianathan V, Raja J, Kavya R, Anuradha N (2010) A 3.1 to 4.85 GHz differential CMOS low noise amplifier for lower band of UWB applications. In: International conference on wireless communication and sensor computing, pp 1–4, Chennai, India
43. Mandai S, Nakura T, Ikeda M, Asada K (2010) Cascaded time difference amplifier using differential logic delay cell. In: Asia and South Pacific design automation conference, pp 355–356, Taipei, Taiwan
44. Popa C (2008) Linearity evaluation technique for CMOS differential amplifier. In: International conference on microelectronics, pp 451–454, Nis, Serbia
45. Popa C (2007) CMOS integrated circuit with improved temperature behavior based on a temperature optimized auto-programmable loop. In: International conference on "computer as a tool", pp 245–249, Warsaw, Poland
46. Dermentzoglou LE, Arapoyanni A, Tsiatouhas Y (2010) A built-in-test circuit for RF differential low noise amplifiers. IEEE Trans Circuits Syst I, Reg Pap 57:1549–1558
47. Figueiredo M, Santin E, et al (2010) Two-stage fully-differential inverter-based self-biased CMOS amplifier with high efficiency. In: International symposium on circuits and systems, pp 2828–2831, Paris, France
48. Enche Ab, Rahim SAE, Ismail MA et al (2010) A wide gain-bandwidth CMOS fully-differential folded cascode amplifier. In: International conference on electronic devices, systems and applications, pp 165–168, Kuala Lumpur, Malaysia

49. Chanapromma C, Daoden K (2010) A CMOS fully differential operational transconductance amplifier operating in sub-threshold region and its application. In: International conference on signal processing systems, pp V2-73–V2-77, Yantai, China
50. Rajput KK, Saini AK, Bose SC (2010) DC offset modeling and noise minimization for differential amplifier in subthreshold operation. In: IEEE computer society annual symposium on VLSI, pp 247–252, Lixouri Kefalonia, Greece
51. Bajaj N, Vermeire B, Bakkaloglu B (2010) A 10 MHz to 100 MHz bandwidth scalable, fully differential current feedback amplifier. In: IEEE international symposium on circuits and systems, pp 217–220, Paris, France
52. Harb A (2010) A rail-to-rail full clock fully differential rectifier and sample-and-hold amplifier. In: IEEE international symposium on circuits and systems, pp 1571–1574, Paris, France
53. C, Zhiqun L et al (2010) A 10-Gb/s CMOS differential transimpedance amplifier for parallel optical receiver. In: International symposium on signals systems and electronics, pp 1–4, Nanjing, China
54. Uhrmann H, Zimmermann H (2009) A fully differential operational amplifier for a low-pass filter in a DVB-H receiver. In: International conference on mixed design of integrated circuits and systems, pp 197–200, Lodz, Poland
55. Popa C (2002) A 0.35um low-power CMOS differential amplifier with improved linearity and extended input range. In: International workshop on symbolic methods and applications to circuit design, pp 61–64, Sinaia, Romania

Chapter 2
Voltage and Current Multiplier Circuits

2.1 Mathematical Analysis for Synthesis of Multipliers

The synthesis of multiplier circuits [1–60] is based on the utilization of some elementary principles, each of them representing the starting point for designing a class of multiplier circuit.

Referring to the input variables, it can be identified two important classes of multiplier circuits:

- Voltage multipliers, having as input variables two single or differential voltages and generating an output current proportional with the product of these input voltages;
- Current multipliers, receiving as inputs two currents and producing an output current proportional with the product of the input currents.

Because the multiplying function uses the characteristic of MOS transistors biased in saturation region, most of mathematical principles are derived from a linear relation between squaring terms (having voltages or currents as variables). The notations used for revealing these principles are: V_1, V_2, V_3 and V_4 represent the input potentials, while, usually, a constant voltage, V_O, is introduced for modeling a voltage shifting; for current multipliers, I_1, I_2 and I_3 are the input currents and I_O represents a reference current. In both cases, I_{OUT} denotes the output current of the multiplier circuit.

C.R. Popa, *Synthesis of Computational Structures for Analog Signal Processing*,
DOI 10.1007/978-1-4614-0403-3_2, © Springer Science+Business Media, LLC 2011

2.1.1   Mathematical Analysis of Voltage Multiplier Circuits

2.1.1.1   First Mathematical Principle (PR 2.1)

The first mathematical principle used for implementing voltage multiplier circuits is based on the following identity:

$$(V_1 + V_2 + V_O)^2 - (V_1 - V_2 + V_O)^2 + (-V_1 - V_2 + V_O)^2$$
$$- (-V_1 + V_2 + V_O)^2 = 2V_2(2V_1 + 2V_O) - 2V_2(-2V_1 + 2V_O) = 8V_1V_2 \quad (2.1)$$

The voltage multipliers based on the previous relation computes a current proportional with the product of two input voltages, V_1 and V_2.

2.1.1.2   Second Mathematical Principle (PR 2.2)

The mathematical relation that models this principle is

$$(V_1 + V_3 + V_O)^2 - (V_2 + V_3 + V_O)^2 + (V_2 + V_4 + V_O)^2$$
$$- (V_1 + V_4 + V_O)^2 = (V_1 - V_2)(V_1 + V_2 + 2V_3 + 2V_O)$$
$$- (V_1 - V_2)(V_1 + V_2 + 2V_4 + 2V_O) = 2(V_1 - V_2)(V_3 - V_4) \quad (2.2)$$

The circuits that use this principle generates a current proportional with the product between two differential input voltages, $V_1 - V_2$ and $V_3 - V_4$.

2.1.1.3   Third Mathematical Principle (PR 2.3)

This principle is illustrated by the following mathematical relation:

$$(V_1 - V_3 + V_O)^2 - (V_1 - V_4 + V_O)^2 + (V_2 - V_4 + V_O)^2$$
$$- (V_2 - V_3 + V_O)^2 = (V_4 - V_3)(2V_1 - V_3 - V_4 + 2V_O)$$
$$- (V_4 - V_3)(2V_2 - V_3 - V_4 + 2V_O) = 2(V_1 - V_2)(V_4 - V_3) \quad (2.3)$$

The multiplier circuits that implement this principle compute, also, an output current proportional with the product between two differential input voltages, $V_1 - V_2$ and $V_4 - V_3$.

2.1.1.4 Fourth Mathematical Principle (PR 2.4)

The fourth mathematical principle is based on the following mathematical relation:

$$I_{OUT} = a\sqrt{I_O}(V_1 - V_2)$$
$$\Rightarrow I_{OUT} = a\sqrt{b}(V_1 - V_2)(V_3 - V_4) \qquad (2.4)$$
$$I_O = b(V_3 - V_4)^2$$

or

$$I_{OUT} = a\left(\sqrt{I_1} - \sqrt{I_2}\right)(V_1 - V_2)$$
$$\Rightarrow I_{OUT} = a\sqrt{b}(V_1 - V_2)(V_3 - V_4) \qquad (2.5)$$
$$\left(\sqrt{I_1} - \sqrt{I_2}\right)^2 = b(V_3 - V_4)^2$$

a and b represent constant coefficients, depending on the particular implementation of the multiplier circuit based on this mathematical principle.

2.1.1.5 Fifth Mathematical Principle (PR 2.5)

The fifth mathematical principle represents the cancellation of a nonlinear dependence of an output current, I_{OUT}, on the input voltage, V_1:

$$I_{OUT} = KV_1\sqrt{\frac{I_O + KV_1^2/4}{K} - \frac{V_1^2}{4}} = \sqrt{KI_O}V_1 \qquad (2.6)$$

2.1.1.6 Sixth Mathematical Principle (PR 2.6)

The mathematical relation that models this principle is

$$\left(\frac{V_1 + V_2}{2} - V_O\right)^2 + V_O^2 - \left(\frac{V_1}{2} - V_O\right)^2 - \left(\frac{V_2}{2} - V_O\right)^2 = V_1 V_2 \qquad (2.7)$$

2.1.1.7 Seventh Mathematical Principle (PR 2.7)

The identity representing the basis of this mathematical relation is

$$(V_1 + V_2)^2 - (V_1 - V_2)^2 = 4V_1 V_2 \qquad (2.8)$$

2.1.1.8 Different Mathematical Principles for Voltage Multipliers (PR 2.Da)

A class of multipliers can be designed starting from different mathematical principles, that are useful for linearizing the behavior of the multiplier circuits.

2.1.2 Mathematical Analysis of Current Multiplier Circuits

2.1.2.1 Ninth Mathematical Principle (PR 2.9)

The ninth mathematical principle uses two square-rooting circuits in order to implement the multiplying function:

$$I_{O1} = a\sqrt{I_O I_{OUT}}$$
$$I_{O2} = a\sqrt{I_1 I_2} \qquad \Rightarrow I_{OUT} = \frac{I_1 I_2}{I_O}$$
$$I_{O1} = I_{O2} \tag{2.9}$$

2.1.2.2 Tenth Mathematical Principle (PR 2.10)

The identity representing the basis of this mathematical relation is

$$\left[I_O + \frac{(I_1 + I_2)^2}{aI_O}\right] - \left[I_O - \frac{(I_1 + I_2)^2}{aI_O}\right] = \frac{4I_1 I_2}{aI_O} \tag{2.10}$$

2.1.2.3 Eleventh Mathematical Principle (PR 2.11)

In order to implement the eleventh mathematical principle, the circuits use only MOS transistors biased in weak inversion region, the translinear loops that contain gate-source voltages generating the product between the input currents.

2.1.2.4 Different Mathematical Principles for Current Multipliers (PR 2.Db)

A class of multipliers can be designed starting from different mathematical principles that are useful for linearizing the behavior of the current multiplier circuits.

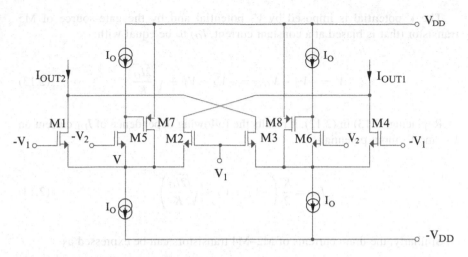

Fig. 2.1 Multiplier circuit (1) based on PR 2.1

2.2 Analysis and Design of Multiplier Circuits

2.2.1 Design of Voltage Multiplier Circuits

Based on the previous presented mathematical analysis, the voltage multipliers can be clustered in eight important functional classes: circuits using PR 2.1–PR 2.7 elementary principle and a class containing multipliers based on different functional relations.

2.2.1.1 Multiplier Circuits Based on the First Mathematical Principle (PR 2.1)

The multiplier structures using as functional basis PR 2.1 present the important advantage of using a symmetrical structure that minimizes the intrinsic linearity error of the designed circuits.

A circuit that implements the product between two input voltages using the PR 2.1 mathematical principle is presented in Fig. 2.1 [1].

The output current of the voltage multiplier can be expressed as a linear function of the currents, I_{OUT1} and I_{OUT2}:

$$I_{OUT} = I_{OUT1} - I_{OUT2} = (I_{D2} + I_{D4}) - (I_{D1} + I_{D3}) \qquad (2.11)$$

The drain current of M1 transistor is

$$I_{D1} = \frac{K}{2}(-V_1 - V - V_T)^2 \qquad (2.12)$$

The V potential is imposed by V_2 potential and by the gate-source of M5 transistor (that is biased at a constant current, I_O) to be equal with:

$$V = -V_2 - V_{GS5} = -V_2 - V_T - \sqrt{\frac{2I_O}{K}} \qquad (2.13)$$

Replacing (2.13) in (2.12), it results the following dependence of I_{D1} current on V_1, and V_2 input potentials:

$$I_{D1} = \frac{K}{2}\left(-V_1 + V_2 + \sqrt{\frac{2I_O}{K}}\right)^2 \qquad (2.14)$$

Similarly, the drain currents of M2–M4 transistors can be expressed as

$$I_{D2} = \frac{K}{2}\left(V_1 + V_2 + \sqrt{\frac{2I_O}{K}}\right)^2 \qquad (2.15)$$

$$I_{D3} = \frac{K}{2}\left(V_1 - V_2 + \sqrt{\frac{2I_O}{K}}\right)^2 \qquad (2.16)$$

$$I_{D4} = \frac{K}{2}\left(-V_1 - V_2 + \sqrt{\frac{2I_O}{K}}\right)^2 \qquad (2.17)$$

The expression of the output current as a function of input potentials can be obtained using relations (2.11) and (2.14)–(2.17):

$$I_{OUT} = \frac{K}{2}2V_1\left(2V_2 + 2\sqrt{\frac{2I_O}{K}}\right) + \frac{K}{2}(-2V_1)\left(-2V_2 + 2\sqrt{\frac{2I_O}{K}}\right) = 4KV_1V_2 \quad (2.18)$$

Another multiplier structure based on the first mathematical principle (PR 2.1) is presented in Fig. 2.2 [2]. Its output current can be expressed as

$$I_{OUT} = (I_{D1} + I_{D3}) - (I_{D2} + I_{D4}) = (I_{D1} - I_{D2}) + (I_{D3} - I_{D4}) \qquad (2.19)$$

For M1–M4 transistors, the gate potentials are imposed by the common-mode voltage V_C and by the differential components $\pm V_1/2$ and $\pm V_2/2$.

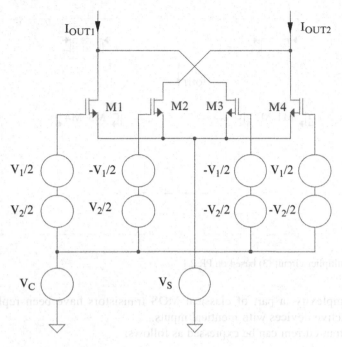

Fig. 2.2 Multiplier circuit (2) based on PR 2.1

Replacing the expressions of the previous drain currents with their quadratic dependence on the gate-source voltages, it results:

$$I_{OUT} = I_{OUT1} - I_{OUT2} = \frac{K}{2}\left(\frac{V_1 + V_2}{2} + V_C - V_S - V_T\right)^2$$

$$+ \frac{K}{2}\left(-\frac{V_1 + V_2}{2} + V_C - V_S - V_T\right)^2 - \frac{K}{2}\left(\frac{V_1 - V_2}{2} + V_C - V_S - V_T\right)^2$$

$$- \frac{K}{2}\left(\frac{V_2 - V_1}{2} + V_C - V_S - V_T\right)^2 \qquad (2.20)$$

So, the output current will be proportional with the product between the differential-mode input voltages:

$$I_{OUT} = \frac{K}{2}V_1(V_2 + 2V_C - 2V_S - 2V_T)$$

$$+ \frac{K}{2}(-V_1)(-V_2 + 2V_C - 2V_S - 2V_T) = KV_1V_2 \qquad (2.21)$$

An alternate realization of a voltage multiplier circuit based on the first mathematical principle (PR 2.1) is presented in Fig. 2.3. In order to reduce the

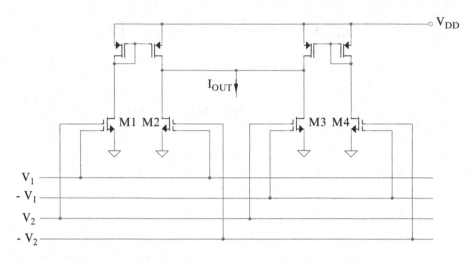

Fig. 2.3 Multiplier circuit (3) based on PR 2.1

circuit complexity, a part of classical MOS transistors have been replaced by
FGMOS active devices with identical inputs.

The output current can be expressed as follows:

$$I_{OUT} = (I_{D1} - I_{D2}) + (I_{D4} - I_{D3}) = \frac{K}{2}\left(\frac{V_1 + V_2}{2} - V_T\right)^2$$

$$-\frac{K}{2}\left(\frac{V_1 - V_2}{2} - V_T\right)^2 + \frac{K}{2}\left(-\frac{V_1 + V_2}{2} - V_T\right)^2$$

$$-\frac{K}{2}\left(\frac{V_2 - V_1}{2} - V_T\right)^2 = \frac{K}{2}V_2(V_1 - 2V_T)\frac{K}{2}V_2(-V_1 - 2V_T) = KV_1V_2 \quad (2.22)$$

2.2.1.2   Multiplier Circuits Based on the Second Mathematical Principle (PR 2.2)

The multipliers based on PR 2.2 can be used in a large area of applications that
require the implementation of the product between two differential voltages.

A voltage multiplier having as functional relation the second mathematical
principle (PR 2.2) is presented in Fig. 2.4 [3].

All MOS transistors are biased in saturation region and V_O represents a constant
voltage that is summed with V_3 and V_4 voltages. The expression of differential
output current is

$$I_{OUT1} - I_{OUT2} = (I_{D5} + I_{D16}) - (I_{D8} + I_{D13}) \quad (2.23)$$

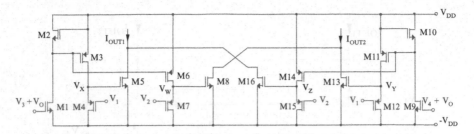

Fig. 2.4 Multiplier circuit (1) based on PR 2.2

where the previous drain currents can be expressed as follows:

$$I_{D5} = \frac{K}{2}(V_X + V_{DD} - V_T)^2 \tag{2.24}$$

$$I_{D16} = \frac{K}{2}(V_Z + V_{DD} - V_T)^2 \tag{2.25}$$

$$I_{D8} = \frac{K}{2}(V_W + V_{DD} - V_T)^2 \tag{2.26}$$

$$I_{D13} = \frac{K}{2}(V_Y + V_{DD} - V_T)^2 \tag{2.27}$$

As a consequence of the circuit configuration, the gate-source voltages of M2 and M3 transistors are equal, resulting $I_{D2} = I_{D3}$. But $I_{D3} = I_{D4}$ and $I_{D1} = I_{D2}$, so $I_{D1} = I_{D4}$. In conclusion, because M1 and M4 transistors are identical, it results $V_{GS1} = V_{SG4}$, equivalent with:

$$V_3 + V_O + V_{DD} = V_X - V_1 \tag{2.28}$$

Thus, the expression of V_X potential will be

$$V_X = V_1 + V_3 + V_O + V_{DD} \tag{2.29}$$

Replacing (2.29) in (2.24), the drain current of M5 transistor will have the following expression:

$$I_{D5} = \frac{K}{2}[(V_1 + V_3) + (2V_{DD} + V_O - V_T)]^2 \tag{2.30}$$

Similarly, the expression of drain currents of M8, M13 and M16 transistor will be

$$I_{D8} = \frac{K}{2}[(V_2 + V_3) + (2V_{DD} + V_O - V_T)]^2 \tag{2.31}$$

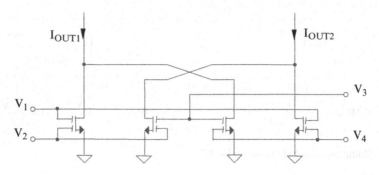

Fig. 2.5 Multiplier circuit (2) based on PR 2.2

$$I_{D13} = \frac{K}{2}[(V_1 + V_4) + (2V_{DD} + V_O - V_T)]^2 \tag{2.32}$$

$$I_{D16} = \frac{K}{2}[(V_2 + V_4) + (2V_{DD} + V_O - V_T)]^2 \tag{2.33}$$

Replacing (2.30)–(2.33) in (2.23), it results:

$$I_{OUT1} - I_{OUT2} = \frac{K}{2}(V_1 - V_2)(V_1 + V_2 + 2V_3 + 4V_{DD} + 2V_O - 2V_T)$$
$$- \frac{K}{2}(V_1 - V_2)(V_1 + V_2 + 2V_4 + 4V_{DD} + 2V_O - 2V_T) \tag{2.34}$$

So

$$I_{OUT1} - I_{OUT2} = K(V_1 - V_2)(V_3 - V_4) \tag{2.35}$$

A possible implementation of a voltage multiplier based on the second mathematical principle (PR 2.2), using FGMOS transistors is presented in Fig. 2.5 [4].

Considering identical inputs for all FGMOS transistors, the differential output current of the voltage multiplier presented in Fig. 2.5 can be expressed as follows:

$$I_{OUT} = I_{OUT1} - I_{OUT2} = \frac{K}{2}\left(\frac{V_1 + V_2}{2} - V_T\right)^2 + \frac{K}{2}\left(\frac{V_3 + V_4}{2} - V_T\right)^2$$
$$- \frac{K}{2}\left(\frac{V_2 + V_3}{2} - V_T\right)^2 - \frac{K}{2}\left(\frac{V_1 + V_4}{2} - V_T\right)^2 = \frac{K}{4}(V_1 - V_3)(V_2 - V_4) \tag{2.36}$$

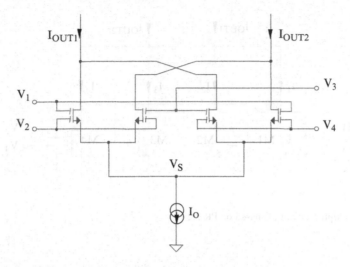

Fig. 2.6 Multiplier circuit (3) based on PR 2.2

An alternate implementation of a voltage multiplier using the second mathematical principle is presented in Fig. 2.6 [4].

The differential output current of the voltage multiplier shown in Fig. 2.6 will have the following expression:

$$I_{OUT} = I_{OUT1} - I_{OUT2} = \frac{K}{2}\left(\frac{V_1 + V_2}{2} - V_S - V_T\right)^2$$

$$+ \frac{K}{2}\left(\frac{V_3 + V_4}{2} - V_S - V_T\right)^2 - \frac{K}{2}\left(\frac{V_2 + V_3}{2} - V_S - V_T\right)^2$$

$$- \frac{K}{2}\left(\frac{V_1 + V_4}{2} - V_S - V_T\right)^2 = \frac{K}{4}(V_1 - V_3)(V_2 - V_4) \qquad (2.37)$$

The multiplier circuit presented in Fig. 2.7 [5] is based on the second mathematical principle (PR 2.2).

The output current of the voltage multiplier is implemented (using an additional current mirror, not shown in Fig. 2.7) to be the difference between I_{OUT2} and I_{OUT1} currents and it can be expressed as follows:

$$I_{OUT} = I_{OUT2} - I_{OUT1} = (I_1 + I_4) - (I_2 + I_3)$$

$$= \frac{K}{2}\left(\frac{V_1 + V_3 + V_G}{3} - V_T\right)^2 + \frac{K}{2}\left(\frac{V_2 + V_4 + V_G}{3} - V_T\right)^2$$

$$- \frac{K}{2}\left(\frac{V_1 + V_4 + V_G}{3} - V_T\right)^2 - \frac{K}{2}\left(\frac{V_2 + V_3 + V_G}{3} - V_T\right)^2 \qquad (2.38)$$

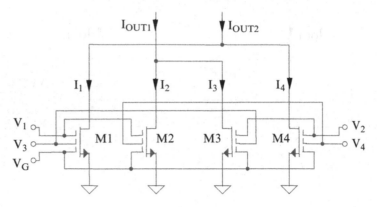

Fig. 2.7 Multiplier circuit (4) based on PR 2.2

resulting:

$$I_{OUT} = \frac{K}{2} \frac{V_3 - V_4}{3} \left(\frac{2V_1 + V_3 + V_4 + 2V_G}{3} - 2V_T \right)$$

$$- \frac{K}{2} \frac{V_3 - V_4}{3} \left(\frac{2V_2 + V_3 + V_4 + 2V_G}{3} - 2V_T \right) \quad (2.39)$$

So, the output current is proportional with the product between the differential input voltages:

$$I_{OUT} = \frac{K}{9} (V_3 - V_4)(V_1 - V_2) \quad (2.40)$$

2.2.1.3 Multiplier Circuits Based on the Third Mathematical Principle (PR 2.3)

Alternative implementations of multiplier circuits designed for differential input voltages uses the mathematical relations described by PR 2.3.

A combination of two differential amplifiers, M1–M2 and M3–M4, can implement the multiplying function (Fig. 2.8), the functional equations of this circuit being obtained using the third mathematical principle (PR 2.3) [6].

The gate-source voltages of M1 and M5 transistors are equal because they are identical and biased at the same drain current, resulting:

$$V_{DD} - V_3 = V - V_1 \quad (2.41)$$

equivalent with:

$$V = V_1 - V_3 + V_{DD} \quad (2.42)$$

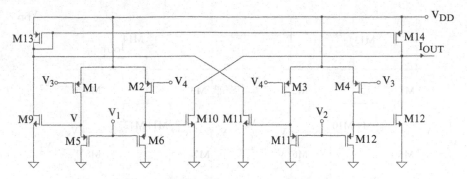

Fig. 2.8 Multiplier circuit (1) based on PR 2.3

The drain current of M9 transistor can be expressed as follows:

$$I_{D9} = \frac{K}{2}(V_{GS9} - V_T)^2 = \frac{K}{2}(V - V_T)^2 = \frac{K}{2}[(V_1 - V_3) + (V_{DD} - V_T)]^2 \quad (2.43)$$

Similarly, the expressions of drain currents for M10, M11 and M12 transistors are:

$$I_{D10} = \frac{K}{2}[(V_1 - V_4) + (V_{DD} - V_T)]^2 \quad (2.44)$$

$$I_{D11} = \frac{K}{2}[(V_2 - V_4) + (V_{DD} - V_T)]^2 \quad (2.45)$$

$$I_{D12} = \frac{K}{2}[(V_2 - V_3) + (V_{DD} - V_T)]^2 \quad (2.46)$$

The output current I_{OUT} of the multiplier will be expressed by

$$I_{OUT} = I_{D9} + I_{D11} - I_{D10} - I_{D12} \quad (2.47)$$

resulting:

$$I_{OUT} = \frac{K}{2}(V_4 - V_3)(2V_1 - V_3 - V_4 + 2V_{DD} - 2V_T)$$
$$+ \frac{K}{2}(V_3 - V_4)(2V_2 - V_3 - V_4 + 2V_{DD} - 2V_T) \quad (2.48)$$

or

$$I_{OUT} = K(V_2 - V_1)(V_3 - V_4) \quad (2.49)$$

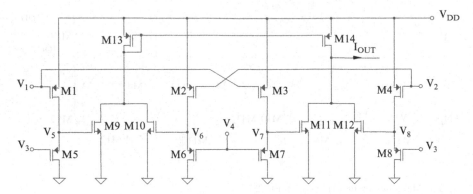

Fig. 2.9 Multiplier circuit (2) based on PR 2.3

The voltage multiplying function can be implemented using the third mathematical principle (PR 2.3) by the structure presented in Fig. 2.9 [7].

The circuit contains four pairs of transistors (M1–M5, M2–M6, M3–M7 and M4–M8) that implement voltage subtraction functions. Because M1 and M5 transistors are identical and biased at the same drain current, their gate-source voltages will be equal, so that:

$$V_{DD} - V_1 = V_5 - V_3 \tag{2.50}$$

resulting that V_5 potential is given by the differential input voltage, $V_3 - V_1$:

$$V_5 = V_{DD} + (V_3 - V_1) \tag{2.51}$$

The V_5 potential is applied on the gate of M9 transistor, its drain current being expressed using the squaring characteristic of the MOS transistor biased in saturation:

$$I_{D9} = \frac{K}{2}(V_5 - V_T)^2 = \frac{K}{2}[(V_{DD} - V_T) + (V_3 - V_1)]^2 \tag{2.52}$$

Similarly, the drain currents of M10, M11 and M12 transistors are

$$I_{D10} = \frac{K}{2}(V_6 - V_T)^2 = \frac{K}{2}[(V_{DD} - V_T) + (V_4 - V_2)]^2 \tag{2.53}$$

$$I_{D11} = \frac{K}{2}(V_7 - V_T)^2 = \frac{K}{2}[(V_{DD} - V_T) + (V_4 - V_1)]^2 \tag{2.54}$$

$$I_{D12} = \frac{K}{2}(V_8 - V_T)^2 = \frac{K}{2}[(V_{DD} - V_T) + (V_3 - V_2)]^2 \tag{2.55}$$

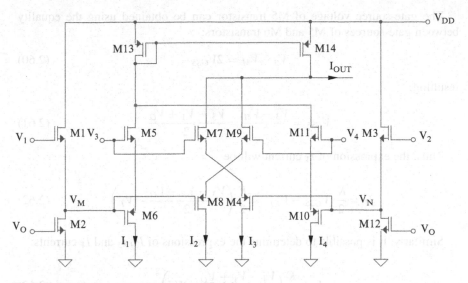

Fig. 2.10 Multiplier circuit (3) based on PR 2.3

The output current of the voltage multiplier presented in Fig. 2.9 can be expressed as a linear relation using the previous currents:

$$I_{OUT} = I_{D9} + I_{D10} - I_{D11} - I_{D12} \tag{2.56}$$

resulting:

$$I_{OUT} = \frac{K}{2}(V_3 - V_4)(2V_{DD} - 2V_T + V_3 + V_4 - 2V_1)$$
$$+ \frac{K}{2}(V_4 - V_3)(2V_{DD} - 2V_T + V_3 + V_4 - 2V_2) \tag{2.57}$$

So, the circuit implements the voltage multiplying function:

$$I_{OUT} = K(V_3 - V_4)(V_2 - V_1) \tag{2.58}$$

A multiplier circuit using exclusively MOS transistors biased in saturation region, based on the third mathematical principle (PR 2.3), is presented in Fig. 2.10 [8].

The M1 and M2 transistors form a difference circuit that generates V_M potential. Considering identical transistors and because they are biased at the same drain current, it results equal gate-source voltages for these transistors. Thus, V_M potential will have the following expression:

$$V_M = V_1 - V_{GS1} = V_1 - V_{GS2} = V_1 - V_O \tag{2.59}$$

The gate-source voltage of M5 transistor can be obtained using the equality between gate-sources of M5 and M6 transistors:

$$V_3 - V_M = 2V_{GS5} \tag{2.60}$$

resulting:

$$V_{GS5} = \frac{V_3 - V_M}{2} = \frac{V_3 - V_1 + V_O}{2} \tag{2.61}$$

Thus, the expression of I_1 current will be

$$I_1 = \frac{K}{2}(V_{GS5} - V_T)^2 = \frac{K}{2}\left(\frac{V_3 - V_1 + V_O}{2} - V_T\right)^2 \tag{2.62}$$

Similarly, it is possible to determine the expressions of I_2, I_3 and I_4 currents:

$$I_2 = \frac{K}{2}\left(\frac{V_4 - V_1 + V_O}{2} - V_T\right)^2 \tag{2.63}$$

$$I_3 = \frac{K}{2}\left(\frac{V_3 - V_2 + V_O}{2} - V_T\right)^2 \tag{2.64}$$

$$I_4 = \frac{K}{2}\left(\frac{V_4 - V_2 + V_O}{2} - V_T\right)^2 \tag{2.65}$$

The output current can be expressed as a linear function of the previous currents:

$$I_{OUT} = I_1 + I_4 - I_2 - I_3 \tag{2.66}$$

resulting:

$$\begin{aligned}I_{OUT} = &\frac{K}{2}\frac{V_2 - V_1}{2}\left(\frac{2V_3 - V_1 - V_2 + 2V_O}{2} - 2V_T\right) \\ &- \frac{K}{2}\frac{V_2 - V_1}{2}\left(\frac{2V_4 - V_1 - V_2 + 2V_O}{2} - 2V_T\right)\end{aligned} \tag{2.67}$$

So

$$I_{OUT} = \frac{K}{4}(V_2 - V_1)(V_3 - V_4) \tag{2.68}$$

The multiplier circuit presented in Fig. 2.11 [9] implements the same mathematical principle PR 2.3 and it is composed from two differential amplifiers, M1–M4

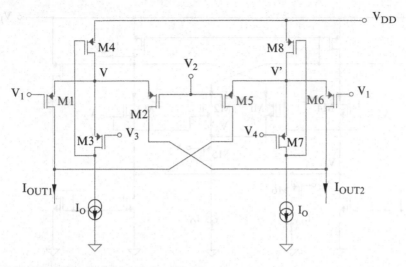

Fig. 2.11 Multiplier circuit (4) based on PR 2.3

and M5–M8. For obtaining a low impedance in the common-source points, each differential amplifier uses a flipped voltage follower (M3–M4 and, M7–M8, respectively). The M4 and M8 transistors absorb current variations of M1–M2 differential pairs and M5–M6.

The differential output current of the first differential amplifier, M1–M2 can be expressed using the squaring law of MOS transistors biased in saturation region:

$$I_{D1} - I_{D2} = \frac{K}{2}(V - V_1 - V_T)^2 - \frac{K}{2}(V - V_2 - V_T)^2$$

$$= \frac{K}{2}(V_2 - V_1)(2V - V_1 - V_2 - 2V_T) \tag{2.69}$$

Similarly, the expression of the differential output current for the second differential amplifier, M5–M6 is

$$I_{D5} - I_{D6} = \frac{K}{2}(V_1 - V_2)(2V' - V_1 - V_2 - 2V_T) \tag{2.70}$$

The differential output current of the entire multiplier circuit presented in Fig. 2.11 will have the following expression:

$$I_{OUT1} - I_{OUT2} = (I_{D1} + I_{D5}) - (I_{D2} + I_{D6}) = (I_{D1} - I_{D2}) + (I_{D5} - I_{D6}) \tag{2.71}$$

Replacing (2.69) and (2.70) in (2.71), it results:

$$I_{OUT1} - I_{OUT2} = K(V_2 - V_1)(V - V') \tag{2.72}$$

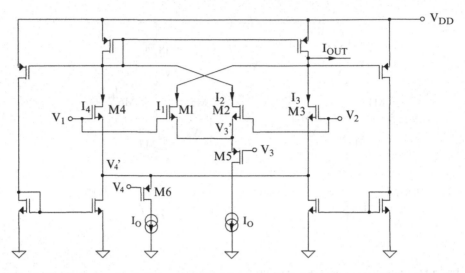

Fig. 2.12 Multiplier circuit (5) based on PR 2.3

Because M3 and M7 transistors are biased at a constant current, I_O, imposed by external current generators, the V and V' potentials can be expressed as follows:

$$V = V_3 + V_{SG3} = V_3 + V_T + \sqrt{\frac{2I_O}{K}} \qquad (2.73)$$

and

$$V' = V_4 + V_{SG7} = V_4 + V_T + \sqrt{\frac{2I_O}{K}} \qquad (2.74)$$

From (2.72), (2.73) and (2.74), it results the multiplying function implemented by the circuit from Fig. 2.11:

$$I_{OUT1} - I_{OUT2} = K(V_2 - V_1)(V_3 - V_4) \qquad (2.75)$$

A possible realization of a voltage multiplier, based on the third mathematical principle (PR 2.3) is presented in Fig. 2.12 [10]. The core of the circuit is represented by the group of M1–M4 transistors. Using identical devices, their drain currents will have the following expressions:

$$I_1 = \frac{K}{2}(V_1 - V_3' - V_T)^2 \qquad (2.76)$$

$$I_2 = \frac{K}{2}(V_2 - V_3' - V_T)^2 \qquad (2.77)$$

$$I_3 = \frac{K}{2}(V_2 - V_4' - V_T)^2 \tag{2.78}$$

$$I_4 = \frac{K}{2}(V_1 - V_4' - V_T)^2 \tag{2.79}$$

As a result of using additional current mirrors, the output current of the differential structure will have a linear variation with respect to $I_1 - I_4$ currents:

$$I_{OUT} = (I_2 - I_1) + (I_4 - I_3) \tag{2.80}$$

resulting:

$$I_{OUT} = \frac{K}{2}(V_2 - V_1)(V_1 + V_2 - 2V_3' - 2V_T)$$
$$+ \frac{K}{2}(V_1 - V_2)(V_1 + V_2 - 2V_4' - 2V_T) = K(V_1 - V_2)(V_3' - V_4') \tag{2.81}$$

Because M5 and M6 transistors are biased at the constant current, I_O, they will introduce a voltage shifting between V_3 and V_3' and, respectively, between V_4 and V_4' potentials, as follows:

$$V_3' = V_3 + V_{SG5} = V_3 + V_T + \sqrt{\frac{2I_O}{K}} \tag{2.82}$$

and

$$V_4' = V_4 + V_{SG6} = V_4 + V_T + \sqrt{\frac{2I_O}{K}} \tag{2.83}$$

From the previous relations, it results the following expression of the output current:

$$I_{OUT} = K(V_1 - V_2)(V_3 - V_4) \tag{2.84}$$

A voltage multiplication that illustrates the third mathematical principle PR 2.3 can be designed using 4 v squaring circuits (Fig. 2.13).

Considering that the output current of the voltage squaring circuits is equal with $K\Delta V^2/2$ ΔV being its differential input voltage, the differential output current of the multiplier presented in Fig. 2.13 will have the following expression:

$$I_{OUT1} - I_{OUT2} = \frac{K}{2}(V_1 - V_4)^2 + \frac{K}{2}(V_2 - V_3)^2$$
$$- \frac{K}{2}(V_1 - V_2)^2 - \frac{K}{2}(V_3 - V_4)^2 \tag{2.85}$$

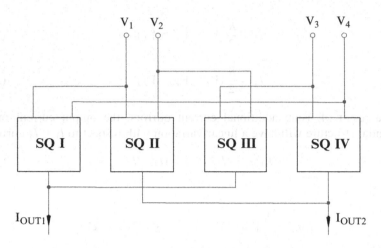

Fig. 2.13 Multiplier circuit (6) based on PR 2.3 – block diagram

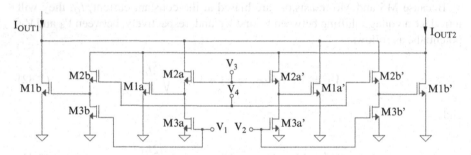

Fig. 2.14 Multiplier circuit (7) based on PR 2.3

resulting:

$$I_{OUT1} - I_{OUT2} = K(V_1 - V_3)(V_2 - V_4) \qquad (2.86)$$

A voltage multiplier that illustrates the third mathematical principle PR 2.3 can be implemented using the symmetrical structure presented in Fig. 2.14 [11].

The circuit is derived from the core shown in Fig. 2.15 [11].

For this circuit core, the $I_{D1} - I_{D2}$ differential output current can be computed replacing the expressions of drain currents by their squaring dependencies on the gate-source voltages (all MOS transistors are supposed to be biased in saturation region).

$$I_{D1} - I_{D2} = \frac{K}{2}(V_{GS1} - V_T)^2 - \frac{K}{2}(V_{GS3} - V_T)^2 \qquad (2.87)$$

Fig. 2.15 The core of the multiplier circuit (7) based on PR 2.3

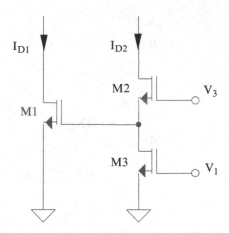

The expression of the gate-source voltage of M1 transistor can be obtained using the equality between the gate-source voltages (if M2 and M3 transistors are identical and biased at the same drain current), resulting:

$$V_{GS1} = V_3 - V_{GS2} = V_3 - V_{GS3} = V_3 - V_1 \tag{2.88}$$

Replacing (2.88) in (2.87), the expression of the output differential current for the circuit presented in Fig. 2.14 can be expressed as follows:

$$I_{D1} - I_{D2} = \frac{K}{2}(V_3 - V_1 - V_T)^2 - \frac{K}{2}(V_1 - V_T)^2$$
$$= \frac{K}{2}(V_3 - 2V_T)(V_3 - 2V_1) \tag{2.89}$$

In order to implement a voltage multiplier circuit, two identical cores from Fig. 2.15 have to be used (Fig. 2.16 [11]), the input voltages for each of them being V_1 and V_3 and, respectively, V_2 and V_3.

For simplifying the analysis of the circuit presented in Fig. 2.16, (2.89) relation can be used, the differential output current of the circuit from Fig. 2.16 being, practically, the difference between two differential output currents of two identical cores, excited using different input voltages:

$$I_L - I_R = (I_{D1}' + I_{D2}') - (I_{D2} + I_{D1}') = (I_{D1} - I_{D2}) - (I_{D1}' - I_{D2}') \tag{2.90}$$

Particularizing (2.89) relation for each circuit core, it results:

$$I_L - I_R = \frac{K}{2}(V_3 - 2V_T)(V_3 - 2V_1) - \frac{K}{2}(V_3 - 2V_T)(V_3 - 2V_2)$$
$$= K(V_2 - V_1)(V_3 - 2V_T) \tag{2.91}$$

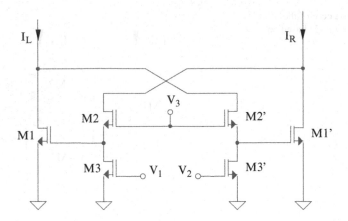

Fig. 2.16 Half-circuit of the multiplier circuit (7) based on PR 2.3

So, for the voltage multiplier presented in Fig. 2.14, the expression of the differential output current will be

$$I_{OUT1} - I_{OUT2} = I_{D1b} + I_{D2a} + I_{D1a}' + I_{D2b}' - I_{D2b} - I_{D1a}$$
$$- I_{D2a}' - I_{D1b}' = [(I_{D1b} - I_{D2b}) - (I_{D1b}' - I_{D2b}')]$$
$$- [(I_{D1a} - I_{D2a}) - (I_{D1b}' - I_{D2a}')] = (I_L - I_R)_a - (I_L - I_R)_b$$
$$= K(V_2 - V_1)(V_4 - 2V_T) - K(V_2 - V_1)(V_3 - 2V_T) = K(V_2 - V_1)(V_4 - V_3)$$

$$(2.92)$$

A modified circuit that implements the multiplication of two differential input voltages is presented in Fig. 2.17 [12] (a four-quadrant multiplier that does not require balanced inputs).

The difference between the gate-source voltages of M2 and M9 transistors can be expressed as follows:

$$V_{GS2} - V_{GS9} = V_4 - V_3 = \sqrt{\frac{2}{K}} \left(\sqrt{I_2} - \sqrt{I_0} \right) \tag{2.93}$$

It results

$$I_2 = I_0 + \frac{K}{2}(V_4 - V_3)^2 + \sqrt{2KI_0}(V_4 - V_3) \tag{2.94}$$

Similarly, for M3–M5, M1–M4, M6–M9, M5–M7 and M1–M8 differential pairs, the differences between their gate-source voltages have the following expressions:

$$V_{GS3} - V_{GS5} = V_3 - V_2 = \sqrt{\frac{2}{K}} \left(\sqrt{I_3} - \sqrt{I_0} \right) \tag{2.95}$$

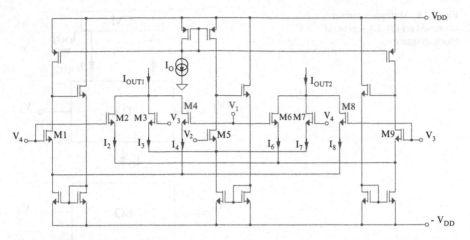

Fig. 2.17 Multiplier circuit (8) based on PR 2.3

$$V_{GS4} - V_{GS1} = V_1 - V_4 = \sqrt{\frac{2}{K}}\left(\sqrt{I_4} - \sqrt{I_O}\right) \tag{2.96}$$

$$V_{GS6} - V_{GS9} = V_1 - V_3 = \sqrt{\frac{2}{K}}\left(\sqrt{I_6} - \sqrt{I_O}\right) \tag{2.97}$$

$$V_{GS7} - V_{GS5} = V_4 - V_2 = \sqrt{\frac{2}{K}}\left(\sqrt{I_7} - \sqrt{I_O}\right) \tag{2.98}$$

and

$$V_{GS8} - V_{GS1} = V_3 - V_4 = \sqrt{\frac{2}{K}}\left(\sqrt{I_8} - \sqrt{I_O}\right) \tag{2.99}$$

resulting:

$$I_3 = I_O + \frac{K}{2}(V_3 - V_2)^2 + \sqrt{2KI_O}\,(V_3 - V_2) \tag{2.100}$$

$$I_4 = I_O + \frac{K}{2}(V_1 - V_4)^2 + \sqrt{2KI_O}\,(V_1 - V_4) \tag{2.101}$$

$$I_6 = I_O + \frac{K}{2}(V_1 - V_3)^2 + \sqrt{2KI_O}\,(V_1 - V_3) \tag{2.102}$$

$$I_7 = I_O + \frac{K}{2}(V_4 - V_2)^2 + \sqrt{2KI_O}\,(V_4 - V_2) \tag{2.103}$$

Fig. 2.18 Multiplier circuit
(1) based on PR 2.4 – general
block diagram

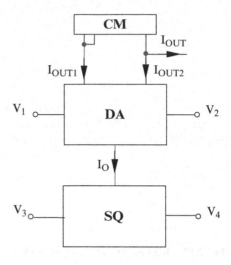

and

$$I_8 = I_O + \frac{K}{2}(V_3 - V_4)^2 + \sqrt{2KI_O}\,(V_3 - V_4) \qquad (2.104)$$

The differential output current for the multiplier circuit presented in Fig. 2.17 will have the following expression:

$$
\begin{aligned}
I_{OUT} = I_{OUT1} - I_{OUT2} &= I_2 + I_3 + I_4 - I_6 - I_7 - I_8 \\
&= K(V_1 - V_2)(V_3 - V_4)
\end{aligned}
\qquad (2.105)
$$

2.2.1.4 Multiplier Circuits Based on the Fourth Mathematical Principle (PR 2.4)

This class of multiplier circuits presents the important advantage of generating, using the same circuit core, multiple circuit functions: amplifying, multiplying, squaring or simulating both positive and negative equivalent resistances.

A method for designing a voltage multiplier using the fourth mathematical principle (PR 2.4) is illustrated by the block diagram presented in Fig. 2.18, the biasing current of the first differential amplifier (with $V_1 - V_2$ differential input voltage) being generated by a voltage squaring circuit having as input another differential voltage $V_3 - V_4$. Supposing that the G_m transconductance of the differential core is proportional with the square-root of the biasing current, I_O (an usual relation for a large class of differential amplifiers), the output current of the multiplier circuit will be proportional with the product between the input voltages of the differential amplifier and voltage squarer circuit.

Fig. 2.19 Multiplier circuit
(1) based on PR 2.4 – block
diagram with SQ circuit
implementation

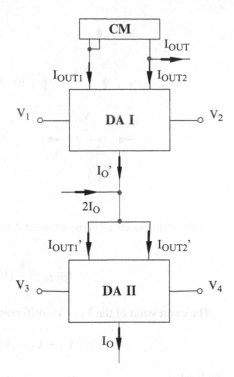

The implementation of a voltage multiplier circuit can be simplified for a particular realization of the differential amplifier, having the transfer characteristic linearized using the method of constant sum of gate-source voltages. In this case, the squarer circuit from Fig. 2.18 can be implemented using the same differential amplifier, the sum of its output currents being proportional with the square of the differential voltage applied on the input pins. The block diagram presented in Fig. 2.18 can be re-drawn replacing the general voltage squarer with its particular implementation based on a differential amplifier (Fig. 2.19). The I_O current from Fig. 2.18 has been replaced with a I_O' current, linearly dependent on the output current of a squaring circuit from Fig. 2.19, having as input the $V_3 - V_4$ differential voltage, as follows:

$$I_O' = I_{out1}' + I_{out2}' - 2I_O \qquad (2.106)$$

A possible realization of the differential amplifier linearized using the previous principle is shown in Fig. 2.20 [13].

Considering a biasing in saturation of MOS transistors, the I_{OUT1} and I_{OUT2} output currents of the differential amplifier from Fig. 2.20 can be expressed as follows:

$$I_{OUT1} = \frac{K}{2}(V_{GS1} - V_T)^2 \qquad (2.107)$$

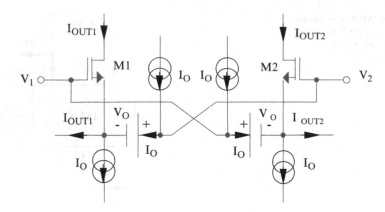

Fig. 2.20 Multiplier circuit (1) based on PR 2.4 – principle implementation of DA block

$$I_{OUT2} = \frac{K}{2}(V_{GS2} - V_T)^2 \tag{2.108}$$

The expression of the $V_1 - V_2$ differential input voltage is

$$V_1 - V_2 = V_O - V_{GS2} = V_{GS1} - V_O \tag{2.109}$$

resulting:

$$V_{GS1} = V_O + (V_1 - V_2) \tag{2.110}$$

and

$$V_{GS2} = V_O - (V_1 - V_2) \tag{2.111}$$

Replacing (2.110) and (2.111) in (2.107) and (2.108), it results:

$$I_{OUT1} = \frac{K}{2}[(V_O - V_T) + (V_1 - V_2)]^2 \tag{2.112}$$

and

$$I_{OUT2} = \frac{K}{2}[(V_O - V_T) - (V_1 - V_2)]^2 \tag{2.113}$$

So, the differential output current of the circuit presented in Fig. 2.20 will be

$$I_{OUT} = I_{OUT1} - I_{OUT2} = 2K(V_O - V_T)(V_1 - V_2) \tag{2.114}$$

In the particular case of implementing each V_O voltage source from Fig. 2.20 using a gate-source voltage of a MOS transistor biased in saturation region, it results:

$$V_O = V_T + \sqrt{\frac{2I_O}{K}} \tag{2.115}$$

so

$$I_{OUT} = \sqrt{8KI_O}\,(V_1 - V_2) \tag{2.116}$$

The sum of the output currents of the differential amplifier will have the following expression:

$$I_{OUT1} + I_{OUT2} = K(V_O - V_T)^2 + K(V_1 - V_2)^2 = 2I_O + K(V_1 - V_2)^2 \tag{2.117}$$

In conclusion, using in the block diagram of the voltage multiplier from Fig. 2.19 the particular implementation of the differential amplifier shown in Fig. 2.20, it is possible to write:

$$I_{OUT} = I_{OUT1} - I_{OUT2} = \sqrt{8KI_O'}\,(V_1 - V_2) \tag{2.118}$$

where I_O' current is linearly dependent on the sum of the output currents of the second differential amplifier:

$$I_O' = I_{OUT1}' + I_{OUT2}' - 2I_O = K(V_3 - V_4)^2 \tag{2.119}$$

Replacing (2.119) in (2.118), it results:

$$I_{OUT} = \sqrt{8K}(V_1 - V_2)\,(V_3 - V_4) \tag{2.120}$$

The complete implementation of the principle illustrated in the block diagram presented in Fig. 2.19 and the utilization of the method of realization shown in Fig. 2.20 allows many possible configurations.

The differential amplifier presented in Fig. 2.21 [14] is realized using M1 and M2 transistors and implements the V_O voltage sources from Fig. 2.20 using the gate-source voltages of M3 and M5 transistors, biased at the same constant current, I_O. So, the differential output current of the differential structure from Fig. 2.21, $I_{OUT} = I_{OUT1} - I_{OUT2}$, will be expressed by (2.116).

The realization of the multiplier circuit based on the block diagram from Fig. 2.19, using the differential amplifier presented in Fig. 2.21 [14] is shown in Fig. 2.22, the expression of the output current of the voltage multiplier shown in Fig. 2.22 being given by (2.120).

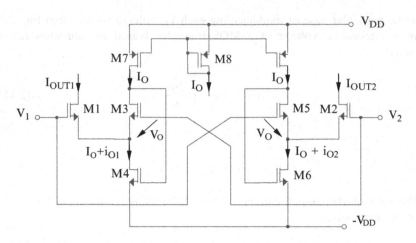

Fig. 2.21 Multiplier circuit (1) based on PR 2.4 – first implementation of DA block

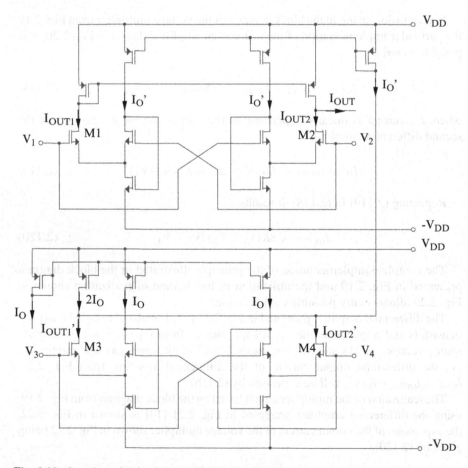

Fig. 2.22 Complete circuit of the multiplier from Fig. 2.19 using the first implementation of DA block

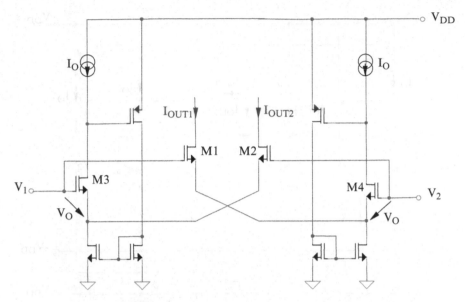

Fig. 2.23 Multiplier circuit (1) based on PR 2.4 – second implementation of DA block

The second possible implementation of the previous presented principle uses as differential amplifier the circuit presented in Fig. 2.23 [14], the V_O voltage sources from Fig. 2.20 being realized using the gate-source voltages of M3 and M4 transistors, biased at the same constant current, I_O, while the differential amplifier is realized with M1 and M2 transistors. The output current of this differential amplifier is expressed by (2.116).

The complete realization of the voltage multiplier is shown in Fig. 2.24, the expression of its output current being given by (2.120).

The third implementation of a differential amplifier based on the principle shown in Fig. 2.20 is presented in Fig. 2.25 [13, 15], the V_O voltage sources from Fig. 2.20 being realized using the gate-source voltages of M3 and M4 transistors, biased at the same constant current, I_O. The output current of this differential amplifier (which is implemented using M1 and M2 transistors) is expressed by (2.116). The disadvantage of this realization of the differential amplifier comparing with the other proposals consists in a biasing of transistors M3 and M4 at variable currents.

The complete realization of the voltage multiplier is shown in Fig. 2.26 [15], the expression of its output current being also given by (2.120).

The fourth implementation of a differential amplifier based on the principle shown in Fig. 2.20 is presented in Fig. 2.27 [16], the V_O voltage sources from Fig. 2.20 being realized using the gate-source voltages of M3a and M3b transistors, biased at the same constant current I_O (because I and I' currents are zero as a result of the circuit configuration). The output current of this differential amplifier is expressed by (2.116).

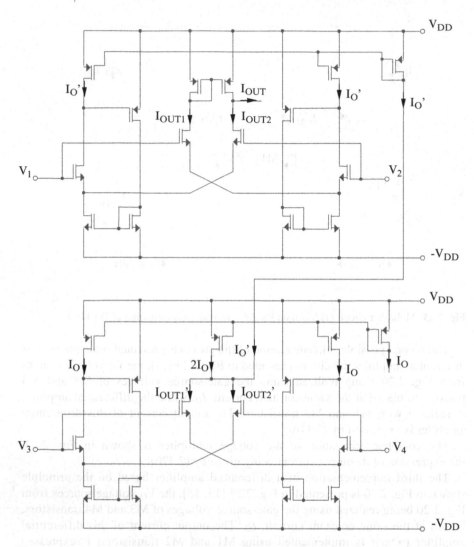

Fig. 2.24 Complete circuit of the multiplier from Fig. 2.19 using the second implementation of DA block

The fifth implementation of a differential amplifier based on the principle shown in Fig. 2.20 is presented in Fig. 2.28 [17], the V_O voltage sources from Fig. 2.20 being realized using the gate-source voltages of M3 and M4 transistors, biased at the same constant current, I_O (because I and I' currents are zero as a result of the circuit configuration). The output current of this differential amplifier (realized with M1 and M2 transistors) is expressed by (2.116).

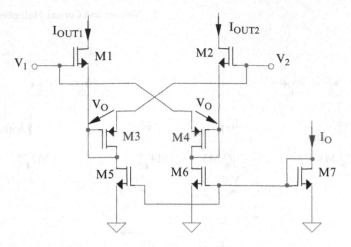

Fig. 2.25 Multiplier circuit (1) based on PR 2.4 – third implementation of DA block

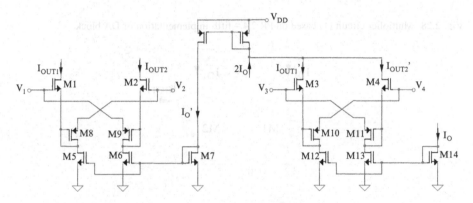

Fig. 2.26 Complete circuit of the multiplier from Fig. 2.19 using the third implementation of DA block

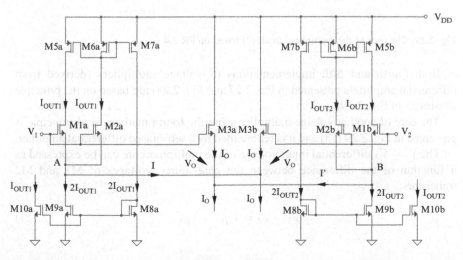

Fig. 2.27 Multiplier circuit (1) based on PR 2.4 – fourth implementation of DA block

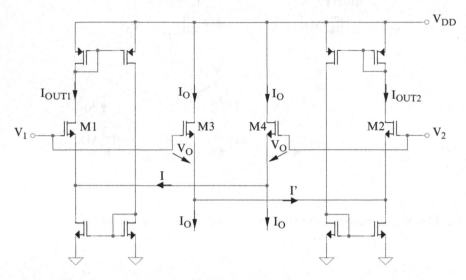

Fig. 2.28 Multiplier circuit (1) based on PR 2.4 – fifth implementation of DA block

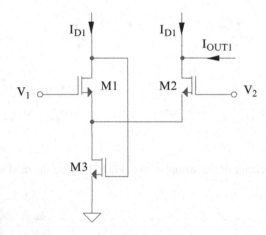

Fig. 2.29 The core of the multiplier circuit (2) based on PR 2.4

Both fourth and fifth implementations of voltage multipliers (derived from differential amplifiers presented in Fig. 2.27 and Fig. 2.28) are based on the principle illustrated in Fig. 2.19.

The core of another voltage multiplier using the fourth mathematical principle is presented in Fig. 2.29 [18] and it is represented by a self-biased differential amplifier.

The $V_1 - V_2$ differential input voltage of the multiplier core can be expressed as a function of the difference between the gate-source voltages of M1 and M2 transistors:

$$V_1 - V_2 = V_{GS1} - V_{GS2} \tag{2.121}$$

Fig. 2.30 Replicated core of
the multiplier circuit (2)
based on PR 2.4

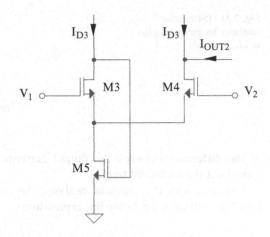

Replacing the square-root dependence of the gate-source voltage on the drain current for a MOS transistor biased in saturation and considering identical transistors, it results:

$$V_1 - V_2 = \sqrt{\frac{2I_{D1}}{K}} - \sqrt{\frac{2I_{D2}}{K}} \tag{2.122}$$

equivalent with:

$$\sqrt{I_{D2}} = \sqrt{I_{D1}} - \sqrt{\frac{K}{2}}(V_1 - V_2) \tag{2.123}$$

Squaring the previous relation, it can be obtained that:

$$I_{D2} = I_{D1} - \sqrt{2KI_{D1}}(V_1 - V_2) + \frac{K}{2}(V_1 - V_2)^2 \tag{2.124}$$

Thus, the output current of the differential core presented in Fig. 2.29, I_{OUT1}, will have the following expression:

$$I_{OUT1} = I_{D2} - I_{D1} = -\sqrt{2KI_{D1}}(V_1 - V_2) + \frac{K}{2}(V_1 - V_2)^2 \tag{2.125}$$

In order to implement the multiplying function, the first linear dependent on the differential input voltage term from the previous relation will be used. The same core permits to realize also the squaring function using the second term from the same relation. The simplest way to remove the last quadratic term is to use a similar structure with the circuit from Fig. 2.29 (presented in Fig. 2.30) [18] and having the I_{D1} current replaced with another current, I_{D3}. As the quadratic term from (2.125) does not depend on I_{D1} and I_{D3} currents, the consideration

Fig. 2.31 Differential
amplifier for generating I_{D3}
and I_{D1} currents

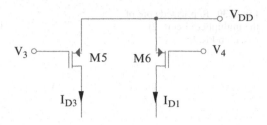

of the difference between the output currents of these similar structures will
cancel out the undesired term.

Similarly with the previous analysis, the output current of the circuit from
Fig. 2.30 will have the following expression:

$$I_{OUT2} = I_{D4} - I_{D3} = -\sqrt{2KI_{D3}}(V_1 - V_2) + \frac{K}{2}(V_1 - V_2)^2 \qquad (2.126)$$

The difference between the output currents I_{OUT1} and I_{OUT2} will be

$$I_{OUT} = I_{OUT1} - I_{OUT2} = \sqrt{2K}\left(\sqrt{I_{D3}} - \sqrt{I_{D1}}\right)(V_1 - V_2) \qquad (2.127)$$

The I_{D3} and I_{D1} currents are generated by another differential amplifier M5–M6,
having $V_3 - V_4$ as differential input voltage (Fig. 2.31).

For this structure, considering that its composing transistors are biased in
saturation, it is possible to write:

$$\sqrt{I_{D3}} - \sqrt{I_{D1}} = \sqrt{\frac{K}{2}[(V_{DD} - V_3 - V_T) - (V_{DD} - V_4 - V_T)]} = \sqrt{\frac{K}{2}}(V_4 - V_3) \quad (2.128)$$

Replacing (2.128) in (2.127) it results the multiplying function:

$$I_{OUT} = K(V_1 - V_2)(V_4 - V_3) \qquad (2.129)$$

The complete circuit of the multiplier is presented in Fig. 2.32 [18]. The differential amplifier from Fig. 2.31 is replaced with two parallel-connected differential
amplifiers M5, M5'–M6, M6' because I_{D1} and I_{D3} currents must be duplicated for
biasing the differential amplifiers, M1–M2 and M3–M4.

Another possible implementation of a voltage multiplier uses the symmetrical
structure presented in Fig. 2.33.

The multiplier is composed from two self-biased differential amplifiers
(M5–M6 and M8–M9, respectively), their active loads being represented by

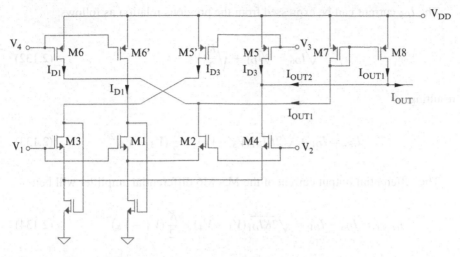

Fig. 2.32 The complete implementation of the multiplier circuit (2) based on PR 2.4

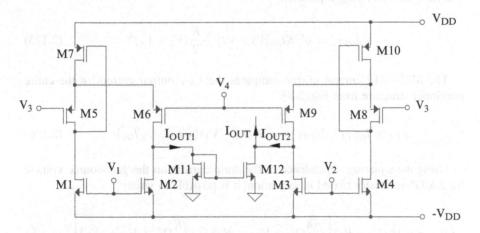

Fig. 2.33 Multiplier circuit (3) based on PR 2.4

M1–M2 and M3–M4 current mirrors. Analyzing the M5–M6 differential amplifier, the $V_3 - V_4$ differential voltage can be expressed as a difference between two gate-source voltages:

$$V_3 - V_4 = V_{SG6} - V_{SG5} \qquad (2.130)$$

Replacing the square-root dependence of the gate-source voltage on the drain current for a MOS transistor biased in saturation and considering identical transistors, it results:

$$V_3 - V_4 = \sqrt{\frac{2I_{D6}}{K}} - \sqrt{\frac{2I_{D1}}{K}} \qquad (2.131)$$

The I_{D6} current can be expressed from the previous relation as follows:

$$\sqrt{I_{D6}} = \sqrt{I_{D1}} + \sqrt{\frac{K}{2}}(V_3 - V_4) \tag{2.132}$$

resulting:

$$I_{D6} = I_{D1} + \sqrt{2KI_{D1}}(V_3 - V_4) + \frac{K}{2}(V_3 - V_4)^2 \tag{2.133}$$

The differential output current of the M5–M6 differential amplifier will be:

$$I_{OUT1} = I_{D6} - I_{D1} = \sqrt{2KI_{D1}}(V_3 - V_4) + \frac{K}{2}(V_3 - V_4)^2 \tag{2.134}$$

Similarly, the differential output current of the M8–M9 differential amplifier will have the following expression:

$$I_{OUT2} = \sqrt{2KI_{D4}}(V_3 - V_4) + \frac{K}{2}(V_3 - V_4)^2 \tag{2.135}$$

The M11–M12 current mirror computes the I_{OUT} output current of the entire multiplier structure from Fig. 2.33:

$$I_{OUT} = I_{OUT2} - I_{OUT1} = \sqrt{2K}(V_3 - V_4)\left(\sqrt{I_{D4}} - \sqrt{I_{D1}}\right) \tag{2.136}$$

Using the squaring dependence of the drain current on the gate-source voltage for a MOS transistor biased in saturation, it is possible to write:

$$I_{OUT} = \sqrt{2K}(V_3 - V_4)\left[\sqrt{\frac{K}{2}}(V_2 + V_{DD} - V_T) - \sqrt{\frac{K}{2}}(V_1 + V_{DD} - V_T)\right] \tag{2.137}$$

$$= K(V_3 - V_4)(V_2 - V_1)$$

A voltage multiplier based on the compensation of the squaring characteristic of the MOS transistor biased in saturation region using two square-root circuits is presented in Fig. 2.34 [19].

The differential input voltage can be expressed as follows:

$$V_1 - V_2 = V_{GS3} - V_{GS4} = \sqrt{\frac{2}{K}}\left(\sqrt{I_2} - \sqrt{I_2'}\right) \tag{2.138}$$

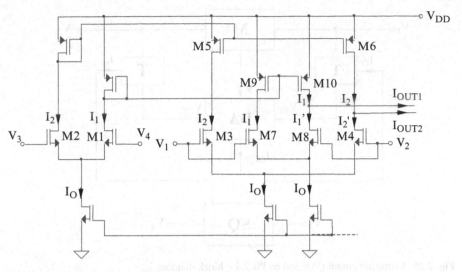

Fig. 2.34 Multiplier circuit (4) based on PR 2.4

resulting:

$$I_2' = I_2 + \frac{K}{2}(V_1 - V_2)^2 - \sqrt{2K}(V_1 - V_2)\sqrt{I_2} \qquad (2.139)$$

The output current of the square-rooting circuit is:

$$I_{OUT2} = I_2 - I_2' = -\frac{K}{2}(V_1 - V_2)^2 + \sqrt{2K}(V_1 - V_2)\sqrt{I_2} \qquad (2.140)$$

Similarly:

$$I_{OUT1} = I_1 - I_1' = -\frac{K}{2}(V_1 - V_2)^2 + \sqrt{2K}(V_1 - V_2)\sqrt{I_1} \qquad (2.141)$$

so:

$$I_{OUT} = I_{OUT2} - I_{OUT1} = \sqrt{2K}(V_1 - V_2)(\sqrt{I_2} - \sqrt{I_1}) \qquad (2.142)$$

The differential input voltage of M1–M2 differential amplifier will have the following expression:

$$V_3 - V_4 = V_{GS2} - V_{GS1} = \sqrt{\frac{2}{K}}(\sqrt{I_2} - \sqrt{I_1}) \qquad (2.143)$$

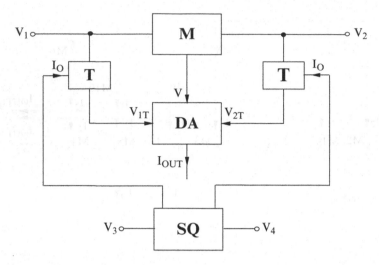

Fig. 2.35 Multiplier circuit (5) based on PR 2.4 – block diagram

resulting:

$$I_{OUT} = K(V_1 - V_2)(V_3 - V_4) \tag{2.144}$$

A possible implementation of a voltage multiplier has the block diagram presented in Fig. 2.35. The "DA" block represents a classical active-load differential amplifier, having the common-sources point biased at a V potential fixed by the circuit "M". This circuit computes the arithmetical mean of input potentials, having the goal of obtaining a very good linearity of the entire structure, with the contribution of "T" blocks (which are used for introducing a translation of input potentials). A squaring circuit, "SQ", is used for generating the biasing current of the two translation blocks, I_O.

The "DA" (Differential Amplifier) block

The "DA" block is implemented as a classical active-load differential amplifier, having the concrete realization presented in Fig. 2.36 [20].

Considering a biasing in saturation of the MOS devices from Fig. 2.36, the output current of the differential amplifier can be expressed as follows:

$$I_{OUT} = I_2 - I_1 = \frac{K}{2}(V_{SG2} - V_T)^2 - \frac{K}{2}(V_{SG1} - V_T)^2 \tag{2.145}$$

Fig. 2.36 Multiplier circuit
(5) based on PR 2.4 –
implementation of DA block

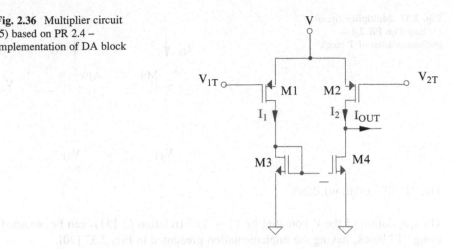

equivalent with:

$$I_{OUT} = \frac{K}{2}(V_{SG2} - V_{SG1})(V_{SG1} + V_{SG2} - 2V_T) \qquad (2.146)$$

Because:

$$V_{SG1} = V - V_{1T} \qquad (2.147)$$

and:

$$V_{SG2} = V - V_{2T} \qquad (2.148)$$

it results:

$$I_{OUT} = \frac{K}{2}(V_{1T} - V_{2T})(2V - V_{1T} - V_{2T} - 2V_T) \qquad (2.149)$$

In order to obtain a linear transfer characteristic $I_{OUT}(V_{1T} - V_{2T})$, it is necessary that the second parenthesis from (2.149) to be constant with respect to the differential input voltage, $V_{1T} - V_{2T}$:

$$2V - V_{1T} - V_{2T} - 2V_T = A = ct. \qquad (2.150)$$

resulting the necessity of implementing a V voltage equal with:

$$V = \frac{V_{1T} + V_{2T}}{2} + V_T + \frac{A}{2} \qquad (2.151)$$

Fig. 2.37 Multiplier circuit
(5) based on PR 2.4 –
implementation of T block

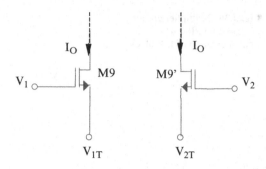

The "T" (Translation) Block

The translation of the V potential by $V_T + A/2$ (relation (2.151)) can be obtained using "T" block, having the implementation presented in Fig. 2.37 [20].

Because the same I_O current is passing through all transistors from Fig. 2.37, it is possible to write that:

$$V_1 = V_{1T} + V_T + \sqrt{\frac{2I_O}{K}} \qquad (2.152)$$

and:

$$V_2 = V_{2T} + V_T + \sqrt{\frac{2I_O}{K}} \qquad (2.153)$$

So, both V_1 and V_2 input potentials are DC shifted with the same amount, $V_T + \sqrt{2I_O/K}$.

The "M" (Arithmetic Mean) Block

In order to obtain the arithmetic mean of input potentials expressed by relation (2.151), the circuit from Fig. 2.38 [20] can be used, having the advantage of using only MOS transistors, biased in saturation region.

The expression of the V potential is:

$$V = \frac{V_1 + V_2}{2} \qquad (2.154)$$

Replacing (2.152) and (2.153) in (2.156), it can be obtained:

$$V = \frac{V_{1T} + V_{2T}}{2} + V_T + \sqrt{\frac{2I_O}{K}} \qquad (2.155)$$

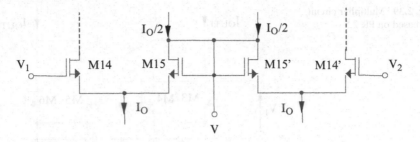

Fig. 2.38 Multiplier circuit (1) based on PR 2.4 – implementation of M block

Comparing relations (2.151) and (2.157), it results that $A = 2\sqrt{2I_O/K}$, so:

$$I_{OUT} = \frac{K}{2}(V_{1T} - V_{2T})2\sqrt{\frac{2I_O}{K}} \tag{2.156}$$

or:

$$I_{OUT} = \sqrt{2KI_O}(V_{1T} - V_{2T}) \tag{2.157}$$

equivalently (using (2.152) and (2.153)) with:

$$I_{OUT} = \sqrt{2KI_O}(V_1 - V_2) = G_m(V_1 - V_2) \tag{2.158}$$

resulting:

$$I_{OUT} = \sqrt{2KI_O}(V_1 - V_2) = G_m(V_1 - V_2) \tag{2.159}$$

$G_m = \sqrt{2KI_O}$ being the equivalent transconductance of the differential amplifier.

Because I_O biasing currents of the translation blocks "T" from Fig. 2.37 are generated by a voltage squaring circuit having as input a differential voltage, $V_3 - V_4$, it results a multiplier circuit with a very good linearity. So, replacing in (2.159) the expression of I_O current:

$$I_O = \frac{K}{4}(V_3 - V_4)^2 \tag{2.160}$$

it results:

$$I_{OUT} = \frac{K}{\sqrt{2}}(V_1 - V_2)(V_3 - V_4) \tag{2.161}$$

Fig. 2.39 Multiplier circuit
(1) based on PR 2.5

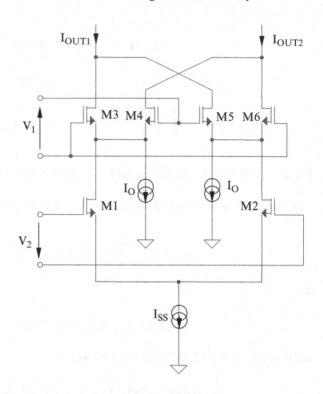

2.2.1.5 Multiplier Circuits Based on the Fifth Mathematical
Principle (PR 2.5)

A method for linearizing the characteristic of a voltage multiplier using the fifth mathematical principle consists in the extension of the linearization technique designed for the CMOS differential amplifier, based on the biasing of the differential structure at a current that is the sum between a constant current and a component proportional with the square of the differential input voltage. The relatively simple implementation of multiplier circuits based on this mathematical principle impose them for a large area of applications in VLSI designs. The expression of the differential output current of the classical CMOS differential amplifier is:

$$I_{OUT} = KV_I \sqrt{\frac{I_{SS}}{K} - \frac{V_I^2}{4}} \qquad (2.162)$$

V_I representing the differential input voltage and I_{SS} being the biasing current of the differential amplifier.

The multiplier with linear characteristic based on the previous presented principle is presented in Fig. 2.39 [21].

The output current of the previous voltage multiplier can be expressed as:

$$I_{OUT} = I_{OUT1} - I_{OUT2} = (I_{D3} + I_{D5}) - (I_{D4} + I_{D6})$$
$$= (I_{D3} - I_{D4}) - (I_{D6} - I_{D5}) \tag{2.163}$$

The differential output currents of the differential amplifiers M3–M4 and M5–M6 can be obtained using the general relation (2.162):

$$I_{D3} - I_{D4} = KV_1\sqrt{\frac{I_{D1} + I_O}{K} - \frac{V_1^2}{4}} \tag{2.164}$$

and:

$$I_{D6} - I_{D5} = K V_1\sqrt{\frac{I_{D2} + I_O}{K} - \frac{V_1^2}{4}} \tag{2.165}$$

The linearization technique is based on the utilization of a current I_O proportional with the squaring of the differential input voltage V_1:

$$I_O = \frac{KV_1^2}{4} \tag{2.166}$$

Replacing (2.164), (2.165) and (2.166) in (2.163), it results the following expression of the output current of the voltage multiplier:

$$I_{OUT} = \sqrt{K}V_1\left(\sqrt{I_{D1}} - \sqrt{I_{D2}}\right) \tag{2.167}$$

Analyzing M1–M2 differential amplifier, the V_2 differential input voltage can be expressed as follows:

$$V_2 = V_{GS1} - V_{GS2} = \left(V_T + \sqrt{\frac{2I_{D1}}{K}}\right) - \left(V_T + \sqrt{\frac{2I_{D2}}{K}}\right)$$
$$= \sqrt{\frac{2}{K}}\left(\sqrt{I_{D1}} - \sqrt{I_{D2}}\right) \tag{2.168}$$

From (2.168) and (2.167), it results the expression of the I_{OUT} output current as a function on the differential input voltages, V_1 and V_2:

$$I_{OUT} = \frac{K}{\sqrt{2}}V_1V_2 \tag{2.169}$$

A similar method, useful for low-voltage operation, is presented in Fig. 40 [21]. In order to reduce the minimal value of the supply voltage, the stacked architecture is replaced with a folded structure.

The $I_{OUT1} - I_{OUT2}$ differential output current of the folded voltage multiplier is:

$$I_{OUT1} - I_{OUT2} = (I_{D3} + I_{D5}) - (I_{D4} + I_{D6})$$
$$= (I_{D3} - I_{D4}) - (I_{D6} - I_{D5}) \tag{2.170}$$

The differential output currents of the differential amplifiers M3–M4 and M5–M6 can be obtained using the general relation (2.162):

$$I_{D3} - I_{D4} = KV_1 \sqrt{\frac{I_{SS} + I_O - I_{D1}}{K} - \frac{V_1^2}{4}} \tag{2.171}$$

and:

$$I_{D6} - I_{D5} = KV_1 \sqrt{\frac{I_{SS} + I_O - I_{D2}}{K} - \frac{V_1^2}{4}} \tag{2.172}$$

The linearization technique is based on the utilization of a I_O current, proportional with the square of the differential input voltage V_1:

$$I_O = \frac{KV_1^2}{4} \tag{2.173}$$

Replacing (2.171), (2.172) and (2.173) in (2.170), it results the following expression of the output current of the voltage multiplier:

$$I_{OUT} = \sqrt{K}V_1 \left(\sqrt{I_{SS} - I_{D1}} - \sqrt{I_{SS} - I_{D2}} \right) \tag{2.174}$$

The M1–M2 differential amplifier is biased at the constant current, I_{SS}, so $I_{D1} + I_{D2} = I_{SS}$. The previous relation can be rewritten as:

$$I_{OUT} = \sqrt{K}V_1 \left(\sqrt{I_{D2}} - \sqrt{I_{D1}} \right) \tag{2.175}$$

Similarly with the previous voltage multiplier, it results the following expression of the I_{OUT} output current as a function on V_1 and V_2 differential input voltages:

$$I_{OUT} = \frac{K}{\sqrt{2}} V_1 V_2 \tag{2.176}$$

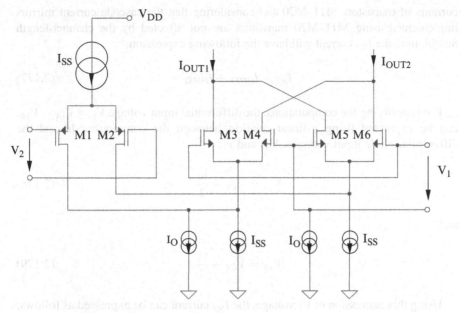

Fig. 2.40 Multiplier circuit (1) based on PR 2.5 – circuit core of the folded version

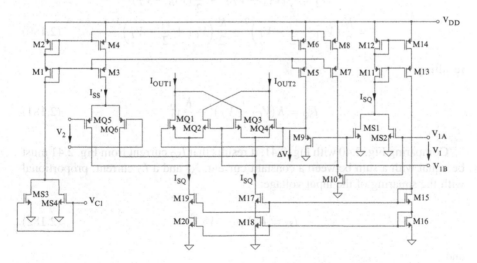

Fig. 2.41 Multiplier circuit (1) based on PR 2.5 – complete implementation of the folded version

The complete implementation of the folded voltage multiplier from Fig. 2.40 is presented in Fig. 2.41 [21].

The M1 and M2 transistors from Fig. 2.40 have been replaced in Fig. 2.41 with MQ5 and MQ6, while the differential amplifiers, M3–M4 and M5–M6 from Fig. 2.40 were renamed MQ1–MQ2 and MQ3–MQ4, respectively. Noting with I_{SQ} the drain currents of transistors M11–M20 and considering that the cascode current mirrors implemented using M11–M20 transistors are not affected by the channel-length modulation, the I_{SQ} current will have the following expression:

$$I_{SQ} = I_{DMS1} + I_{DMS2} \qquad (2.177)$$

For simplifying the computations, the differential input voltage $V_1 = V_{1A} - V_{1B}$, can be expressed using a linear relation between the common-mode and the differential-mode input voltages, V_{C1} and v_1:

$$V_{1A} = V_{C1} - \frac{v_1}{2} \qquad (2.178)$$

and:

$$V_{1B} = V_{C1} + \frac{v_1}{2} \qquad (2.179)$$

Using this expression of V_1 voltage, the I_{SQ} current can be expressed as follows:

$$
\begin{aligned}
I_{SQ} &= \frac{K}{2}(V_{1A} - V_T)^2 + \frac{K}{2}(V_{1B} - V_T)^2 \\
&= \frac{K}{2}\left(V_{C1} - \frac{v_1}{2} - V_T\right)^2 + \frac{K}{2}\left(V_{C1} + \frac{v_1}{2} - V_T\right)^2 \qquad (2.180)
\end{aligned}
$$

resulting:

$$I_{SQ} = K(V_{C1} - V_T)^2 + \frac{K}{4}v_1^2 \qquad (2.181)$$

Comparing Fig. 2.40 with Fig. 2.41, it results that I_{SQ} current from Fig. 2.41 must be equal with a sum between a constant current, I_{SS} and a I_O current, proportional with the squaring of the input voltage:

$$I_{SS} = K(V_{C1} - V_T)^2 \qquad (2.182)$$

and:

$$I_O = \frac{K}{4}v_1^2 \qquad (2.183)$$

For a proper operation of the folded multiplier, the I_{SS}' biasing current of the differential amplifier MQ5–MQ6 must be equal with I_{SS}. This I_{SS}' current is

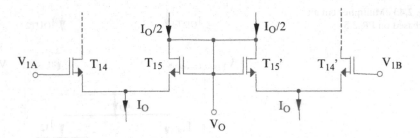

Fig. 2.42 Arithmetical mean circuit

generated by the MS3–MS4 pair, each transistor being biased at the common-mode component of the input voltage V_1:

$$I_{SS}' = I_{DMS3} + I_{DMS4} \tag{2.184}$$

resulting:

$$I_{SS}' = 2\frac{K}{2}(V_{C1} - V_T)^2 = K(V_{C1} - V_T)^2 = I_{SS} \tag{2.185}$$

The M9 and M10 transistors are used for transferring the differential input voltage, V_1, on the input of cross-connected differential amplifiers MQ1–MQ2 and MQ3–MQ4:

$$\Delta V = (V_{1B} + V_{SG9}) - (V_{1A} + V_{SG10}) \tag{2.186}$$

The M9 and M10 transistors being identical and working at the same drain current, it results $V_{SG9} = V_{SG10}$. So:

$$\Delta V = V_{1B} - V_{1A} = -V_1 \tag{2.187}$$

Concluding that the circuits presented in Fig. 2.40 and Fig. 2.41 are functionally identical, the I_{OUT} output current of the complete implementation of the folded voltage multiplier circuit can be obtained replacing in (2.176) V_1 with $-V_1$:

$$I_{OUT} = I_{OUT1} - I_{OUT2} = -\frac{K}{\sqrt{2}}V_1 V_2 \tag{2.188}$$

An arithmetical mean circuit (Fig. 2.42) [22] must be used for extracting the common-mode component V_{C1} of the input voltage V_1.

The V_O output voltage for this circuit will be:

$$V_O = \frac{V_{1A} + V_{1B}}{2} = V_{C1} \tag{2.189}$$

Fig. 2.43 Multiplier circuit
(2) based on PR 2.5

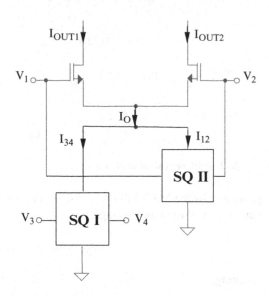

The voltage multiplier presented in Fig. 2.43 [23], using the fifth mathematical principle (PR 2.5) is derived from a differential amplifier with linear transfer characteristic.

The differential output current, I_{OUT}, for the circuit presented in Fig. 2.43 will present a strong nonlinear dependence on the $V_1 - V_2$ differential input voltage, that can be expressed as:

$$I_{OUT} = I_{OUT1} - I_{OUT2} = I_0 \sqrt{\frac{K(V_1 - V_2)^2}{I_0} - \frac{K^2(V_1 - V_2)^4}{4I_0^2}} \tag{2.190}$$

equivalent with:

$$I_{OUT} = \frac{V_1 - V_2}{2} \sqrt{4KI_0 - K^2(V_1 - V_2)^2} \tag{2.191}$$

I_0 being the biasing current of the differential structure. So, superior-order distortions will characterize the behavior of this structure, imposing the design of a linearization technique for removing the superior-order terms from the transfer characteristic. The method for obtaining a linear transfer characteristic is to obtain the I_0 bias current of the entire differential structure as a sum of two terms: I_{12}, proportional with the squaring of the $V_1 - V_2$ differential input voltage and I_{34}, proportional with the squaring of another differential voltage, $V_3 - V_4$:

$$I_0 = I_{12} + I_{34} = \frac{K}{4}(V_1 - V_2)^2 + \frac{K}{4}(V_3 - V_4)^2 \tag{2.192}$$

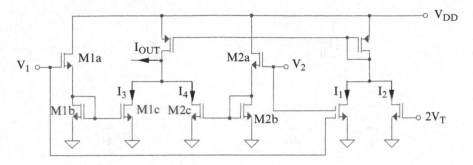

Fig. 2.44 Multiplier circuit (1) based on PR 2.6

resulting, in this case, a perfect proportionality of the output current on the differential input voltages:

$$I_{OUT} = \frac{K}{2}(V_1 - V_2)(V_3 - V_4) \tag{2.193}$$

2.2.1.6 Multiplier Circuits Based on the Sixth Mathematical Principles (PR 2.6)

The area of applications of multiplier structures based on PR 2.6 is restricted to the circuits that do not require the multiplication of differential input voltages.

The following voltage multiplier circuit is based on the sixth mathematical principle. In order to obtain a linear characteristic of the circuit, a perfect symmetrical structure with respect to the two input potentials is presented in Fig. 2.44 [24, 25].

Considering an operation in saturation for all MOS devices, the output current can be expressed as:

$$I_{OUT} = (I_1 + I_2) - (I_3 + I_4) \tag{2.194}$$

where:

$$I_1 = \frac{K}{2}\left(\frac{V_1 + V_2}{2} - V_T\right)^2 \tag{2.195}$$

$$I_2 = \frac{K}{2}V_T^2 \tag{2.196}$$

$$I_3 = \frac{K}{2}\left(\frac{V_1}{2} - V_T\right)^2 \tag{2.197}$$

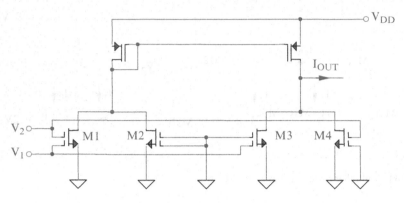

Fig. 2.45 Multiplier circuit (2) based on PR 2.6

$$I_4 = \frac{K}{2}\left(\frac{V_2}{2} - V_T\right)^2 \tag{2.198}$$

For the previous mathematical relations, it results a perfect linear dependence of the output current on the input voltages:

$$I_{OUT} = \frac{K}{2}V_1V_2 \tag{2.199}$$

Another possible implementation of a voltage multiplier circuit using the sixth mathematical principle (PR 2.6) is presented in Fig. 2.45.

The output current of the voltage multiplier can be expressed as follows:

$$I_{OUT} = (I_{D1} + I_{D2}) - (I_{D3} + I_{D4}) \tag{2.200}$$

where:

$$I_{D1} = \frac{K}{2}\left(\frac{V_1 + V_2}{2} - V_T\right)^2 \tag{2.201}$$

$$I_{D2} = \frac{K}{2}V_T^2 \tag{2.202}$$

$$I_{D3} = \frac{K}{2}\left(\frac{V_1}{2} - V_T\right)^2 \tag{2.203}$$

$$I_{D4} = \frac{K}{2}\left(\frac{V_2}{2} - V_T\right)^2 \tag{2.204}$$

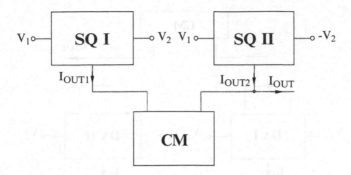

Fig. 2.46 Multiplier circuit (1) based on PR 2.7 – block diagram

From the previous mathematical relations it results a perfect linear dependence of the output current on the input voltages:

$$I_{OUT} = \frac{K}{2}V_1V_2 \qquad (2.205)$$

2.2.1.7 Multiplier Circuits Based on the Seventh Mathematical Principles (PR 2.7)

Practically derived from the implementation of voltage squaring circuits, the multiplier structures that use as functional basis PR 2.7 find many applications in analog signal processing.

A similar approach of a voltage multiplier uses 2 v squaring circuit, connected as it is shown in Fig. 2.46.

Considering that the squaring circuits have an output current proportional with $K\Delta V^2/2$, ΔV being the differential input voltage, the output current of the circuit presented in Fig. 2.46 will have the following expression:

$$I_{OUT} = I_{OUT2} - I_{OUT1} = \frac{K}{2}(V_1 + V_2)^2 - \frac{K}{2}(V_1 - V_2)^2 = 2KV_1V_2 \qquad (2.206)$$

In order to obtain the multiplying function using two squaring circuits, a similar method is proposed in Fig. 2.47. The squaring circuits from Fig. 2.46 have been replaced in Fig. 2.47 with two particular implementations of a differential amplifier (presented in Fig. 2.20).

The implementation of the voltage multiplier is shown in Fig. 2.28 [13]. The I_{OUT} output current will have the following expression:

$$I_{OUT} = \left[2I_O + K(V_1 + V_2)^2\right] - \left[2I_O + K(V_1 - V_2)^2\right] = 4KV_1V_2 \qquad (2.207)$$

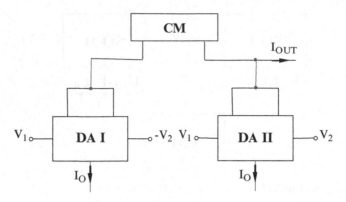

Fig. 2.47 Multiplier circuit (2) based on PR 2.7 – block diagram

Two complete realizations of the previous circuit, using specific implementations of V_O sources from Fig. 2.48, are shown in Fig. 2.49 [13] and Fig. 2.50 [13], respectively.

A multiplier circuit can be designed starting from a squaring characteristic implemented using a classical differential amplifier (Fig. 2.51) [18].

The differential input voltage for the circuit presented in Fig. 2.51 can be expressed as follows:

$$V_1 - V_2 = \sqrt{\frac{2}{K}}\left(\sqrt{I} - \sqrt{I_O}\right) \tag{2.208}$$

resulting:

$$I = I_O + \frac{K}{2}(V_1 - V_2)^2 + \sqrt{2KI_O}(V_1 - V_2) \tag{2.209}$$

In order to obtain the multiplying function, four differential amplifiers from Fig. 2.51 can be connected as it is shown in Fig. 2.52.

The output current of the multiplier circuit will have the following expression:

$$I_{OUT} = I_1 + I_2 - I_3 - I_4 \tag{2.210}$$

resulting:

$$I_{OUT} = \frac{K}{2}(-V_1 - V_2)^2 + \frac{K}{2}(V_1 + V_2)^2$$
$$- \frac{K}{2}(-V_1 + V_2)^2 - \frac{K}{2}(V_1 - V_2)^2 = 4KV_1V_2 \tag{2.211}$$

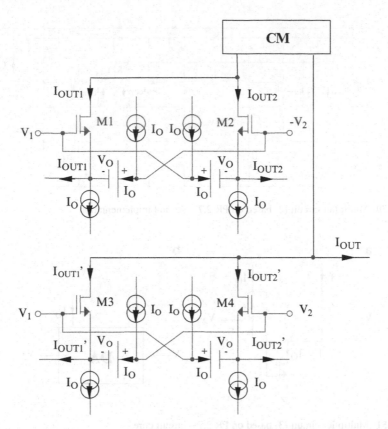

Fig. 2.48 Multiplier circuit (2) based on PR 2.7 – principle implementation

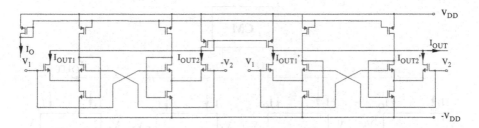

Fig. 2.49 Multiplier circuit (2) based on PR 2.7 – first implementation

The circuit presented in Fig. 2.53 [12] represents a four-quadrant multiplier with balanced inputs.

The difference between the gate-source voltages of M1 and M3 transistors can be expressed as follows:

$$V_1 - V_2 = V_{GS1} - V_{GS3} \tag{2.212}$$

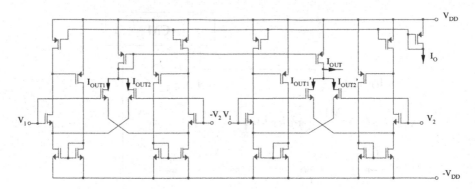

Fig. 2.50 Multiplier circuit (2) based on PR 2.7 – second implementation

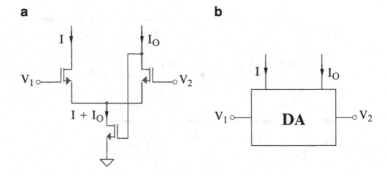

Fig. 2.51 Multiplier circuit (3) based on PR 2.7 – circuit core

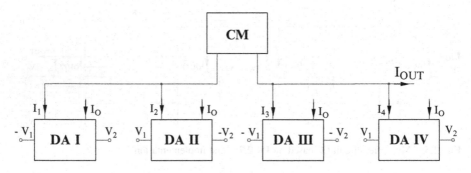

Fig. 2.52 Multiplier circuit (3) based on PR 2.7 – block diagram

For a biasing in saturation of all MOS transistors from Fig. 2.53, it results:

$$V_1 - V_2 = \sqrt{\frac{2}{K}}\left(\sqrt{I_1} - \sqrt{I_O}\right) \qquad (2.213)$$

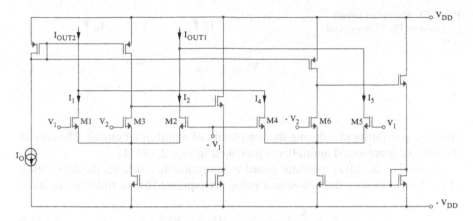

Fig. 2.53 Multiplier circuit (4) based on PR 2.7

So, the expression of I_1 current will be:

$$I_1 = I_O + \frac{K}{2}(V_1 - V_2)^2 + \sqrt{2KI_O}(V_1 - V_2) \tag{2.214}$$

Similarly, computing the difference between the gate-source voltages of M2–M3 transistors, it results:

$$I_2 = I_O + \frac{K}{2}(-V_1 - V_2)^2 + \sqrt{2KI_O}(-V_1 - V_2) \tag{2.215}$$

The $I_1 - I_2$ differential current will have the following expression:

$$I_1 - I_2 = 2V_1\sqrt{2KI_O} - 2KV_1V_2 \tag{2.216}$$

Similarly, for the structure implemented using M4–M6 transistors, the differential output current can be expressed as follows:

$$I_4 - I_5 = -2V_1\sqrt{2KI_O} - 2KV_1V_2 \tag{2.217}$$

The differential output current for the entire multiplier structure presented in Fig. 2.53 will be:

$$I_{OUT} = I_{OUT1} - I_{OUT2} = I_2 + I_5 - I_1 - I_4 = 4KV_1V_2 \tag{2.218}$$

2.2.1.8 Multiplier Circuits Based on Different Mathematical Principles (PR 2.Da)

Alternate implementations of the previous presented multiplier circuits are based on different mathematical principles. The utilization of the bulk as an active terminal

Fig. 2.54 Multiplier circuit
(1) based on PR 2.Da – circuit
core

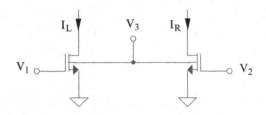

gives the possibility of reducing the complexity of a multiplier circuit. The core of
the following presented multiplier is presented in Fig. 2.54 [11].

A model of the MOS transistor biased in saturation that includes the dependence
of the drain current on the bulk-source voltage is expressed by the following relation:

$$I_D = \frac{K}{2}\left(V_{GS} - V_T - AV_{BS} - BV_{BS}^2\right)^2 \tag{2.219}$$

A and B being constants. The differential output current of the multiplier core
from Fig. 2.54, $I_L - I_R$, will be:

$$I_L - I_R = \frac{K}{2}\left(V_1 - V_T - AV_3 - BV_3^2\right)^2 - \frac{K}{2}\left(V_2 - V_T - AV_3 - BV_3^2\right)^2 \tag{2.220}$$

resulting:

$$I_L - I_R = \frac{K}{2}(V_1 - V_2)\left(V_1 + V_2 - 2V_T - 2AV_3 - 2BV_3^2\right) \tag{2.221}$$

In order to obtain the multiplying function, two circuits from Fig. 2.54 can be
cross-connected, resulting the multiplier presented in Fig. 2.55 [11].

For this circuit, the differential output current can be expressed as the difference
between the differential output currents of each core:

$$I_{OUT1} - I_{OUT2} = (I_{L2} + I_{R1}) - (I_{L1} + I_{R2}) = (I_{L2} - I_{R2}) - (I_{L1} - I_{R1}) \tag{2.222}$$

Replacing (2.221) in (2.222), it results:

$$I_{OUT1} - I_{OUT2} = \frac{K}{2}(V_1 - V_2)\left(V_1 + V_2 - 2V_T - 2AV_4 - 2BV_4^2\right)$$
$$- \frac{K}{2}(V_1 - V_2)\left(V_1 + V_2 - 2V_T - 2AV_3 - 2BV_3^2\right) \tag{2.223}$$

equivalent with:

$$I_{OUT1} - I_{OUT2} = K(V_1 - V_2)\left[A(V_3 - V_4) + B\left(V_3^2 - V_4^2\right)\right]$$
$$= K(V_1 - V_2)(V_3 - V_4)[A + B(V_3 + V_4)] \tag{2.224}$$

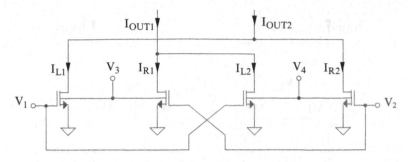

Fig. 2.55 Multiplier circuit (1) based on PR 2.Da – complete implementation

If the V_1, V_2, V_3 and V_4 input voltages contain common-mode terms (V_{12} and V_{34}) and differential-mode terms (v_{12} and v_{34}) as follows:

$$V_1 = V_{12} + \frac{v_{12}}{2} \tag{2.225}$$

$$V_2 = V_{12} - \frac{v_{12}}{2} \tag{2.226}$$

$$V_3 = V_{34} + \frac{v_{34}}{2} \tag{2.227}$$

$$V_4 = V_{34} - \frac{v_{34}}{2} \tag{2.228}$$

it results:

$$I_{OUT1} - I_{OUT2} = K v_{12} v_{34}(A + 2BV_{34}) \tag{2.229}$$

so, the differential output current of the voltage multiplier presented in Fig. 2.55 will be proportional with the product of the differential-mode components of input voltages.

An alternate approach of a voltage multiplier, based on bulk-driven MOS devices using another model for the dependence of the threshold voltage V_T on the biasing of the bulk (V_{BS}), is shown in Fig. 2.56 [26].

The differential output current of this multiplier can be expressed as follows:

$$I_{OUT1} - I_{OUT2} = (I_{D1} + I_{D3}) - (I_{D2} + I_{D4}) = (I_{D1} - I_{D2}) + (I_{D3} - I_{D4}) \tag{2.230}$$

where it is considered that the drain current of a MOS transistor depends on the gate-source voltage following a quadratic law (2.231) and on the bulk-source voltage as a consequence of the bulk effect using the mathematical relation (2.232):

$$I_D = \frac{K}{2}(V_{GS} - V_T)^2 \tag{2.231}$$

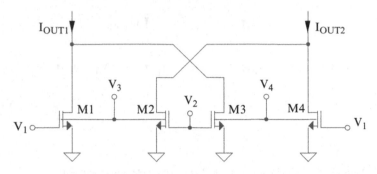

Fig. 2.56 Multiplier circuit (2) based on PR 2.Da

$$V_T = V_{T0} + \gamma\left(\sqrt{2\Phi_F - V_{BS}} - \sqrt{2\Phi_F}\right) \qquad (2.232)$$

V_T being the threshold voltage of the MOS transistor biased at a bulk-source voltage equal with V_{BS}, γ is a model parameter and Φ_F represents the Fermi potential. Replacing (2.232) in (2.231) and using the fact that $V_{T1} = V_{T2}$ and $V_{T3} = V_{T4}$ (because $V_{BS1} = V_{BS2} = V_3$ and $V_{BS3} = V_{BS4} = V_4$), it results the following expression of the output current:

$$I_{OUT1} - I_{OUT2} = \frac{K}{2}(V_1 - V_2)(V_1 + V_2 - 2V_{T1}) + \frac{K}{2}(V_2 - V_1)(V_1 + V_2 - 2V_{T3})$$

$$(2.233)$$

so:

$$I_{OUT1} - I_{OUT2} = K(V_1 - V_2)(V_{T3} - V_{T1}) \qquad (2.234)$$

Using the (2.232) relation that models the bulk effect, the previous expression of the output current can be rewritten as follows:

$$I_{OUT1} - I_{OUT2} = K(V_1 - V_2)\gamma\left(\sqrt{2\Phi_F - V_4} - \sqrt{2\Phi_F - V_3}\right)$$

$$= K(V_1 - V_2)\gamma\sqrt{2\Phi_F}\left(\sqrt{1 - \frac{V_4}{2\Phi_F}} - \sqrt{1 - \frac{V_3}{2\Phi_F}}\right) \qquad (2.235)$$

For V_3 and V_4 input signals much smaller than the Fermi potential, it is possible to use the first-order Taylor expansion $\sqrt{1+x} \cong 1 + x/2$, for $x \ll 1$, the expression of the output current becoming proportional with the product between the differential input voltages:

$$I_{OUT1} - I_{OUT2} = K(V_1 - V_2)\gamma\sqrt{2\Phi_F}\left(\frac{V_3}{4\Phi_F} - \frac{V_4}{4\Phi_F}\right)$$

$$= \frac{\gamma K}{2\sqrt{2\Phi_F}}(V_1 - V_2)(V_3 - V_4) \qquad (2.236)$$

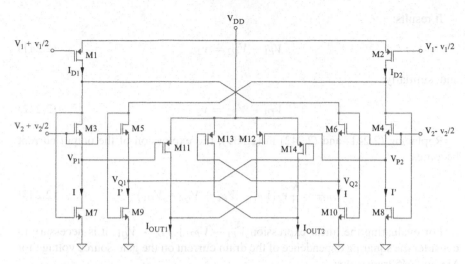

Fig. 2.57 Multiplier circuit (3) based on PR 2.Da

A symmetrical implementation of a voltage multiplier circuit is presented in Fig. 2.57 [27]. In order to improve the frequency response, all MOS transistors are biased in saturation region.

The differential output current of the voltage multiplier can be expressed as follows:

$$I_{OUT} = I_{OUT1} - I_{OUT2} = (I_{D11} + I_{D12}) - (I_{D13} + I_{D14}) \qquad (2.237)$$

resulting:

$$
\begin{aligned}
I_{OUT} &= \frac{K}{2}(V_{DD} - V_{P1} - V_T)^2 - \frac{K}{2}(V_{DD} - V_{Q1} - V_T)^2 \\
&\quad + \frac{K}{2}(V_{DD} - V_{P2} - V_T)^2 - \frac{K}{2}(V_{DD} - V_{Q2} - V_T)^2
\end{aligned} \qquad (2.238)
$$

or:

$$
\begin{aligned}
I_{OUT} &= \frac{K}{2}(V_{Q2} - V_{P1})(2V_{DD} - V_{P1} - V_{Q2} - 2V_T) \\
&\quad + \frac{K}{2}(V_{Q1} - V_{P2})(2V_{DD} - V_{P2} - V_{Q1} - 2V_T)
\end{aligned} \qquad (2.239)
$$

As a result of the connections between circuit transistors, the gate-source voltages of M7 and M10 transistors are equal. Additionally, because the same current I is passing through M3, M6, M7 and M10 transistors, all their gate-source voltages will be also equal. The identity $V_{GS3} = V_{GS6}$ can be written as:

$$V_2 + \frac{v_2}{2} - V_{P1} = V_2 - \frac{v_2}{2} - V_{Q2} \qquad (2.240)$$

It results:

$$V_{P1} - V_{Q2} = v_2 \tag{2.241}$$

and, similarly:

$$V_{Q1} - V_{P2} = v_2 \tag{2.242}$$

Replacing (2.241) and (2.242) in (2.239), the expression of the output current becomes:

$$I_{OUT} = \frac{K}{2} v_2 (V_{P1} - V_{P2} + V_{Q2} - V_{Q1}) \tag{2.243}$$

For evaluating the linear expression $V_{P1} - V_{P2} + V_{Q2} - V_{Q1}$, it is necessary to consider the squaring dependence of the drain current on the gate-source voltage for M3 and M5 transistors:

$$I = \frac{K}{2} (V_{GS3} - V_T)^2 = \frac{K}{2} \left(V_2 + \frac{v_2}{2} - V_{P1} - V_T\right)^2 \tag{2.244}$$

and:

$$I' = \frac{K}{2} (V_{GS5} - V_T)^2 = \frac{K}{2} \left(V_2 + \frac{v_2}{2} - V_{Q1} - V_T\right)^2 \tag{2.245}$$

resulting:

$$V_2 + \frac{v_2}{2} - V_{P1} - V_T = \sqrt{\frac{2I}{K}} \tag{2.246}$$

and:

$$V_2 + \frac{v_2}{2} - V_{Q1} - V_T = \sqrt{\frac{2I'}{K}} \tag{2.247}$$

Summing (2.246) and (2.247), it results:

$$2V_2 + v_2 - V_{P1} - V_{Q1} - 2V_T = \sqrt{\frac{2}{K}} \left(\sqrt{I} + \sqrt{I'}\right) \tag{2.248}$$

A similar analysis for M4 and M6 transistors will conclude to:

$$2V_2 - v_2 - V_{P2} - V_{Q2} - 2V_T = \sqrt{\frac{2}{K}} \left(\sqrt{I} + \sqrt{I'}\right) \tag{2.249}$$

Because $I_{D1} = I_{D3} + I_{D6}$ and $I_{D3} = I_{D6} = I$, it results $I_{D1} = 2I$ and, similarly, $I_{D2} = 2I'$. So:

$$\sqrt{\frac{2}{K}}\left(\sqrt{I} + \sqrt{I'}\right) = \sqrt{\frac{1}{K}}\left(\sqrt{I_{D1}} + \sqrt{I_{D2}}\right) = \sqrt{\frac{1}{K}}\sqrt{\frac{K}{2}}(V_{SG1} + V_{SG2} - 2V_T) \quad (2.250)$$

The source-gate voltages of M1 and M2 transistors can be expressed as $V_{SG1} = V_{DD} - V_1 - v_1/2$, and $V_{SG2} = V_{DD} - V_1 + v_1/2$, so:

$$\sqrt{\frac{2}{K}}\left(\sqrt{I} + \sqrt{I'}\right) = \sqrt{2}(V_{DD} - V_1 - V_T) \quad (2.251)$$

Replacing (2.251) in (2.248) and (2.249), it results:

$$2V_2 + v_2 - V_{P1} - V_{Q1} - 2V_T = 2V_2 - v_2 - V_{P2} - V_{Q2} - 2V_T$$
$$= \sqrt{2}(V_{DD} - V_1 - V_T) \quad (2.252)$$

The M1 and M2 transistors form a differential amplifier, its differential output current having the following expression:

$$I_{D1} - I_{D2} = \frac{K}{2}\left(V_{DD} - V_1 - \frac{v_1}{2} - V_T\right)^2 - \frac{K}{2}\left(V_{DD} - V_1 + \frac{v_1}{2} - V_T\right)^2$$
$$= -Kv_1(V_{DD} - V_1 - V_T) \quad (2.253)$$

As function on V_2 and v_2 input voltages, I_{D1} and I_{D2} currents could be expressed as follows:

$$I_{D1} = I_{D3} + I_{D6} = \frac{K}{2}\left(V_2 + \frac{v_2}{2} - V_{P1} - V_T\right)^2 + \frac{K}{2}\left(V_2 - \frac{v_2}{2} - V_{Q2} - V_T\right)^2 \quad (2.254)$$

and:

$$I_{D2} = I_{D4} + I_{D5} = \frac{K}{2}\left(V_2 - \frac{v_2}{2} - V_{P2} - V_T\right)^2 + \frac{K}{2}\left(V_2 + \frac{v_2}{2} - V_{Q1} - V_T\right)^2 \quad (2.255)$$

resulting:

$$I_{D1} - I_{D2} = \frac{K}{2}\left(V_{Q1} - V_{P1}\right)\left(2V_2 + v_2 - V_{P1} - V_{Q1} - 2V_T\right)$$
$$+ \frac{K}{2}\left(V_{P2} - V_{Q2}\right)\left(2V_2 - v_2 - V_{P2} - V_{Q2} - 2V_T\right) \quad (2.256)$$

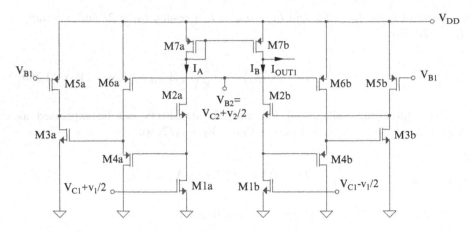

Fig. 2.58 Multiplier circuit (4) based on PR 2.Da – half circuit

Replacing (2.252) in (2.250), it results:

$$I_{D1} - I_{D2} = \frac{K}{\sqrt{2}} (V_{DD} - V_1 - V_T)(V_{Q1} - V_{P1} + V_{P2} - V_{Q2}) \qquad (2.257)$$

Comparing (2.253) with (2.257), it can write:

$$V_{Q1} - V_{P1} + V_{P2} - V_{Q2} = -\sqrt{2}v_1 \qquad (2.258)$$

Replacing (2.258) in (2.243), the output current of the multiplier will depend on the product between v_1 and v_2 voltages:

$$I_{OUT} = \frac{K}{\sqrt{2}} v_1 v_2 \qquad (2.259)$$

A voltage multiplier using MOS transistors biased in linear region is presented in Fig. 2.58 [28].

The symmetrical structure shown in Fig. 2.58 has the following expression of the output current:

$$I_{OUT1} = I_A - I_B = \frac{K}{2} \left[2(V_{GS1a} - V_T)V_{DS1a} - V_{DS1a}^2 \right]$$
$$- \frac{K}{2} \left[2(V_{GS1b} - V_T)V_{DS1b} - V_{DS1b}^2 \right] \qquad (2.260)$$

Drain-source voltages of M1a and M1b transistors are equal because M3a–M3b and M4a–M4b pairs contains identical transistors, biased at equal drain currents (for each pair):

$$V_{DS1a} = V_{DS1b} \overset{not.}{=} V_{DS1} = V_{GS3a} - V_{SG4a} = V_{GS3b} - V_{SG4b} \qquad (2.261)$$

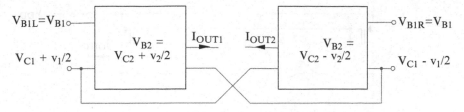

Fig. 2.59 Multiplier circuit (4) based on PR 2.Da – block diagram

As M3a and M5a and, respectively, M4a and M6a transistors are identical and biased at the same drain current, their gate-source voltages are equal, $V_{GS3a} = V_{SG5a}$ and $V_{SG4a} = V_{SG6a}$, so:

$$V_{DS1} = V_{SG5a} - V_{SG6a} = (V_{DD} - V_{B1}) - (V_{DD} - V_{B2})$$
$$= V_{B2} - V_{B1} = V_{C2} + \frac{v_2}{2} - V_{B1} \qquad (2.262)$$

From (2.260), (2.261) ad (2.262), it results the following expression of the output current for the circuit presented in Fig. 2.58:

$$I_{OUT1} = K(V_{GS1a} - V_{GS1b})V_{DS1} = Kv_1 V_{DS1} = Kv_1\left(V_{C2} + \frac{v_2}{2} - V_{B1}\right) \qquad (2.263)$$

because $V_{GS1a} = V_{C1} + v_1/2$, $V_{GS1b} = V_{C1} - v_1/2$.

In order to obtain the multiplying function, two similar circuits from Fig. 2.58 must be used, the difference between them being the value of V_{B2} potential: $V_{B2} = V_{C2} + v_2/2$ for the left structure from Fig. 2.59 [28] and $V_{B2} = V_{C2} - v_2/2$ for the right structure.

The expression of the output current of the multiplier circuit presented in Fig. 2.59 will be:

$$I_{OUT1} - I_{OUT2} = Kv_1\left(V_{C2} + \frac{v_2}{2} - V_{B1}\right) - Kv_1\left(V_{C2} - \frac{v_2}{2} - V_{B1}\right) = Kv_1 v_2 \qquad (2.264)$$

Another possible realization of the voltage multiplier circuit is designed with the main goal of reducing the total harmonic distortions coefficient (THD).

Because of the quadratic characteristic of a MOS transistor biased in saturation, the linearity of the basic multiplier presented in Fig. 2.60 [29, 31] is rather poor. The core of this circuit is a modified Gilbert cell, extended to implement with good linearity the multiplication function.

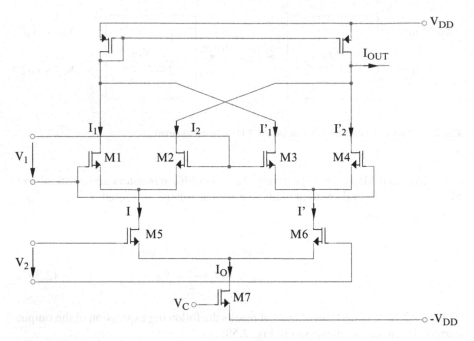

Fig. 2.60 Multiplier circuit (5) based on PR 2.Da

Considering a biasing in saturation of all transistors from Fig. 2.60, the expressions of the drain currents for M1–M4 transistors are:

$$I_{1,2} = \frac{I}{2}\left(1 \pm \sqrt{\frac{KV_1^2}{I} - \frac{K^2 V_1^4}{4I^2}}\right); \quad I'_{2,1} = \frac{I'}{2}\left(1 \pm \sqrt{\frac{KV_1^2}{I'} - \frac{K^2 V_1^4}{4(I')^2}}\right) \quad (2.265)$$

The output current expression is:

$$I_{OUT} = (I_1 - I_2) + (I'_1 - I'_2) = V_1\left(\sqrt{KI - \frac{K^2}{4}V_1^2} - \sqrt{KI' - \frac{K^2}{4}V_1^2}\right) \quad (2.266)$$

Similarly, the drain currents of M5 and M6 transistors will have the following expressions:

$$I = \frac{I_O}{2}\left(1 + \sqrt{\frac{KV_2^2}{I_O} - \frac{K^2 V_2^4}{4I_O^2}}\right) \quad (2.267)$$

$$I' = \frac{I_O}{2}\left(1 - \sqrt{\frac{KV_2^2}{I_O} - \frac{K^2 V_2^4}{4I_O^2}}\right) \quad (2.268)$$

All transistors from Fig. 2.60 are supposed to be identical. Considering the limited expansion $\sqrt{1+x} \cong 1 + x/2$, from the previous relations, it results the approximate expression of the basic multiplier output current:

$$I_{OUT} \cong \frac{K}{\sqrt{2}} V_1 V_2 - \frac{K^2}{8\sqrt{2}I_O} V_1 V_2^3 \qquad (2.269)$$

Thus, the total harmonic distortions coefficient introduced by the circuit (approximated with the third-order one) will be expressed as:

$$THD_3 \cong \frac{KV_2^2}{8I_O} = \frac{1}{4}\left(\frac{V_2}{V_C + V_{DD} - V_T}\right)^2 \qquad (2.270)$$

In conclusion, THD_3 is directly proportional with the ratio between V_2 input signal amplitude and the effective gate-source voltage of the biasing transistor.

In order to improve the circuit linearity, the modified multiplier presented in Fig. 2.61 [29] replaces the differential amplifier M5–M6 from Fig. 2.60 with a cross-connected one, M5–M8 from Fig. 2.61. The output current expression for the multiplier with improved linearity has the same form (2.266), but the expressions of I and I' currents become:

$$I = I_{p1} + I'_{p1} \qquad (2.271)$$

$$I' = I_{p2} + I'_{p2} \qquad (2.272)$$

where:

$$I_{p_{1,2}} = \frac{I_{O1}}{2}\left(1 \pm \sqrt{\frac{K_1 V_2^2}{I_{O1}} - \frac{K_1^2 V_2^4}{4I_{O1}^2}}\right) \qquad (2.273)$$

$$I'_{p_{1,2}} = \frac{I_{O2}}{2}\left(1 \mp \sqrt{\frac{K_2 V_2^2}{I_{O2}} - \frac{K_2^2 V_2^4}{4I_{O2}^2}}\right) \qquad (2.274)$$

Similarly with the basic circuit analysis, considering the more accurate expansion $\sqrt{1+x} \cong 1 + x/2 - x^2/4$, the output current of the multiplier from Fig. 2.61 will have the following expression:

$$I_{OUT} \cong V_1 \sqrt{\frac{K}{a}}(bV_2 + cV_2^3 + dV_2^5) \qquad (2.275)$$

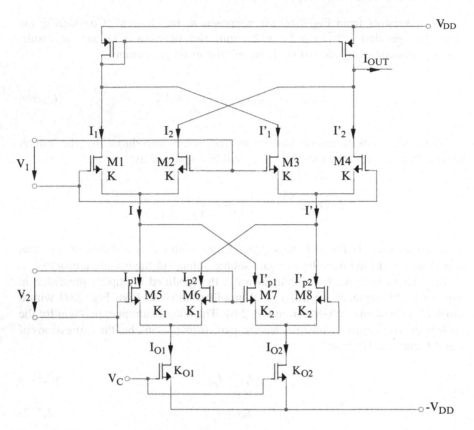

Fig. 2.61 Multiplier circuit (6) based on PR 2.Da

where b, c and d are constants, expressed as follows:

$$b = \frac{K_1^{1/2}I_{O1}^{1/2} - K_2^{1/2}I_{O2}^{1/2}}{2} \qquad (2.276)$$

$$c = \frac{K_2^{3/2}I_{O2}^{-1/2} - K_1^{3/2}I_{O1}^{-1/2}}{16} \qquad (2.277)$$

$$d = \frac{K_1^{5/2}I_{O1}^{-3/2} - K_2^{5/2}I_{O2}^{-3/2}}{128} \qquad (2.278)$$

Because the main nonlinearity from the output current expression is caused by the third-order term of relation (2.242), the proposed linearization technique is referring to the cancellation of the third-order distortions ($c = 0$), equivalent with the following design condition:

$$\frac{I_{O2}}{I_{O1}} = \left(\frac{K_2}{K_1}\right)^3 \tag{2.279}$$

and, in consequence:

$$b = \frac{(K_1 I_{O1})^{1/2}}{2}\left[1 - \left(\frac{K_2}{K_1}\right)^2\right] \tag{2.280}$$

$$d = \frac{K_1^{5/2} I_{O1}^{-3/2}}{128}\left[1 - \left(\frac{K_1}{K_2}\right)^2\right] \tag{2.281}$$

The total harmonic distortions for the multiplier circuit with improved linearity (Fig. 2.61), approximated with the fifth-order one, will be:

$$THD_5 = \left(\frac{K_1^2}{8K_2 I_{O1}}\right)^2 V_2^4 = \left(\frac{K_1^2}{4K_2 K_{O1}}\right)^2 \left(\frac{V_2}{V_C + V_{DD} - V_T}\right)^4 \tag{2.282}$$

Considering the particular case that $K_2 = K_{O1}$ and $K_1/K_2 = 1/2$, it results:

$$THD_5 = \frac{1}{256}\left(\frac{V_2}{V_C + V_{DD} - V_T}\right)^4 \tag{2.283}$$

Thus, the linearity improvement from the circuit presented in Fig. 2.61 with respect to the basic multiplier presented in Fig. 2.60 is about two orders of magnitude:

$$\frac{THD_3}{THD_5} = 64\left(\frac{V_C + V_{DD} - V_T}{V_2}\right)^2 \tag{2.284}$$

For the multiplier circuit presented in Fig. 2.62 [32], the expression of the output current is:

$$I_{OUT} = I_1 + I_2 - I_3 - I_4 \tag{2.285}$$

or:

$$I_{OUT} = \frac{K}{2}(V_4 - V_X - V_T)^2 + \frac{K}{2}(V_3 - V_Y - V_T)^2$$
$$- \frac{K}{2}(V_3 - V_X - V_T)^2 - \frac{K}{2}(V_4 - V_Y - V_T)^2 \tag{2.286}$$

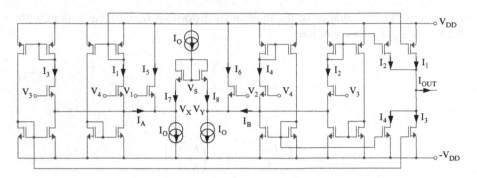

Fig. 2.62 Multiplier circuit (7) based on PR 2.Da

resulting:

$$I_{OUT} = \frac{K}{2}(V_4 - V_3)(V_3 + V_4 - 2V_X - 2V_T)$$
$$+ \frac{K}{2}(V_3 - V_4)(V_3 + V_4 - 2V_Y - 2V_T) \tag{2.287}$$

So:

$$I_{OUT} = K(V_3 - V_4)(V_X - V_Y) \tag{2.288}$$

Because of the current mirrors from the circuit, $I_A = I_B = 0$, so between the currents from the circuit will exist the following linear relations:

$$I_7 + I_5 = I_O \tag{2.289}$$

$$I_7 + I_8 = I_O \tag{2.290}$$

and:

$$I_8 + I_6 = I_O \tag{2.291}$$

resulting $I_5 = I_8$ and $I_6 = I_7$. So:

$$V_1 - V_X = V_S - V_Y \tag{2.292}$$

and:

$$V_2 - V_Y = V_S - V_X \tag{2.293}$$

equivalent with:

$$V_X - V_Y = V_1 - V_S \tag{2.294}$$

Fig. 2.63 Asymmetrical
differential structure

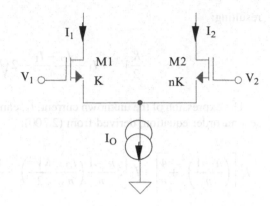

and:

$$V_X - V_Y = V_S - V_2 \tag{2.295}$$

It results:

$$V_S = \frac{V_1 + V_2}{2} \tag{2.296}$$

and:

$$V_X - V_Y = V_1 - \frac{V_1 + V_2}{2} = \frac{V_1 - V_2}{2} \tag{2.297}$$

Replacing (2.297) in (2.288), the output current will be proportional with the square of the differential input voltage:

$$I_{OUT} = \frac{K}{2}(V_1 - V_2)(V_3 - V_4) \tag{2.298}$$

A possible realization of a voltage multiplier circuit is based on a particular implementation of a voltage squaring circuit. The method for designing such a voltage squaring circuit uses a differential amplifier (Fig. 2.63) [17, 33] having a controllable asymmetry between the geometries of its two composing transistors. This difference between the aspect ratios of MOS transistors will introduce in the output currents of the differential amplifier a term proportional with the square of the differential input voltage.

Noting with $V = V_1 - V_2$ the differential input voltage, it can be expressed as follows:

$$V = V_{GS1} - V_{GS2} = \sqrt{\frac{2I_1}{K}} - \sqrt{\frac{2(I_O - I_1)}{nK}} \tag{2.299}$$

resulting:

$$\frac{K}{2}V^2 = I_1 + \frac{I_O - I_1}{n} - 2\sqrt{\frac{I_1(I_O - I_1)}{n}} \qquad (2.300)$$

The expression of the unknown current, I_1, can be obtained solving the following second-order equation, derived from (2.300):

$$I_1^2\left[\left(\frac{n-1}{n}\right)^2 + \frac{4}{n}\right] + I_1\left[2\frac{n-1}{n}\left(\frac{I_O}{n} - \frac{KV^2}{2}\right) - \frac{4I_O}{n}\right] + \left(\frac{I_O}{n} - \frac{KV^2}{2}\right)^2 = 0 \quad (2.301)$$

So:

$$I_1 = \frac{I_O}{n+1} + \frac{n(n-1)}{2(n+1)^2}KV^2 + \frac{nV}{(n+1)^2}\sqrt{2KI_O(n+1) - K^2nV^2} \qquad (2.302)$$

and:

$$I_2 = I_O - I_1 = \frac{nI_O}{n+1} - \frac{n(n-1)}{2(n+1)^2}KV^2 - \frac{nV}{(n+1)^2}\sqrt{2KI_O(n+1) - K^2nV^2} \quad (2.303)$$

The complete realization of a voltage squaring circuit uses a cross-coupling of two differential amplifier having controllable asymmetries between their geometries, M1–M2 and M3–M4 (Fig. 2.64) [17, 33].

Using (2.302) and (2.303), it results:

$$I_{OUT} = I_{D2} + I_{D4} = \left[\frac{nI_O}{n+1} - \frac{n(n-1)}{2(n+1)^2}KV^2 - \frac{nV}{(n+1)^2}\sqrt{2KI_O(n+1) - K^2nV^2}\right]$$

$$+ \left[\frac{nI_O}{n+1} - \frac{n(n-1)}{2(n+1)^2}KV^2 + \frac{nV}{(n+1)^2}\sqrt{2KI_O(n+1) - K^2nV^2}\right]$$

$$= \frac{2nI_O}{n+1} - \frac{n(n-1)}{(n+1)^2}KV^2 \qquad (2.304)$$

An application of the previous presented voltage squarer circuit is represented by a voltage multiplier with linear characteristic (Fig. 2.65).

The method of designing the multiplying function is to use two cross-connected M1–M2 and M3–M4 differential amplifiers, each of them being biased at a drain

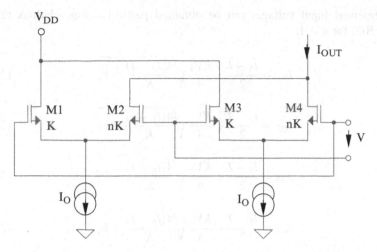

Fig. 2.64 Voltage squaring circuit

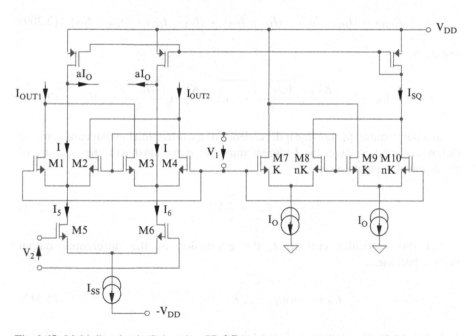

Fig. 2.65 Multiplier circuit (8) based on PR 2.Da

current equal with the difference between a drain current of another differential amplifier, M5–M6 (I_5 and, respectively, I_6) and a current, I, which must have a term proportional with the square of the first differential input voltage, in order to compensate the intrinsic nonlinearity of the differential amplifiers. Because M1–M4 transistors are identical, the dependencies of their drain currents on

V_1 differential input voltage, can be obtained particularizing relations (2.302) and (2.303) for $n = 1$:

$$I_{D1} = \frac{I_5 - I}{2} - \frac{KV_1}{4}\sqrt{\frac{4(I_5 - I)}{K} - V_1^2} \tag{2.305}$$

$$I_{D2} = \frac{I_5 - I}{2} + \frac{KV_1}{4}\sqrt{\frac{4(I_5 - I)}{K} - V_1^2} \tag{2.306}$$

$$I_{D3} = \frac{I_6 - I}{2} + \frac{KV_1}{4}\sqrt{\frac{4(I_6 - I)}{K} - V_1^2} \tag{2.307}$$

$$I_{D4} = \frac{I_6 - I}{2} - \frac{KV_1}{4}\sqrt{\frac{4(I_6 - I)}{K} - V_1^2} \tag{2.308}$$

The expression of the differential output current of the entire structure will be:

$$I_{OUT1} - I_{OUT2} = (I_{D1} + I_{D3}) - (I_{D2} + I_{D4}) = (I_{D1} - I_{D2}) - (I_{D4} - I_{D3}) \tag{2.309}$$

resulting:

$$I_{OUT1} - I_{OUT2} = -\frac{KV_1}{2}\sqrt{\frac{4(I_5 - I)}{K} - V_1^2} + \frac{KV_1}{2}\sqrt{\frac{4(I_6 - I)}{K} - V_1^2} \tag{2.310}$$

In order to cancel the nonlinear dependence of the differential output current on the differential input voltage, the I current must be proportional with the squaring of the differential input voltage:

$$I = -\frac{K}{4}V_1^4 \tag{2.311}$$

For this particular current, I, the expression of the differential output current becomes:

$$I_{OUT1} - I_{OUT2} = \sqrt{K}V_1\left(\sqrt{I_6} - \sqrt{I_5}\right) \tag{2.312}$$

Considering the squaring dependencies of the drain currents of M5–M6 differential amplifier on the gate-source voltages of their composing transistors, it results:

$$I_{OUT1} - I_{OUT2} = \sqrt{K}V_1\left[\sqrt{\frac{K}{2}}(V_{GS6} - V_{GS5})\right] = \frac{1}{\sqrt{2}}KV_1V_2 \tag{2.313}$$

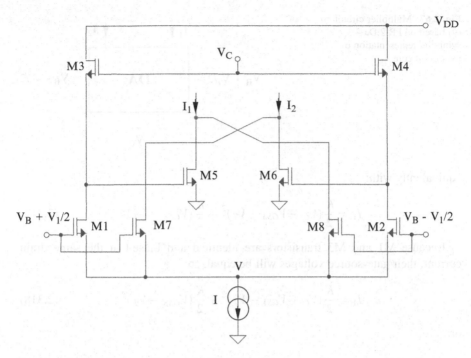

Fig. 2.66 Multiplier circuit (9) based on PR 2.Da – circuit core

So, the differential output current is proportional with the product between the input voltages. The implementation of the I current having the (2.311) expression is realized using M7–M10 transistors, representing a voltage squaring circuit, similar with the structure presented in Fig. 2.64. Comparing (2.304) with (2.311), it results the following condition that must be imposed to the constant n:

$$\frac{n(n-1)}{(n+1)^2} = \frac{1}{4}$$

(2.314)

So $n = 2.15$ and:

$$I_{SQ} = 1.36I_O - \frac{K}{4}V_1^2$$

(2.315)

Because $I = I_{SQ} - aI_O$, the expression (2.311) of I current can be obtained for $a = 1.36$.

Another possible implementation of a multiplier circuit is based on the differential amplifier presented in Fig. 2.66 [34].

The expression of the first output current is:

$$I_1 = \frac{K}{2}(V_{GS5} - V_T)^2 + \frac{K}{2}(V_{GS8} - V_T)^2$$

(2.316)

Fig. 2.67 Multiplier circuit
(9) based on PR 2.Da –
symbolic representation of
the circuit core

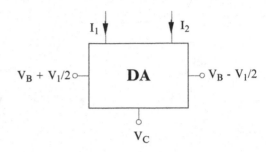

equivalently with:

$$I_1 = \frac{K}{2}(V_C - V_{GS3} - V_T)^2 + \frac{K}{2}(V_{GS8} - V_T)^2 \qquad (2.317)$$

Because M1 and M3 transistors are identical and biased at the same drain current, their gate-source voltages will be equal, so:

$$I_1 = \frac{K}{2}(V_C - V_{GS1} - V_T)^2 + \frac{K}{2}(V_{GS8} - V_T)^2 \qquad (2.318)$$

or:

$$I_1 = \frac{K}{2}\left(V_C - V_B + V - V_T - \frac{V_1}{2}\right)^2 + \frac{K}{2}\left(V_B - V - V_T - \frac{V_1}{2}\right)^2 \qquad (2.319)$$

Similarly, the second output current, I_{OUT2}, will have the following expression:

$$I_2 = \frac{K}{2}\left(V_C - V_B + V - V_T + \frac{V_1}{2}\right)^2 + \frac{K}{2}\left(V_B - V - V_T + \frac{V_1}{2}\right)^2 \qquad (2.320)$$

The differential output current can be expressed as follows:

$$I_2 - I_1 = KV_1(V_C - V_B + V - V_T) + KV_1(V_B - V - V_T) = K(V_C - 2V_T)V_1 \quad (2.321)$$

The symbolic representation of the circuit is shown in Fig. 2.67.

A voltage multiplier can be realized using two previous presented differential amplifiers having the V_C voltage sources replaced by two voltages that are dependent on a second input voltage, V_2. The complete multiplier circuit is shown in Fig. 2.68.

The differential output current of the voltage multiplier presented in Fig. 2.68 can be expressed as follows:

$$I_{OUT1} - I_{OUT2} = (I_1 + I_3) - (I_2 + I_4) = (I_1 - I_2) + (I_3 - I_4) \qquad (2.322)$$

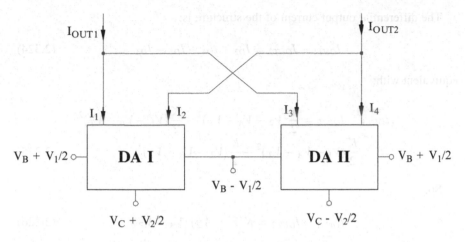

Fig. 2.68 Multiplier circuit (9) based on PR 2.Da – block diagram

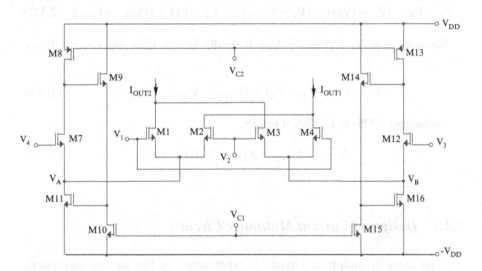

Fig. 2.69 Multiplier circuit (10) based on PR 2.Da

resulting, using (2.321):

$$I_{OUT1} - I_{OUT2} = K\left[\left(V_C + \frac{V_2}{2}\right) - 2V_T\right]V_1$$

$$+ K\left[\left(V_C - \frac{V_2}{2}\right) - 2V_T\right](-V_1) = KV_1V_2 \qquad (2.323)$$

A possible realization of a voltage multiplier circuit is presented in Fig. 2.69 [35].

The differential output current of the structure is:

$$I_{OUT1} - I_{OUT2} = I_{D2} + I_{D4} - I_{D1} - I_{D3} \tag{2.324}$$

equivalent with:

$$I_{OUT1} - I_{OUT2} = \frac{K}{2}(V_2 - V_A - V_T)^2 + \frac{K}{2}(V_1 - V_B - V_T)^2$$
$$- \frac{K}{2}(V_1 - V_A - V_T)^2 - \frac{K}{2}(V_2 - V_B - V_T)^2 \tag{2.325}$$

So:

$$I_{OUT1} - I_{OUT2} = K(V_1 - V_2)(V_A - V_B) \tag{2.326}$$

The $V_A - V_B$ differential voltage can be expressed as follows:

$$V_A - V_B = (V_4 - V_{GS7}) - (V_3 - V_{GS12}) = (V_4 - V_3) + (V_{GS12} - V_{GS7}) \tag{2.327}$$

Because $I_{D7} = I_{D8}$ and $I_{D12} = I_{D13}$, it results $V_{GS7} = V_{SG8}$ and $V_{GS12} = V_{SG13}$. So:

$$V_A - V_B = (V_4 - V_3) + (V_{SG13} - V_{SG8}) = V_4 - V_3 \tag{2.328}$$

Replacing (2.328) in (2.326), it results:

$$I_{OUT1} - I_{OUT2} = K(V_1 - V_2)(V_4 - V_3) \tag{2.329}$$

2.2.2 Design of Current Multiplier Circuits

Another class of multiplier circuits is represented by the current multipliers, the output current of these circuits being proportional with the product between two input currents. The current-mode operation of the multiplier circuits has the important advantage of increasing the frequency response of the designed structures.

2.2.2.1 Multiplier Circuits Based on the Ninth Mathematical Principle (PR 2.9)

A possibility of designing a current multiplier uses as circuit cores two current square-root circuits (Fig. 2.70) [36].

Fig. 2.70 CMOS square-root circuit for the multiplier circuit (1) based on PR 2.9

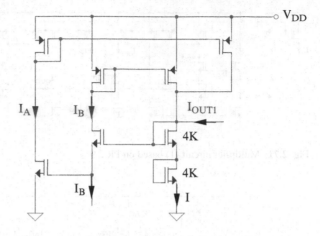

For a biasing in saturation of all the MOS transistors from Fig. 2.70, it is possible to write that:

$$\left(V_T + \sqrt{\frac{2I_A}{K}}\right) + \left(V_T + \sqrt{\frac{2I_B}{K}}\right) = 2\left(V_T + \sqrt{\frac{2I}{4K}}\right) \tag{2.330}$$

equivalent with:

$$I = I_A + I_B + 2\sqrt{I_A I_B} \tag{2.331}$$

Implementing the following linear relation between the previous current of the square-root circuit:

$$I_{OUT1} = I - I_A - I_B \tag{2.332}$$

the output current of the circuit from Fig. 2.70 will be proportional with the square-root of the input current:

$$I_{OUT1} = 2\sqrt{I_A I_B} \tag{2.333}$$

The multiplier circuit can be obtained using two square-root circuits from Fig. 2.70 connected as it is shown in Fig. 2.71. The computed functions are $I_{OUT1} = 2\sqrt{I_{OUT}I_O}$ and $I_{OUT2} = 2\sqrt{I_1 I_2}$. Because $I_{OUT1} = I_{OUT2}$, the function implemented by the circuit from Fig. 2.71 [36] will be:

$$I_{OUT} = \frac{I_1 I_2}{I_O} \tag{2.334}$$

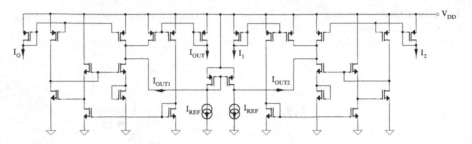

Fig. 2.71 Multiplier circuit (1) based on PR 2.9

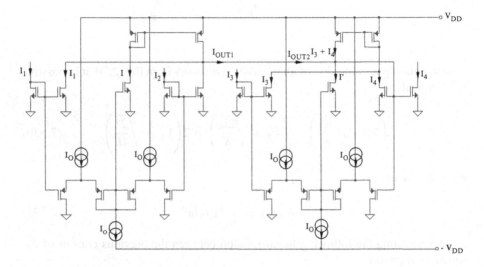

Fig. 2.72 Multiplier circuit (2) based on PR 2.9

The design of a multiplier/divider circuit can be done using two square-root circuits, as it is shown in Fig. 2.72 [37].

The square-root circuits have the implementation presented in Fig. 2. The output currents of these circuits have the following expressions:

$$I_{OUT1} = 2\sqrt{I_1 I_2} \tag{2.335}$$

and:

$$I_{OUT2} = 2\sqrt{I_3 I_4} \tag{2.336}$$

Imposing by design $I_{OUT1} = I_{OUT2}$, it results:

$$I_4 = \frac{I_1 I_2}{I_3} \tag{2.337}$$

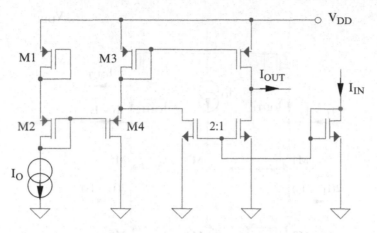

Fig. 2.73 CMOS squaring circuit for the multiplier circuit (1) based on PR 2.10

2.2.2.2 Multiplier circuits based on the tenth mathematical principle (PR 2.10)

The following multiplier structure (Fig. 2.74) is derived from the squaring circuit presented in Fig. 2.73 [38].

The characteristic equation of the translinear loop including M1–M4 transistors is:

$$V_{SG1} + V_{SG2} = V_{SG3} + V_{SG4} \tag{2.338}$$

resulting:

$$2\sqrt{I_{D1}} = \sqrt{I_{D3}} + \sqrt{I_{D4}} \tag{2.339}$$

equivalent with:

$$2\sqrt{I_O} = \sqrt{I_{OUT} + I_{IN}} + \sqrt{I_{OUT} - I_{IN}} \tag{2.340}$$

The expression of the output current will be:

$$I_{OUT} = I_O + \frac{I_{IN}^2}{4I_O} \tag{2.341}$$

Using the squaring circuit presented in Fig. 2.73, it is possible to design a current multiplier structure (Fig. 2.74) [38].

The output current of the multiplier circuit can be expressed as follows:

$$I_{OUT} = I_{OUT2} - I_{OUT1} = \left[I_O + \frac{(I_1 + I_2)^2}{4I_O}\right] - \left[I_O - \frac{(I_1 + I_2)^2}{4I_O}\right] = \frac{I_1 I_2}{I_O} \tag{2.342}$$

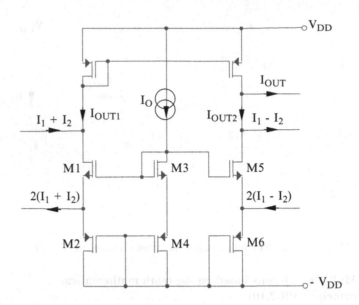

Fig. 2.75 Multiplier circuit (2) based on PR 2.10

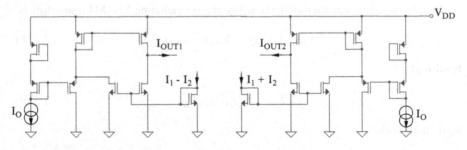

Fig. 2.74 Multiplier circuit (1) based on PR 2.10

The multiplier/divider circuit presented in Fig. 2.75 [39] uses two translinear loops implemented with M1–M4 and M3–M6 transistors, respectively.

The characteristic equation of the first translinear loop is:

$$V_{GS3} + V_{SG4} = V_{GS1} + V_{SG2} \qquad (2.343)$$

Considering a biasing in saturation of all MOS transistors from Fig. 2.75, it results:

$$2\sqrt{I_O} = \sqrt{I_{OUT1} + (I_1 + I_2)} + \sqrt{I_{OUT1} - (I_1 + I_2)} \qquad (2.344)$$

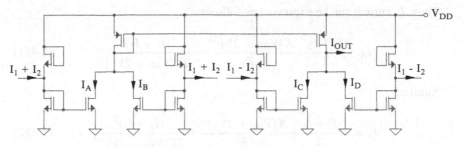

Fig. 2.76 Multiplier circuit (2) based on PR 2.10

The I_{OUT1} current will be expressed as follows:

$$I_{OUT1} = I_O + \frac{(I_1 + I_2)^2}{4I_O} \tag{2.345}$$

Similarly, analyzing the characteristic equation for the second translinear loop, it results:

$$2\sqrt{I_O} = \sqrt{I_{OUT1} + (I_1 - I_2)} + \sqrt{I_{OUT1} - (I_1 - I_2)} \tag{2.346}$$

So, the expression of I_{OUT2} current will be:

$$I_{OUT2} = I_O + \frac{(I_1 - I_2)^2}{4I_O} \tag{2.347}$$

The output current of the multiplier/divider circuit presented in Fig. 2.75 will have the following expression:

$$I_{OUT} = I_{OUT1} - I_{OUT2} = \frac{I_1 I_2}{I_O} \tag{2.348}$$

The most important advantage of the multiplier is the independence of the circuit performances on technological errors.

A current multiplier circuit (Fig. 2.76) [40] can be designed using four current squaring circuits.

Because:

$$V_{DD} = V_{GS}(I_A) + V_{GS}(I_A - I_1 - I_2) \tag{2.349}$$

it results:

$$\sqrt{I_A} + \sqrt{I_A - (I_1 + I_2)} = \sqrt{\frac{K}{2}}(V_{DD} - 2V_T) \tag{2.350}$$

Thus, I_A current can be expressed as follows:

$$I_A = \frac{I_1 + I_2}{2} + \frac{K(V_{DD} - 2V_T)^2}{8} + \frac{(I_1 + I_2)^2}{2K(V_{DD} - 2V_T)^2} \qquad (2.351)$$

Similarly:

$$I_B = -\frac{I_1 + I_2}{2} + \frac{K(V_{DD} - 2V_T)^2}{8} + \frac{(I_1 + I_2)^2}{2K(V_{DD} - 2V_T)^2} \qquad (2.352)$$

$$I_C = \frac{I_1 - I_2}{2} + \frac{K(V_{DD} - 2V_T)^2}{8} + \frac{(I_1 - I_2)^2}{2K(V_{DD} - 2V_T)^2} \qquad (2.353)$$

$$I_D = -\frac{I_1 - I_2}{2} + \frac{K(V_{DD} - 2V_T)^2}{8} + \frac{(I_1 - I_2)^2}{2K(V_{DD} - 2V_T)^2} \qquad (2.354)$$

The output current of the multiplier circuit presented in Fig. 2.76 will have the following expression:

$$I_{OUT} = I_A + I_B - I_C - I_D = \frac{4I_1 I_2}{K(V_{DD} - 2V_T)^2} \qquad (2.355)$$

Comparing with the previous circuit, the operation of this multiplier structure is affected by technological errors (both K and V_T technological parameters appear in the expression of the output current).

The current multiplier circuit presented in Fig. 2.77 [41] contains two current squaring circuits.

The expressions of the output current for these circuits are:

$$I_{OUT1} = 2I_O + \frac{(I_1 + I_2)^2}{8I_O} \qquad (2.356)$$

and:

$$I_{OUT2} = 2I_O + \frac{(I_1 - I_2)^2}{8I_O} \qquad (2.357)$$

The output current of the multiplier circuit presented in Fig. 2.77 will have the following expression:

$$I_{OUT} = I_{OUT1} - I_{OUT2} = \frac{I_1 I_2}{2I_O} \qquad (2.358)$$

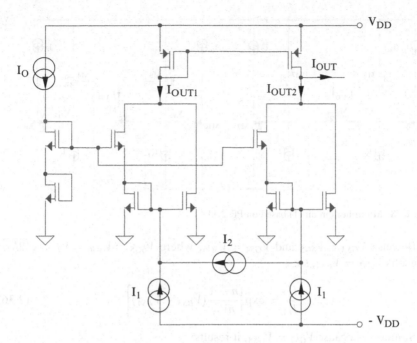

Fig. 2.77 Multiplier circuit (4) based on PR 2.10

2.2.2.3 Multiplier Circuits Based on the Eleventh Mathematical Principle (PR 2.11)

The following multiplier structures are designed for low-power applications, the reducing of their current consumptions being obtained by a biasing in weak inversion of all MOS active devices.

A current multiplier/divider using bulk-driven subthreshold-operated MOS transistors is presented in Fig. 2.78 [42, 43]. The double drive of the MOS devices (on gate and on bulk) allows the reduction of the computational circuit complexity. Unfortunately, the possibility of implementation in silicon is limited to CMOS technologies with independent wells. Imposing a weak inversion of all MOS transistors from Fig. 2.78, it is possible to write:

$$I_{OUT1} = I_{DO} \exp\left(\frac{V_{GS1} + (n-1)V_{BS1}}{nV_{th}}\right) \qquad (2.359)$$

$$I_{OUT2} = I_{DO} \exp\left(\frac{V_{GS2} + (n-1)V_{BS2}}{nV_{th}}\right) \qquad (2.360)$$

So:

$$\frac{I_{OUT1}}{I_{OUT2}} = \exp\left(\frac{V_{X2} - V_{Y2}}{nV_{th}}\right) \exp\left[\frac{n-1}{nV_{th}}(V_{X1} - V_{Y1})\right] \qquad (2.361)$$

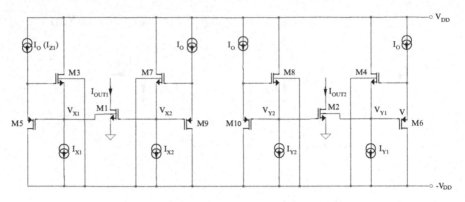

Fig. 2.78 Multiplier circuit (1) based on PR 2.11

Because $V_{GS3} = V_{SG5}$ and $V_{GS4} = V_{SG6}$, where $V_{SG5} = V_{SG6} = V_T + \sqrt{2I_O/K}$ it results $V_{GS3} = V_{GS4}$, so:

$$\frac{I_{X1}}{I_{Y1}} = \exp\left[\frac{n-1}{nV_{th}}(V_{BS3} - V_{BS4})\right] \tag{2.362}$$

Similarly, because $V_{GS7} = V_{GS8}$, it results:

$$\frac{I_{X2}}{I_{Y2}} = \exp\left[\frac{n-1}{nV_{th}}(V_{BS7} - V_{BS8})\right] \tag{2.363}$$

Knowing that $V_{X1} - V_{Y1} = V_{BS4} - V_{BS3}$, $V_{X2} - V_{Y2} = V_{BS8} - V_{BS7}$ and using (2.361) and (2.362), it can write that:

$$\frac{I_{OUT1}}{I_{OUT2}} = \frac{I_{Y1}}{I_{X1}}\left(\frac{I_{Y2}}{I_{X2}}\right)^{\frac{1}{n-1}} \tag{2.364}$$

A generalization of (2.364) relation for different values of I_O current sources (I_{Z1}, I_{W1} and I_{Z2}, I_{W2}, respectively) allows to obtain a more complex relation between the circuit currents:

$$\frac{I_{OUT1}}{I_{OUT2}} = \frac{I_{Y1}}{I_{X1}}\frac{I_{Z1}}{I_{W1}}\left(\frac{I_{Y2}}{I_{X2}}\frac{I_{Z2}}{I_{W2}}\right)^{\frac{1}{n-1}} \tag{2.365}$$

Another implementation of a current-mode multiplier/divider is referred to a bulk-driven active-load differential amplifier from Fig. 2.79 [42]. Considering a weak inversion operation of all transistors from Fig. 2.79, the ratio of I_1 and I_2 currents will be:

$$\frac{I_{OUT1}}{I_{OUT2}} = \exp\left(\frac{V_{GS1} - V_{GS2}}{nV_{th}}\right) \tag{2.366}$$

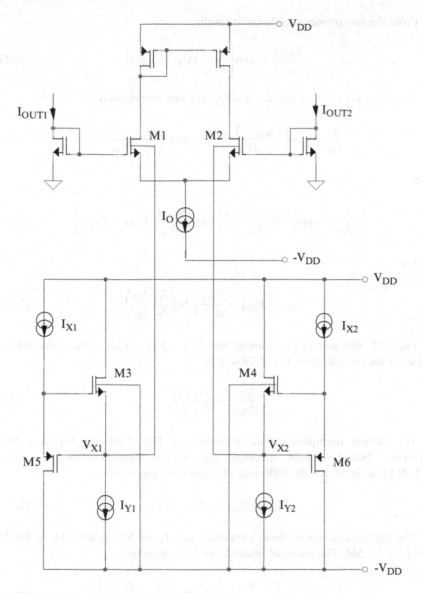

Fig. 2.79 Multiplier circuit (2) based on PR 2.11

For the active-load differential amplifier, it can write:

$$\frac{I_{D1}}{I_{D2}} = 1 = \exp\left(\frac{V_{GS1} - V_{GS2}}{nV_{th}}\right) \exp\left[\frac{n-1}{nV_{th}}(V_{BS1} - V_{BS2})\right] \tag{2.367}$$

From the two previous relations, it results:

$$\frac{I_{OUT1}}{I_{OUT2}} = \exp\left[\frac{1-n}{nV_{th}}(V_{X1} - V_{X2})\right]$$ (2.368)

Similarly, the ratios of I_{X1}, I_{X2} and I_{Y1}, I_{Y2} are, respectively:

$$\frac{I_{X1}}{I_{X2}} = \exp\left(\frac{V_{SG6} - V_{SG5}}{nV_{th}}\right) = \exp\left(\frac{V_{GS3} - V_{GS4}}{nV_{th}}\right)$$ (2.369)

and:

$$\frac{I_{Y1}}{I_{Y2}} = \exp\left(\frac{V_{GS3} - V_{GS4}}{nV_{th}}\right) \exp\left[\frac{n-1}{nV_{th}}(V_{BS3} - V_{BS4})\right]$$ (2.370)

So:

$$V_{BS3} - V_{BS4} = \frac{nV_{th}}{n-1} \ln\left(\frac{I_{Y1}}{I_{Y2}}\frac{I_{X2}}{I_{X1}}\right)$$ (2.371)

From (2.368) and (2.371), using that $V_{X1} - V_{X2} = V_{BS4} - V_{BS3}$, the relation between the currents from Fig. 2.79 will be:

$$\frac{I_{OUT1}}{I_{OUT2}} = \frac{I_{Y1}}{I_{Y2}}\frac{I_{X2}}{I_{X1}}$$ (2.372)

The current multiplier circuit presented in Fig. 2.80 [44–46] uses MOS transistors biased in weak inversion region. The translinear loop containing M1–M4 transistors has the following characteristic equation:

$$V_{GS1} + V_{GS2} = V_{GS3} + V_{GS4}$$ (2.373)

The biasing currents of these transistors are: I_1 for M1, I_2 for M2, I_O for M3 and I_{OUT} for M4. The previous relation can be written as:

$$nV_{th} \ln\left[\frac{I_1}{(W/L)I_{D0}}\right] + nV_{th} \ln\left[\frac{I_2}{(W/L)I_{D0}}\right]$$
$$= nV_{th} \ln\left[\frac{I_O}{(W/L)I_{D0}}\right] + nV_{th} \ln\left[\frac{I_{OUT}}{(W/L)I_{D0}}\right]$$ (2.374)

resulting:

$$I_{OUT} = \frac{I_1 I_2}{I_O}$$ (2.375)

Fig. 2.80 Multiplier circuit (3) based on PR 2.11

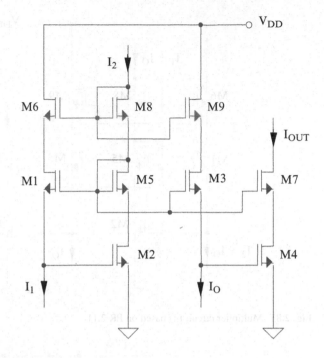

A four-quadrant multiplier derived from the circuit presented in Fig. 2.80 is shown in Fig. 2.81 [44].

Similarly with the previous circuit:

$$I_{D7} = \frac{(I_O + I_1)(I_O + I_2)}{I_O} = I_O + I_1 + I_2 + \frac{I_1 I_2}{I_O} \qquad (2.376)$$

resulting the following expression of the output current:

$$I_{OUT} = I_{D7} - I_O - I_1 - I_2 = \frac{I_1 I_2}{I_O} \qquad (2.377)$$

The circuit presented in Fig. 2.82 [47, 48] represents a current multiplier implemented using MOS transistors biased in weak inversion region. Using the exponential dependence of the drain current on the gate-source and bulk-source voltages for a subthreshold-operated MOS device, the ratio between I_2 and I_{OUT} currents can be expressed as follows:

$$\frac{I_2}{I_{OUT}} = \frac{I_{DO} \exp\left[\frac{V_{SG2} + (n-1)V_{SB2}}{nV_{th}}\right]}{I_{DO} \exp\left[\frac{V_{SG4} + (n-1)V_{SB4}}{nV_{th}}\right]} = \exp\left(\frac{V_{SG2} - V_{SG4}}{nV_{th}}\right) \qquad (2.378)$$

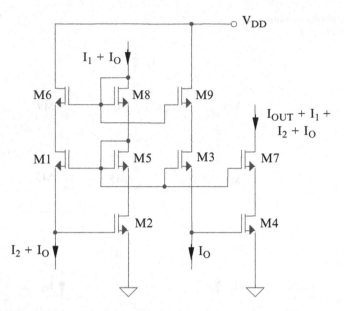

Fig. 2.81 Multiplier circuit (4) based on PR 2.11

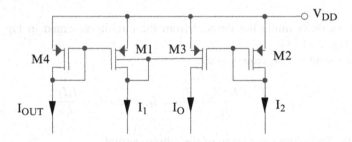

Fig. 2.82 Multiplier circuit (5) based on PR 2.11

Using the exponential dependence of the drain current on the gate-source voltage for a MOS transistor biased in weak inversion region, it can be obtained:

$$V_{SG2} = V_{SG3} = nV_{th} \ln\left(\frac{I_O}{I_{DO}}\right) - (n-1)V_{SB3} \tag{2.379}$$

and:

$$V_{SG1} = V_{SG4} = nV_{th} \ln\left(\frac{I_1}{I_{DO}}\right) - (n-1)V_{SB1} \tag{2.380}$$

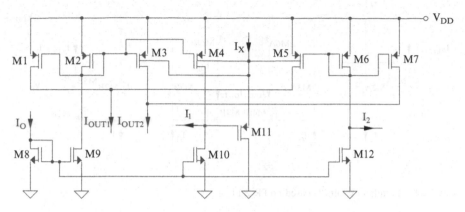

Fig. 2.83 Multiplier circuit (6) based on PR 2.11

Because $V_{SB3} = V_{SB1}$, from the previous relations, it results:

$$I_{OUT} = \frac{I_1 I_2}{I_O} \tag{2.381}$$

A four-quadrant multiplier and two-quadrant divider is presented in Fig. 2.83 [47]. The M2, M4, M5 and M6 transistors form the multiplier presented in Fig. 2.82. The output currents I_{OUT1} and I_{OUT2} can be expressed as follows:

$$I_{OUT1} = I_{D1} + I_{D5} = I_O + \frac{I_{D6} I_{D4}}{I_{D2}} = I_O + \frac{(I_O + I_1)(I_O + I_2)}{I_O} \tag{2.382}$$

and:

$$I_{OUT2} = I_{D3} + I_{D7} = I_{D4} + I_{D6} = (I_O + I_1) + (I_O + I_2) \tag{2.383}$$

resulting:

$$I_{OUT1} - I_{OUT2} = \frac{I_1 I_2}{I_O} \tag{2.384}$$

The circuit presented in Fig. 2.84 [49] implements the current multiplying function.

The characteristic equation of the translinear loop realized using the gate-source voltages of M1A–M5A transistors can be written as follows:

$$V_{GS1A} + V_{GS2A} = V_{GS3A} + V_{GS4A} \tag{2.385}$$

For a weak inversion operation for all MOS devices, it results the following expression of the first output current:

$$I_{OUT1} = \frac{I_2}{I_O}(I_{REF} + I_1) \tag{2.386}$$

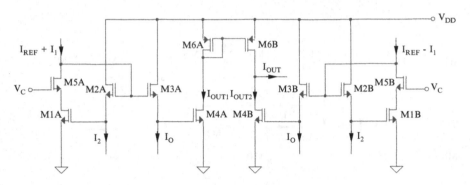

Fig. 2.84 Multiplier circuit (7) based on PR 2.11

Fig. 2.85 Multiplier circuit
(8) based on PR 2.11

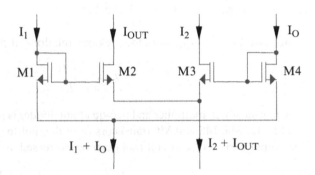

and, similarly, for the translinear loop implemented using the gate-source voltages of M1B–M5B transistors:

$$I_{OUT2} = \frac{I_2}{I_O}(I_{REF} - I_1) \tag{2.387}$$

The expression of the differential output current of the entire structure will be:

$$I_{OUT} = I_{OUT1} - I_{OUT2} = 2\frac{I_1 I_2}{I_O} \tag{2.388}$$

A possible realization of a current multiplier is presented in Fig. 2.85 [50].

The translinear loop implemented using M1–M4 transistors has the following characteristic equation:

$$V_{GS1} + V_{GS3} = V_{GS2} + V_{GS4} \tag{2.389}$$

For a biasing in weak inversion of the circuit transistors, the previous relation becomes:

$$I_{OUT} = \frac{I_1 I_2}{I_O} \tag{2.390}$$

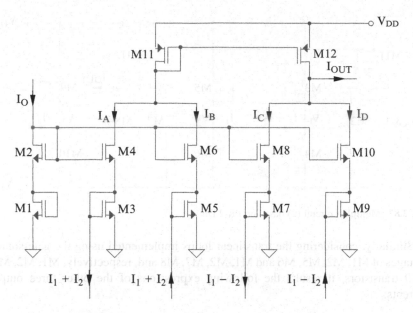

Fig. 2.86 Multiplier circuit (1) based on PR 2.Db

2.2.2.4 Multiplier circuits based on different mathematical principle (PR 2.Db)

The circuit presented in Fig. 2.86 [51] represents a current multiplier, having a principle of operation based on four translinear loops. The first loop contains M1–M4 transistors, M1–M2 pair being biased at the reference current, I_O, M3 transistor – at $I_A - I_1 - I_2$ drain current, while M3 transistor is working at I_A current.

The characteristic equation of the translinear loop can be written as follows:

$$V_{GS1} + V_{GS2} = V_{GS3} + V_{GS4} \tag{2.391}$$

resulting, for a biasing in saturation of all MOS transistors:

$$2\sqrt{I_O} = \sqrt{I_A} + \sqrt{I_A - I_1 - I_2} \tag{2.392}$$

Squaring the previous relation, it can be obtained:

$$4I_O = 2I_A - I_1 - I_2 - 2\sqrt{I_A(I_A - I_1 - I_2)} \tag{2.393}$$

So, the I_A current will have the following expression:

$$I_A = I_O + \frac{I_1}{2} + \frac{I_2}{2} + \frac{I_1 I_2}{8I_O} + \frac{I_1^2}{16I_O} + \frac{I_2^2}{16I_O} \tag{2.394}$$

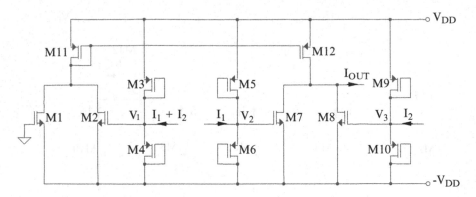

Fig. 2.87 Multiplier circuit (2) based on PR 2.Db

Similarly, considering the translinear loops implemented using the gate-source voltages of M1, M2, M5, M6 and M1, M2, M7, M8 and, respectively, M1, M2, M9, M10 transistors, it results the following expressions of the other three output currents:

$$I_B = I_O - \frac{I_1}{2} - \frac{I_2}{2} + \frac{I_1 I_2}{8 I_O} + \frac{I_1^2}{16 I_O} + \frac{I_2^2}{16 I_O} \tag{2.395}$$

$$I_C = I_O + \frac{I_1}{2} - \frac{I_2}{2} - \frac{I_1 I_2}{8 I_O} + \frac{I_1^2}{16 I_O} + \frac{I_2^2}{16 I_O} \tag{2.396}$$

$$I_D = I_O - \frac{I_1}{2} + \frac{I_2}{2} - \frac{I_1 I_2}{8 I_O} + \frac{I_1^2}{16 I_O} + \frac{I_2^2}{16 I_O} \tag{2.397}$$

The output current can be expressed as follows:

$$I_{OUT} = (I_A + I_B) - (I_C + I_D) = \frac{I_1 I_2}{2 I_O} \tag{2.398}$$

The independence of the circuit performances on technological parameters and, as a result, their immunity with respect to technological errors, represents an important characteristic of the previously presented multiplier circuit.

The circuit presented in Fig. 2.87 [52] implements the current multiplying function. Transistors M5 and M6 form a voltage divider with an input current, I_1, that will modify the divided potential V_2. The current balance in this point can be quantitatively evaluated by the following relation:

$$\frac{K}{2}(V_{DD} - V_2 - V_T)^2 + I_1 = \frac{K}{2}(V_2 + V_{DD} - V_T)^2 \tag{2.399}$$

resulting:

$$V_2 = \frac{I_1}{2K(V_{DD} - V_T)}$$ (2.400)

The output current of the circuit will linearly depend on the following drain currents:

$$I_{OUT} = I_{D1} + I_{D2} - I_{D7} - I_{D8}$$ (2.401)

the previous currents having the following expressions:

$$I_{D1} = \frac{K}{2}(V_{DD} - V_T)^2$$ (2.402)

$$I_{D2} = \frac{K}{2}(V_1 + V_{DD} - V_T)^2 = \frac{K}{2}\left[\frac{I_1 + I_2}{2K(V_{DD} - V_T)} + V_{DD} - V_T\right]^2$$ (2.403)

$$I_{D7} = \frac{K}{2}(V_2 + V_{DD} - V_T)^2 = \frac{K}{2}\left[\frac{I_1}{2K(V_{DD} - V_T)} + V_{DD} - V_T\right]^2$$ (2.404)

and:

$$I_{D8} = \frac{K}{2}(V_3 + V_{DD} - V_T)^2 = \frac{K}{2}\left[\frac{I_2}{2K(V_{DD} - V_T)} + V_{DD} - V_T\right]^2$$ (2.405)

So, the output current will be proportional with the product between the input currents:

$$I_{OUT} = \frac{I_1 I_2}{4K(V_{DD} - V_T)^2}$$ (2.406)

The disadvantage of this circuit is represented by the dependence of the output current on the supply voltage and on the threshold voltage.

2.3 Conclusion

Chapter describes the principle of operation of CMOS multiplier circuits and, starting from their functional principle of operation, it presents many implementations in CMOS technology of these computational structures. There were analyzed multiplier circuits having both current-input and voltage-input variables. Depending on the power consumption imposed to the designs, two important classes of multiplier structures have been presented. For improving the frequency response, circuits using exclusively MOS transistors biased in saturation region have been used, while low-power multipliers have been designed based on subthreshold-operated MOS active devices. The linearity of multiplier circuits has been improved by applying specific design techniques.

References

1. Kim YH, Park SB (1992) Four-quadrant CMOS analogue multiplier. Electron Lett 28:649–650
2. Wallinga H, Bult K (1989) Design and analysis of CMOS analog signal processing circuits by means of a graphical MOST model. IEEE J Solid-State Circuits 24:672–680
3. Shen-Iuan L, Chen-Chieh C (1997) Low-voltage CMOS four-quadrant multiplier. Electron Lett 33:207–208
4. Gunhee H, Sanchez-Sinencio E (1998) CMOS transconductance multipliers: a tutorial. IEEE Trans Circuits Syst II: Analog Digit Signal Process 12:1550–1563
5. Chen JJ, Liu SI, Hwang YS (1998) Low-voltage single power supply four-quadrant multiplier using floating-gate MOSFETs. IEE proceedings on circuits, devices and systems, pp 40–43
6. Sawigun C, Mahattanakul J (2008) A 1.5 V, wide-input range, high-bandwidth, CMOS four-quadrant analog multiplier. IEEE international symposium on circuits and systems, pp 2318–2321, Washington, USA
7. Sawigun C, Demosthenous A, Pal D (2007) A low-voltage, low-power, high-linearity CMOS four-quadrant analog multiplier. European conference on circuit theory and design, pp 751–754, Seville, Spain
8. Liu SI, Hwang YS (1993) CMOS four-quadrant multiplier using bias offset crosscoupled pairs. Electron Lett 29:1737–1738
9. Ramirez-Angulo J, Carvajal RG, Martinez-Heredia J (2000) 1.4 V supply, wide swing, high frequency CMOS analogue multiplier with high current efficiency. IEEE international symposium on circuits and systems, pp 533–536, Geneva, Switzerland
10. Popa C (2006) Improved linearity active resistor with controllable negative resistance. IEEE international conference on integrated circuit design and technology, pp 1–4, Padova, Italy
11. Langlois PJ (1990) Comments on "A CMOS four-quadrant multiplier": effects of threshold voltage. IEEE J Solid-State Circuits 25:1595–1597
12. Zarabadi SR, Ismail M, Chung-Chih H (1998) High performance analog VLSI computational circuits. IEEE J Solid-State Circuits 33:644–649
13. Popa C (2009) High accuracy CMOS multifunctional structure for analog signal processing. International semiconductor conference, pp 427–430, Sinaia, Romania
14. De La Cruz Blas CA, Feely O (2008) Limit cycle behavior in a class-AB second-order square root domain filter. IEEE International conference on electronics, circuits and systems, pp 117–120, St. Julians, Malta
15. Popa C (2001) Low-power rail-to-rail CMOS linear transconductor. International semiconductor conference, pp 557–560, Sinaia, Romania
16. Sakurai S, Ismail M (1992) A CMOS square-law programmable floating resistor independent of the threshold voltage. IEEE Trans Circuits Syst II: Analog Digit Signal Process 39:565–574
17. Jong-Kug S, Charlot J (2000) A CMOS inverse trigonometric function circuit. IEEE midwest symposium on circuits and systems, pp 474–477, Michigan, USA
18. Popa C (2010) CMOS multifunctional computational structure with improved performances. International semiconductors conference, pp 471–474, Sinaia, Romania

19. Popa C (2002) CMOS transconductor with extended linearity range. IEEE international conference on automation, quality and testing, robotics, pp 349–354, Cluj, Romania
20. Manolescu AM, Popa C (2009) Low-voltage low-power improved linearity CMOS active resistor circuits. Springer J Analog Integr Circuits Signal Process 62:373–387
21. Babanezhad JN, Temes GC (1985) A 20-V four-quadrant CMOS analog multiplier. IEEE J Solid-State Circuits 20:1158–1168
22. Popa C, Manolescu AM (2007) CMOS differential structure with improved linearity and increased frequency response. International semiconductor conference, pp 517–520, Sinaia, Romania
23. Popa C (2008) Programmable CMOS active resistor using computational circuits. International semiconductor conference, pp 389–392, Sinaia, Romania
24. Popa C (2009) Multiplier circuit with improved linearity using FGMOS transistors. International symposium ELMAR, pp 159–162, Zadar, Croatia
25. Seng YK, Rofail SS (1998) Design and analysis of a ±1 V CMOS four-quadrant analogue multiplier. IEE proceedings on circuits, devices and systems, pp 148–154, Florida, USA
26. Kathiresan G, Toumazou C (1999) A low voltage bulk driven downconversion mixer core. IEEE international symposium on circuits and systems, pp 598–601, Florido, USA
27. Szczepanski S, Koziel S (2004) 1.2 V low-power four-quadrant CMOS transconductance multiplier operating in saturation region. International symposium on circuits and systems, pp 1016–1019, Vancouver, Canada
28. Coban AL, Allen PE (1994) A 1.5 V four-quadrant analog multiplier. Midwest symposium on Sch of Electr and Comput Eng, pp 117–120, La Fayette, USA
29. Manolescu AM, Popa C (2011) A 2.5 GHz CMOS mixer with improved linearity. J Circuits Syst Comp 20:233–242
30. Popa C, Coada D (2003) A new linearization technique for a CMOS differential amplifier using bulk-driven weak-inversion MOS transistors. International symposium on circuits and systems, pp 589–592, Iasi, Romania
31. Akshatha BC, Akshintala VK (2009) Low voltage, low power, high linearity, high speed CMOS voltage mode analog multiplier. International conference on emerging trends in engineering and technology, pp 149–154, Nagpur, India
32. Shen-Iuan L, Yuh-Shyan H (1995) CMOS squarer and four-quadrant multiplier. IEEE Trans Circuits Syst I: Fundam Theory Appl 42:119–122
33. Xiang-Luan Jia WH, Shi-Cai Q (1995) A new CMOS analog multiplier with improved input linearity. IEEE region 10 international conference on microelectronics and VLSI, pp 135–136, Hong Kong
34. Szczepanski S, Koziel S (2002) A 3.3 V linear fully balanced CMOS operational transconductance amplifier for high-frequency applications. IEEE international conference on circuits and systems for communications, pp 38–41, St. Petersburg, Russia
35. Mahmoud SA (2009) Low voltage low power wide range fully differential CMOS four-quadrant analog multiplier. IEEE international midwest symposium on circuits and systems, pp 130–133, Cancun, Mexico
36. Popa C (2010) Improved linearity CMOS active resistor based on complementary computational circuits. IEEE international conference on electronics, circuits, and systems, pp 455–458, Athens, Greece
37. Psychalinos C, Vlassis S (2002) A systematic design procedure for square-root-domain circuits based on the signal flow graph approach. IEEE Trans Circuits Syst I: Fundam Theory Appl 49:1702–1712
38. Naderi A et al (2009) Four-quadrant CMOS analog multiplier based on new current squarer circuit with high-speed. IEEE international conference on "Computer as a tool", pp 282–287, St. Petersburg, Russia
39. Naderi A, Khoei A, Hadidi K (2007) High speed, low power four-quadrant CMOS current-mode multiplier. IEEE international conference on electronics, circuits and systems, pp 1308–1311, Marracech, Morocco

40. Arthansiri T, Kasemsuwan V (2006) Current-mode pseudo-exponential-control variable-gain amplifier using fourth-order Taylor's series approximation. Electron Lett 42:379–380
41. Bult K, Wallinga H (1987) A class of analog CMOS circuits based on the square-law characteristic of an MOS transistor in saturation. IEEE J Solid-State Circuits 22:357–365
42. Popa C (2003) Low-power CMOS bulk-driven weak-inversion accurate current-mode multiplier/divider circuits. International conference on electrical and electronics engineering, pp 66–73, Bursa, Turkey
43. Popa C (2009) Computational circuits using bulk-driven MOS devices. IEEE international conference on "Computer as a tool", pp 246–251, St. Petersburg, Russia
44. Wilamowski BM (1998) VLSI analog multiplier/divider circuit. IEEE international symposium on industrial electronics, pp 493–496, Pretoria, South Africa
45. De La Cruz-Blas CA, Lopez-Martin A, Carlosena A (2003) 1.5-V MOS translinear loops with improved dynamic range and their applications to current-mode signal processing. IEEE Trans Circuits Syst II: Analog Digit Signal Process 50:918–927
46. Popa C (2003) A new curvature-corrected voltage reference based on the weight difference of gate-source voltages for subthreshold-operated MOS transistors. International symposium on circuits and systems, pp 585–588, Iasi, Romania
47. Cheng-Chieh C, Li S-I (1998) Weak inversion four-quadrant multiplier and two-quadrant divider. Electron Lett 34:2079–2080
48. Khateb F, Biolek D, Khatib N, Vavra J (2010) Utilizing the bulk-driven technique in analog circuit design. IEEE international symposium on design and diagnostics of electronic circuits and systems, pp 16–19, Vienna, Austria
49. Shu-Xiang S, Guo-Ping Y, Hua C (2007) A new CMOS electronically tunable current conveyor based on translinear circuits. International conference on ASIC, pp 569–572, Guilin, China
50. Gravati M, Valle M, Ferri G, Guerrini N, Reyes N (2005) A novel current-mode very low power analog CMOS four quadrant multiplier. Solid-state circuits conference, pp 495–498, Grenoble, France
51. Oliveira VJS, Oki N (2005) Low voltage analog synthesizer of orthogonal signals: a current-mode approach. IEEE international symposium on circuits and systems, pp 3708–3712, Kobe, Japan
52. Gravati M, Valle M, et al (2005) A novel current-mode very low power analog CMOS four quadrant multiplier. Solid-state circuits conference, pp 495–498, Grenoble, France
53. Cheng-Chieh C, Shen-Iuan L (1998) Weak inversion four-quadrant multiplier and two-quadrant divider. Electron Lett 34:2079–2080
54. Sawigun C, Serdijn WA (2009) Ultra-low-power, class-AB, CMOS four-quadrant current multiplier. Electron Lett 45:483–484
55. Hidayat R, Dejhan K, Moungnoul P, Miyanaga Y (2008) OTA-based high frequency CMOS multiplier and squaring circuit. International symposium on intelligent signal processing and communications systems, pp 1–4, Bangkok, Thailand
56. Machowski W, Kuta S, Jasielski J, Kolodziejski W (2010) Quarter-square analog four-quadrant multiplier based on CMOS inverters and using low voltage high speed control circuits. International conference on mixed design of integrated circuits and systems, pp 333–336, Wroclaw, Poland
57. Ehsanpour M, Moallem P, Vafaei A (2010) Design of a novel reversible multiplier circuit using modified full adder. International conference on computer design and applications, pp V3-230–V3-234, Hebei, China
58. Parveen T, Ahmed MT (2009) OFC based versatile circuit for realization of impedance converter, grounded inductance, FDNR and component multipliers. International multimedia, signal processing and communication technologies, pp 81–84, Aligarh, India
59. Feldengut T, Kokozinski R, Kolnsberg S (2009) A UHF voltage multiplier circuit using a threshold-voltage cancellation technique. Research in microelectronics and electronics, pp 288–291, Cork, Ireland
60. Naderi A, Mojarrad H, Ghasemzadeh H, Khoei A, Hadidi K (2009) Four-quadrant CMOS analog multiplier based on new current squarer circuit with high-speed. IEEE international conference on "Computer as a tool", pp 282–287, St. Petersburg, Russia

Chapter 3
Squaring Circuits

3.1 Mathematical Analysis for Synthesis of Squaring Circuits

The synthesis of squaring circuits [1–55] is based on the utilization of some elementary principle, each of them representing the starting point for designing a class of squarers. For voltage squarers, the notations used are: V_1 and V_2 represent the input potentials, while, usually, a constant voltage, V_O is introduced for modeling a voltage shifting. For current squarers, I_O is a reference current and I_{IN} denotes the input current. The notation I_{OUT} is used for the output current of both voltage and current squaring circuits.

3.1.1 Mathematical Analysis of Voltage Squaring Circuits

3.1.1.1 First Mathematical Principle (PR 3.1)

The first mathematical principle used for implementing squaring circuits is based on the following relation:

$$V_1 - V_2 = \left(V_T + \sqrt{\frac{2I_{OUT}}{K_1}} \right) - \left(V_T + \sqrt{\frac{2I_{OUT}}{K_2}} \right)$$

$$= \sqrt{I_{OUT}} \left(\frac{1}{\sqrt{K_1}} - \frac{1}{\sqrt{K_2}} \right) \Rightarrow I_{OUT} = \left(\frac{1}{\sqrt{K_1}} - \frac{1}{\sqrt{K_2}} \right)^{-2} (V_1 - V_2)^2 \qquad (3.1)$$

3.1.1.2 Second Mathematical Principle (PR 3.2)

The mathematical relation that models this principle is:

$$\left[A(V_1 - V_2)^2 + B(V_1 - V_2) + C\right] + \left[A(V_1 - V_2)^2 - B(V_1 - V_2) + C\right]$$
$$- 2C = 2A(V_1 - V_2)^2 \tag{3.2}$$

3.1.1.3 Third Mathematical Principle (PR 3.3)

The third mathematical principle is illustrated by the following relation:

$$2\left(\frac{V_1 + V_2}{2} - V_O\right)^2 - (V_1 - V_O)^2 - (V_2 - V_O)^2$$
$$= \frac{V_2 - V_1}{2}\left(\frac{3V_1 + V_2}{2} - 2V_O\right) + \frac{V_1 - V_2}{2}\left(\frac{3V_2 + V_1}{2} - 2V_O\right) = -\frac{1}{2}(V_1 - V_2)^2 \tag{3.3}$$

3.1.1.4 Different Mathematical Principles for Voltage Squaring Circuits (PR 3.Da)

A class of voltage squaring circuits is based on different mathematical principles.

3.1.2 Mathematical Analysis of Current Squaring Circuits

3.1.2.1 Fifth Mathematical Principle (PR 3.5)

The mathematical relation that models this principle is:

$$2\sqrt{I_O} = \sqrt{aI_O + bI_{OUT} + cI_{IN}} + \sqrt{aI_O + bI_{OUT} - cI_{IN}}$$
$$\Rightarrow 4I_O = 2aI_O + 2bI_{OUT} + 2\sqrt{(aI_O + bI_{OUT})^2 - c^2I_{IN}^2}$$
$$\Rightarrow 4(2 - a)^2 I_O^2 + 4b^2 I_{OUT}^2 - 8(2 - a)bI_O I_{OUT}$$
$$= 4a^2 I_O^2 + 4b^2 I_{OUT}^2 + 8abI_O I_{OUT} - 4c^2 I_{IN}^2$$
$$\Rightarrow I_{OUT} = \frac{1 - a}{b}I_O + \frac{c^2}{4b}\frac{I_{IN}^2}{I_O} \tag{3.4}$$

3.1.2.2 Sixth Mathematical Principle (PR 3.6)

The utilization of MOS transistors biased in weak inversion region allows to implement the following mathematical relation:

$$2\ln(I_{IN}) = \ln(I_O) + \ln(I_{OUT}) \Rightarrow I_{OUT} = \frac{I_{IN}^2}{I_O} \tag{3.5}$$

3.1.2.3 Seventh Mathematical Principle (PR 3.7)

The mathematical relation that models this principle is:

$$\frac{9}{4}I_O + \frac{3}{2}I_{IN} + I_{OUT} = \left(\sqrt{I_O} + \sqrt{I_{OUT} + \frac{I_O}{4} + \frac{I_{IN}}{2}}\right)^2$$

$$\Rightarrow I_O + I_{IN} = 2\sqrt{I_O\left(I_{OUT} + \frac{I_O}{4} + \frac{I_{IN}}{2}\right)} \Rightarrow I_{OUT} = \frac{I_{IN}^2}{4I_O} \tag{3.6}$$

3.1.2.4 Eighth Mathematical Principle (PR 3.8)

This principle is illustrated by the following mathematical relation:

$$(I_O - I_{IN})^2 + (I_O + I_{IN})^2 = 2I_O^2 + 2I_{IN}^2 \tag{3.7}$$

3.1.2.5 Ninth Mathematical Principle (PR 3.9)

This mathematical principle uses a current square-rooting circuit for implementing the squaring function.

3.1.2.6 Different Mathematical Principles for Current Squaring Circuits (PR 3.Db)

A class of squaring circuits is based on different mathematical principles.

3.2   Analysis and Design of Squaring Circuits

Based on the previous presented mathematical analysis, it is possible to design different types of squaring circuits, concentrated in two important classes: voltage squaring and current squaring circuits.

3.2.1   Design of Voltage Squaring Circuits

The voltage squaring circuits are grouped in four classes, corresponding to the first four mathematical principles (PR 3.1 – PR 3.Da).

3.2.1.1   Squaring Circuits Based on the First Mathematical Principle (PR 3.1)

A method for obtaining a voltage squaring circuit using the first mathematical principle (PR 3.1) is based on the utilization of an unbalanced MOS differential amplifier (M1–M7 in Fig. 3.1) [1].

All MOS transistors from Fig. 3.1 are identical, excepting M6 that has an aspect ratio nth times greater than other transistors. This controllable asymmetry will be equivalent with a nonzero differential input voltage in the equilibrium state.

The drain currents of M1 and M5 transistors are equal as a result of the M2–M3 and M4–M5 currents mirrors. Because $I_{D1} + I_{D7} = I_{D5} + I_{D6}$, it can be obtained $I_{D6} = I_{D7}$. The current mirror M4–M6 with different transistors imposes $I_{D6} = nI_{D4} = nI_{D5}$, so $I_{D7} = nI_{D1}$, equivalent with different gate-source voltages for M1 and M7 transistors (that compose the differential stage):

$$V_{GS7} = \sqrt{\frac{2I_{D7}}{K}} + V_T = \sqrt{\frac{2nI_{D1}}{K}} + V_T \tag{3.8}$$

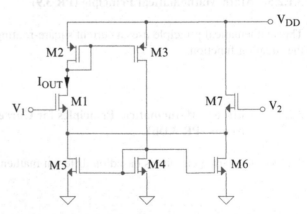

Fig. 3.1 Squaring circuit (1) based on PR 3.1

where:

$$I_{D1} = \frac{K}{2}(V_{GS1} - V_T)^2 \tag{3.9}$$

It results the following dependence of V_{GS7} on V_{GS1}:

$$V_{GS7} = \sqrt{n}(V_{GS1} - V_T) + V_T \tag{3.10}$$

The $V_1 - V_2$ differential input voltage quantitatively evaluates the asymmetry of the differential stage:

$$V_1 - V_2 = V_{GS1} - V_{GS7} = V_{GS1}(1 - \sqrt{n}) + V_T(\sqrt{n} - 1) \tag{3.11}$$

So, the gate-source voltage of M1 transistor can be expressed as a function of the differential input voltage, as follows:

$$V_{GS1} = \frac{V_1 - V_2}{1 - \sqrt{n}} + V_T \tag{3.12}$$

Using the (3.9) squaring dependence, the output current of the circuit presented in Fig. 3.1 will have the following expression:

$$I_{OUT} = I_{D1} = \frac{K}{2(1 - \sqrt{n})^2}(V_1 - V_2)^2 \tag{3.13}$$

Based on the same mathematical principle, a stacked stage, M3–M4 (Fig. 3.2) [2] can implement the squaring function:

$$V_1 = V_{GS4} - V_{GS3} = \sqrt{\frac{2I_{OUT}}{K_4}} - \sqrt{\frac{2I_{OUT}}{K_3}} = \sqrt{2I_{OUT}}\left(\frac{1}{\sqrt{K_4}} - \frac{1}{\sqrt{K_3}}\right) \tag{3.14}$$

resulting:

$$I_{OUT} = \frac{V_1^2}{2\left(\frac{1}{\sqrt{K_4}} - \frac{1}{\sqrt{K_3}}\right)^2} = AV_1^2 \tag{3.15}$$

A being a constant that models the squaring dependence of the output current, I_{OUT}, on the input voltage, V_1.

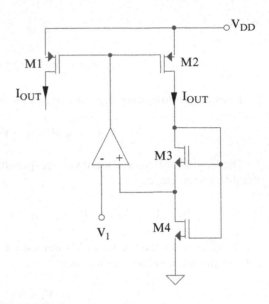

Fig. 3.2 Squaring circuit (2) based on PR 3.1

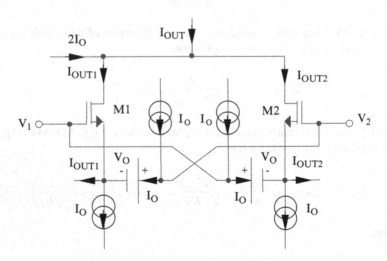

Fig. 3.3 Squaring circuit (1) based on PR 3.2 – principle circuit

3.2.1.2   Squaring Circuits Based on the Second Mathematical Principle (PR 3.2)

A voltage squaring circuit can be implemented using the second mathematical principle (PR 3.2), starting from the differential amplifier with the transfer characteristic linearized using a method based on the constant sum of gate-source voltages (Fig. 3.3) [3].

Considering a biasing in saturation of MOS transistors, I_{OUT1} and I_{OUT2} output currents of the circuit from Fig. 3.3 can be expressed as follows:

$$I_{OUT1} = \frac{K}{2}(V_{GS1} - V_T)^2 \tag{3.16}$$

$$I_{OUT2} = \frac{K}{2}(V_{GS2} - V_T)^2 \tag{3.17}$$

The $V_1 - V_2$ differential input voltage has the following expressions:

$$V_1 - V_2 = V_O - V_{GS2} = V_{GS1} - V_O \tag{3.18}$$

resulting:

$$V_{GS1} = V_O + (V_1 - V_2) \tag{3.19}$$

and:

$$V_{GS2} = V_O - (V_1 - V_2) \tag{3.20}$$

Replacing (3.19) and (3.20) in (3.16) and (3.17). Thus:

$$I_{OUT1} = \frac{K}{2}[(V_O - V_T) + (V_1 - V_2)]^2 \tag{3.21}$$

and:

$$I_{OUT2} = \frac{K}{2}[(V_O - V_T) - (V_1 - V_2)]^2 \tag{3.22}$$

So, the sum of the output currents for the circuit presented in Fig. 3.3 will be:

$$I_{OUT1} + I_{OUT2} = K(V_O - V_T)^2 + K(V_1 - V_2)^2 \tag{3.23}$$

In order to avoid a dependence of the output current on the threshold voltage of the MOS devices, V_O voltage sources from Fig. 3.3 are implemented using gate-source voltages of MOS transistors biased in saturation region. Considering that the current passing through these current-controlled voltage sources is a constant current, I_O, the V_O voltage will have the following expression:

$$V_O = V_T + \sqrt{\frac{2I_O}{K}} \tag{3.24}$$

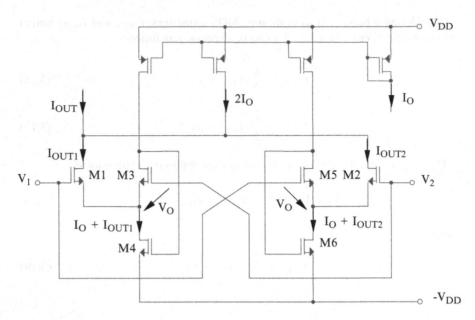

Fig. 3.4 Squaring circuit (1) based on PR 3.2 – first implementation

Replacing (3.24) in (3.23), it results:

$$I_{OUT1} + I_{OUT2} = 2I_O + K(V_1 - V_2)^2 \qquad (3.25)$$

The output current of the voltage squaring circuit, I_{OUT}, can be expressed using a linear relation between the currents from the circuit:

$$I_{OUT} = I_{OUT1} + I_{OUT2} - 2I_O = K(V_1 - V_2)^2 \qquad (3.26)$$

It exists many possibilities of implementing this principle (Fig. 3.4–3.7) [4, 5]. The V_O voltage sources are realized in Fig. 3.4 using M3 and M5 transistors, in Fig. 3.5 and in Fig. 3.6 – using M3 and M4 transistors, while in Fig. 3.7 – using M3 and M4 transistors. The expressions of the output current are given by (3.26) for all four squaring circuits.

Starting from the general circuit presented in Fig. 3.3, it is possible to design a current squaring circuit, implementing proper linear relations between the currents from the circuit and using multiple current mirrors, as it is shown in Fig. 3.8 [3]:

$$I_{OUT1} + I_{OUT2} = 8I_A + 8I_{IN} + 2I_O \qquad (3.27)$$

The sum of the output currents of the differential amplifier (the "Linear DA" block from Fig. 3.8 is realized using the differential amplifier presented in Fig. 3.3) is:

$$I_{OUT1} + I_{OUT2} = 2I_O + K(V_A - V_B)^2 \qquad (3.28)$$

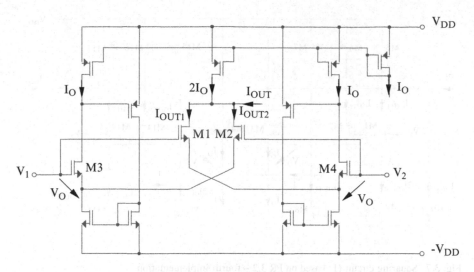

Fig. 3.5 Squaring circuit (1) based on PR 3.2 – second implementation

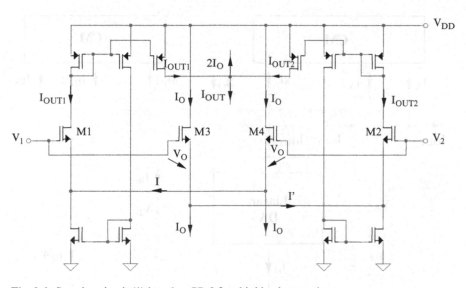

Fig. 3.6 Squaring circuit (1) based on PR 3.2 – third implementation

The $V_A - V_B$ differential voltage, that is fixed by I_A and I_B currents, can be expressed as a function on the gate-source voltages as follows:

$$V_A - V_B = 2V_{GS}(I_A) - 2V_{GS}(I_B) \qquad (3.29)$$

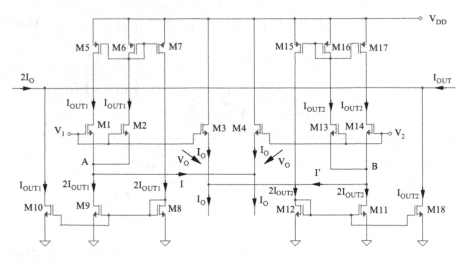

Fig. 3.7 Squaring circuit (1) based on PR 3.2 – fourth implementation

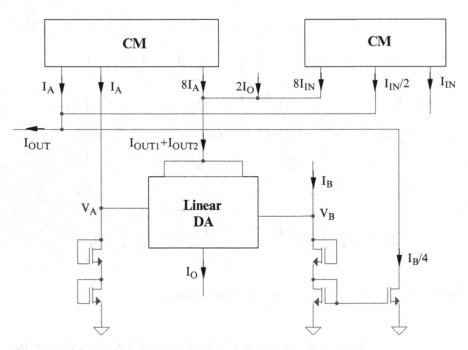

Fig. 3.8 Squaring circuit (2) based on PR 3.2 – block diagram

Replacing (3.29) in (3.28) and using the square-root dependence of the drain current on its gate-source voltage for a MOS transistor biased in saturation, it can be obtained:

$$I_{OUT1} + I_{OUT2} = 2I_O + K\left(2\sqrt{\frac{2I_A}{K}} - 2\sqrt{\frac{2I_B}{K}}\right)^2 \tag{3.30}$$

equivalent with:

$$I_{OUT1} + I_{OUT2} = 2I_O + 8I_A + 8I_B - 16\sqrt{I_A I_B} \tag{3.31}$$

From (3.27) and (3.31), it results:

$$I_A = \frac{(I_B - I_{IN})^2}{4I_B} = \frac{I_B}{4} - \frac{I_{IN}}{2} + \frac{I_{IN}^2}{4I_B} \tag{3.32}$$

The expression of the output current of the current squaring circuit will be:

$$I_{OUT} = I_A + \frac{I_{IN}}{2} - \frac{I_B}{4} = \frac{I_{IN}^2}{4I_B} \tag{3.33}$$

where I_{IN} is the input current and I_B represents the reference current. For simplicity, I_B current can be considered to be equal with the other reference current, I_O, that biases the differential amplifier, resulting:

$$I_{OUT} = \frac{I_{IN}^2}{4I_O} \tag{3.34}$$

The complete implementation of the current squaring circuit, having the principle shown in Fig. 3.8 is presented in Fig. 3.9 [3]. The "Linear DA" block is realized using the differential amplifier from Fig. 3.5.

The squaring circuit presented in Fig. 3.10 [6] is also based on the same mathematical principles.

Noting with $V_{GS}(I)$ the gate-source voltage of a MOS transistor having the drain current equal with I, the differential input voltage can be expressed as follows:

$$V_1 - V_2 = 2V_{GS}(I_O) - 2V_{GS}(I_{OUT1}) = 2\sqrt{\frac{2}{K}}\left(\sqrt{I_O} - \sqrt{I_{OUT1}}\right) \tag{3.35}$$

$$V_1 - V_2 = 2V_{GS}(I_{OUT2}) - 2V_{GS}(I_O) = 2\sqrt{\frac{2}{K}}\left(\sqrt{I_{OUT2}} - \sqrt{I_O}\right) \tag{3.36}$$

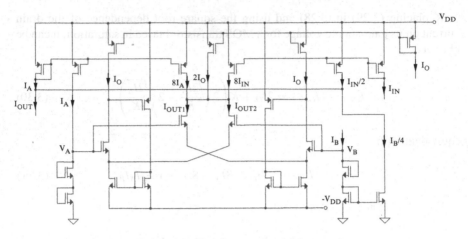

Fig. 3.9 Squaring circuit (2) based on PR 3.2 – complete implementation

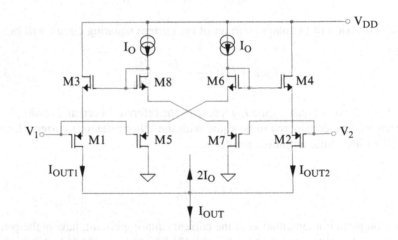

Fig. 3.10 Squaring circuit (3) based on PR 3.2

resulting:

$$\sqrt{I_{OUT1}} = \sqrt{I_O} - \frac{V_1 - V_2}{2}\sqrt{\frac{K}{2}} \qquad (3.37)$$

and:

$$\sqrt{I_{OUT2}} = \sqrt{I_O} + \frac{V_1 - V_2}{2}\sqrt{\frac{K}{2}} \qquad (3.38)$$

equivalent with:

$$I_{OUT1} = I_O - \sqrt{\frac{KI_O}{2}}(V_1 - V_2) + \frac{K}{8}(V_1 - V_2)^2 \qquad (3.39)$$

Fig. 3.11 Squaring circuit
(4) based on PR 3.2 –
principle circuit

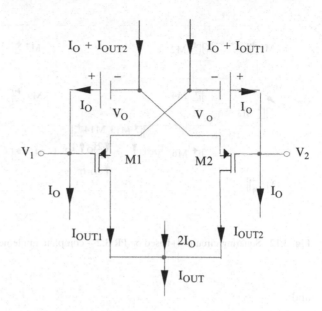

and:

$$I_{OUT2} = I_O + \sqrt{\frac{KI_O}{2}}(V_1 - V_2) + \frac{K}{8}(V_1 - V_2)^2 \qquad (3.40)$$

The output current for the structure presented in Fig. 3.10 will be:

$$I_{OUT} = I_{OUT1} + I_{OUT2} - 2I_O = \frac{K}{4}(V_1 - V_2)^2 \qquad (3.41)$$

Another possible realization of a voltage squaring circuit using the second mathematical principle (PR 3.2) is illustrated in Fig. 3.11, while the complete implementation of the circuit is shown in Fig. 3.12. The circuit represents a complementary approach of the structure presented in Fig. 3.3. The M1 and M2 transistors from Fig. 3.11 are replaced in Fig. 3.12 [7] by M13 and M14 transistors, V_O voltage sources being implemented using the source-gate voltages of M3 and M10 transistors. The other devices from Fig. 3.12 are used for mirroring the currents in the circuit. As M3, M5, M10, M12 and M15 transistors are identical and biased at the same drain current, I_O, their gate-source voltages are equal, so:

$$V_O = V_T + \sqrt{\frac{2I_O}{K}} \qquad (3.42)$$

For the circuit presented in Fig. 3.11, it is possible to write:

$$V_1 - V_2 = V_{SG2} - V_O \qquad (3.43)$$

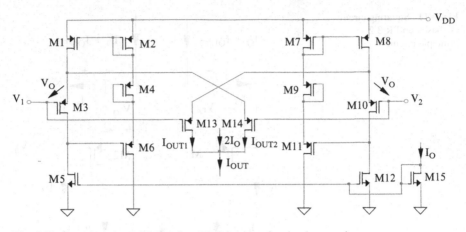

Fig. 3.12 Squaring circuit (4) based on PR 3.2 – complete implementation

and:

$$V_1 - V_2 = V_O - V_{SG1} \tag{3.44}$$

resulting:

$$V_{SG1} = V_O - (V_1 - V_2) \tag{3.45}$$

and:

$$V_{SG2} = V_O + (V_1 - V_2) \tag{3.46}$$

The output current can be expressed as follows:

$$I_{OUT} = I_{OUT1} + I_{OUT2} - 2I_O = \frac{K}{2}(V_{SG1} - V_T)^2 + \frac{K}{2}(V_{SG2} - V_T)^2 - 2I_O \tag{3.47}$$

So:

$$I_{OUT} = \frac{K}{2}[(V_O - V_T) - (V_1 - V_2)]^2 + \frac{K}{2}[(V_O - V_T) + (V_1 - V_2)]^2 - 2I_O$$
$$= K(V_O - V_T)^2 + K(V_1 - V_2)^2 - 2I_O \tag{3.48}$$

Replacing (3.42) in (3.48), it results:

$$I_{OUT} = K(V_1 - V_2)^2 \tag{3.49}$$

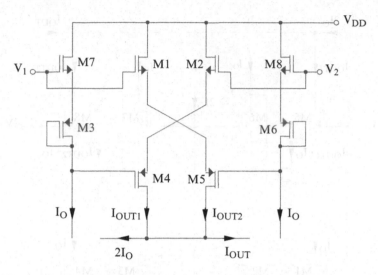

Fig. 3.13 Squaring circuit (5) based on PR 3.2

A current squaring circuit based on the same mathematical principle, using a translinear loop, is presented in Fig. 3.13 [8].

The translinear loop containing M2, M3, M4 and M7 transistors has the following characteristic equation:

$$V_1 - V_2 = 2V_{GS}(I_O) - 2V_{GS}(I_1) = 2\sqrt{\frac{2}{K}}\left(\sqrt{I_O} - \sqrt{I_1}\right) \qquad (3.50)$$

resulting:

$$I_{OUT1} = I_O + \frac{K}{8}(V_1 - V_2)^2 - \sqrt{\frac{KI_O}{2}}(V_1 - V_2) \qquad (3.51)$$

Similarly, analyzing the translinear loop containing M1, M5, M6 and M8 transistors, it results:

$$I_{OUT2} = I_O + \frac{K}{8}(V_1 - V_2)^2 + \sqrt{\frac{KI_O}{2}}(V_1 - V_2) \qquad (3.52)$$

The output current of the circuit presented in Fig. 3.13 will have the following expression:

$$I_{OUT} = I_{OUT1} + I_{OUT2} - 2I_O = \frac{K}{4}(V_1 - V_2)^2 \qquad (3.53)$$

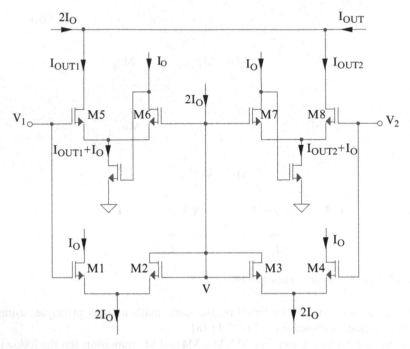

Fig. 3.14 Squaring circuit (6) based on PR 3.2

A realization of a voltage squaring circuit using the computation of the input potentials arithmetical mean is presented in Fig. 3.14.

As M1–M4 transistors implement an arithmetical mean circuit, the expression of V potential will be:

$$V = \frac{V_1 + V_2}{2} \qquad (3.54)$$

For M5–M6 differential amplifier, the differential input voltage can be expressed as follows:

$$V_1 - V = V_{GS5} - V_{GS6} = \sqrt{\frac{2}{K}}\left(\sqrt{I_{OUT1}} - \sqrt{I_O}\right) \qquad (3.55)$$

Replacing (3.54) in (3.55), it results:

$$I_{OUT1} = I_O + \sqrt{\frac{KI_O}{2}}(V_1 - V_2) + \frac{K}{8}(V_1 - V_2)^2 \qquad (3.56)$$

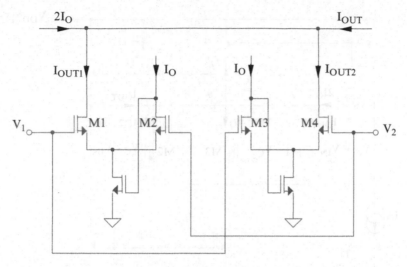

Fig. 3.15 Squaring circuit (7) based on PR 3.2

Similarly, for M7–M8 differential amplifier, it can be obtained:

$$I_{OUT2} = I_O - \sqrt{\frac{KI_O}{2}}(V_1 - V_2) + \frac{K}{8}(V_1 - V_2)^2 \tag{3.57}$$

The output current of the squaring circuit presented in Fig. 3.14 will be:

$$I_{OUT} = I_{OUT1} + I_{OUT2} - 2I_O = \frac{K}{4}(V_1 - V_2)^2 \tag{3.58}$$

A voltage squaring circuit containing two differential amplifiers is presented in Fig. 3.15 [9].

For M1–M2 differential amplifier, the differential input voltage can be expressed as follows:

$$V_1 - V_2 = \sqrt{\frac{2}{K}}\left(\sqrt{I_{OUT1}} - \sqrt{I_O}\right) \tag{3.59}$$

resulting:

$$I_{OUT1} = I_O + \sqrt{2KI_O}(V_1 - V_2) + \frac{K}{2}(V_1 - V_2)^2 \tag{3.60}$$

Similarly, for M3–M4 differential amplifier, the expression of I_2 current will be:

$$I_{OUT2} = I_O - \sqrt{2KI_O}(V_1 - V_2) + \frac{K}{2}(V_1 - V_2)^2 \tag{3.61}$$

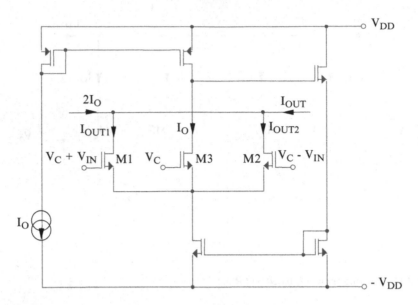

Fig. 3.16 Squaring circuit (8) based on PR 3.2

The output current of the squaring circuit presented in Fig. 3.15 is:

$$I_{OUT} = I_{OUT1} + I_{OUT2} - 2I_O = K(V_1 - V_2)^2 \qquad (3.62)$$

The circuit presented in Fig. 3.16 [5] is used for obtaining the squaring of a differential input voltage, V_{IN}.

The difference between the gate-source voltages of M1 and M3 transistors can be expressed as follows:

$$V_{GS1} - V_{GS3} = (V_C + V_{IN}) - V_C \qquad (3.63)$$

For a biasing in saturation of all MOS transistors from Fig. 3.16, it results:

$$V_{IN} = \sqrt{\frac{2}{K}} \left(\sqrt{I_{OUT1}} - \sqrt{I_O} \right) \qquad (3.64)$$

So, the expression of I_{OUT1} current will be:

$$I_{OUT1} = I_O + \frac{K}{2} V_{IN}^2 + \sqrt{2KI_O}\, V_{IN} \qquad (3.65)$$

Similarly, computing the difference between the gate-source voltages of M2–M3 transistors, it results:

$$I_{OUT2} = I_O + \frac{K}{2} V_{IN}^2 - \sqrt{2KI_O}\, V_{IN} \qquad (3.66)$$

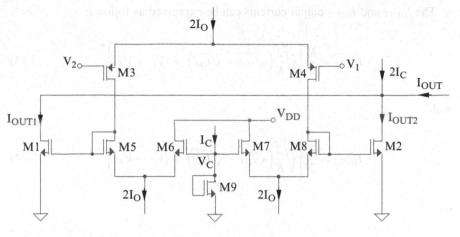

Fig. 3.17 Squaring circuit (9) based on PR 3.2

The output current for the circuit presented in Fig. 3.16 is:

$$I_{OUT} = I_{OUT1} + I_{OUT2} - 2I_O = KV_{IN}^2 \qquad (3.67)$$

The circuit presented in Fig. 3.17 [10] implements the squaring function using PR 3.2.

The current sources and the circuit's connections impose the following relation between the currents from the circuit:

$$I_{D3} + I_{D6} = I_{D4} + I_{D7} = I_{D3} + I_{D4} = 2I_O \qquad (3.68)$$

resulting $I_{D6} = I_{D4}$ and $I_{D7} = I_{D3}$. The translinear loops containing M1, M5, M6 and M2, M7, M8 transistors have the following characteristic equations:

$$V_{GS1} - V_C = V_{GS5} - V_{GS6} \qquad (3.69)$$

and:

$$V_{GS2} - V_C = V_{GS8} - V_{GS7} \qquad (3.70)$$

resulting:

$$V_T + \sqrt{\frac{2I_{OUT1}}{K}} - V_C = \sqrt{\frac{2}{K}}\left(\sqrt{I_{D3}} - \sqrt{I_{D4}}\right) \qquad (3.71)$$

and:

$$V_T + \sqrt{\frac{2I_{OUT2}}{K}} - V_C = \sqrt{\frac{2}{K}}\left(\sqrt{I_{D4}} - \sqrt{I_{D3}}\right) \qquad (3.72)$$

The I_{OUT1} and I_{OUT2} output currents can be expressed as follows:

$$I_{OUT1} = \frac{K}{2}\left[\sqrt{\frac{2}{K}}\left(\sqrt{I_{D3}} - \sqrt{I_{D4}}\right) + (V_C - V_T)\right]^2 \qquad (3.73)$$

and:

$$I_{OUT2} = \frac{K}{2}\left[\sqrt{\frac{2}{K}}\left(\sqrt{I_{D4}} - \sqrt{I_{D3}}\right) + (V_C - V_T)\right]^2 \qquad (3.74)$$

or:

$$I_{OUT1} = \left(\sqrt{I_{D3}} - \sqrt{I_{D4}}\right)^2 + \sqrt{2K}\left(\sqrt{I_{D3}} - \sqrt{I_{D4}}\right)(V_C - V_T) + \frac{K}{2}(V_C - V_T)^2 \quad (3.75)$$

and:

$$I_{OUT2} = \left(\sqrt{I_{D4}} - \sqrt{I_{D3}}\right)^2 + \sqrt{2K}\left(\sqrt{I_{D4}} - \sqrt{I_{D3}}\right)(V_C - V_T) + \frac{K}{2}(V_C - V_T)^2 \quad (3.76)$$

The differential input voltage of the circuit is equal with the difference between two source-gate voltages:

$$V_1 - V_2 = V_{SG3} - V_{SG4} = \sqrt{\frac{2}{K}}\left(\sqrt{I_{D3}} - \sqrt{I_{D4}}\right) \qquad (3.77)$$

From (3.75), (3.76) and (3.77), the expressions of the output currents become:

$$I_{OUT1} = \frac{K}{2}(V_1 - V_2)^2 + K(V_1 - V_2)(V_C - V_T) + \frac{K}{2}(V_C - V_T)^2 \qquad (3.78)$$

and:

$$I_{OUT2} = \frac{K}{2}(V_1 - V_2)^2 - K(V_1 - V_2)(V_C - V_T) + \frac{K}{2}(V_C - V_T)^2 \qquad (3.79)$$

As V_C voltage is equal with the gate-source voltage of M9 transistor, which is biased at the constant current, I_C, the previous relations become:

$$I_{OUT1} = \frac{K}{2}(V_1 - V_2)^2 + \sqrt{2KI_C}(V_1 - V_2) + I_C \qquad (3.80)$$

Fig. 3.18 Squaring circuit (10) based on PR 3.2

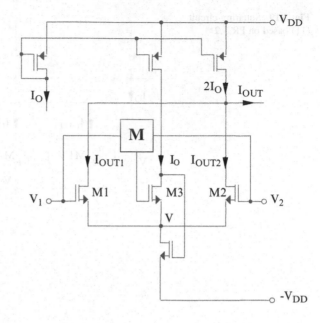

and:

$$I_{OUT2} = \frac{K}{2}(V_1 - V_2)^2 - \sqrt{2KI_C}(V_1 - V_2) + I_C \qquad (3.81)$$

The sum of the output currents will be:

$$I_{OUT1} + I_{OUT2} = K(V_1 - V_2)^2 + 2I_C \qquad (3.82)$$

So, the output current of the circuit presented in Fig. 3.17 will be proportional with the square of the differential input voltage:

$$I_{OUT} = I_{OUT1} + I_{OUT2} - 2I_C = K(V_1 - V_2)^2 \qquad (3.83)$$

A simple realization of a voltage squaring circuit exploits the second mathematical principle, being based on a differential amplifier having the sources of transistors connected to a potential fixed by a constant current. The realization of the circuit is presented in Fig. 3.18 [11]. The "M" block represents a circuit that computes the arithmetical mean of V_1 and V_2 input potentials. A possible implementation of this block is presented in Fig. 3.19 [11]. The FGMOS transistor is functionally equivalent with the "M" block and M3 transistor

The output currents of the differential amplifier can be expressed as follows:

$$I_{OUT1} = \frac{K}{2}(V_1 - V - V_T)^2 \qquad (3.84)$$

Fig. 3.19 Squaring circuit
(11) based on PR 3.2

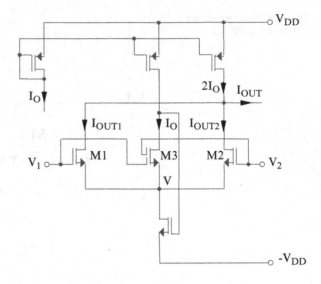

and:

$$I_{OUT2} = \frac{K}{2}(V_2 - V - V_T)^2 \qquad (3.85)$$

while V potential is imposed by V_1 and V_2 potentials and by M3 transistor, biased at
a constant drain current, I_O:

$$I_O = \frac{K}{2}\left(\frac{V_1 + V_2}{2} - V - V_T\right)^2 \Rightarrow V = \frac{V_1 + V_2}{2} - V_T - \sqrt{\frac{2I_O}{K}} \qquad (3.86)$$

The expressions of the differential amplifier output currents becomes:

$$I_{OUT1} = \frac{K}{2}\left(\frac{V_1 - V_2}{2} + \sqrt{\frac{2I_O}{K}}\right)^2 \qquad (3.87)$$

and:

$$I_{OUT2} = \frac{K}{2}\left(-\frac{V_1 - V_2}{2} + \sqrt{\frac{2I_O}{K}}\right)^2 \qquad (3.88)$$

The sum of the output currents will contain a term proportional with the square
of the differential input voltage:

$$I_{OUT1} + I_{OUT2} = 2I_O + \frac{K}{4}(V_1 - V_2)^2 \qquad (3.89)$$

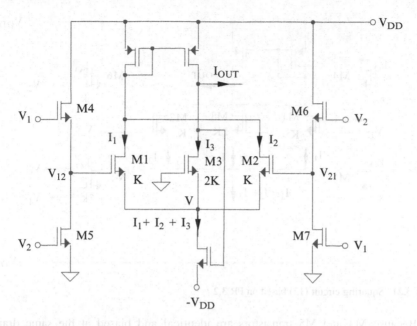

Fig. 3.20 Squaring circuit (12) based on PR 3.2

The output current of the entire circuit from Fig. 3.19 will be proportional with $(V_1 - V_2)^2$:

$$I_{OUT} = I_{OUT1} + I_{OUT2} - 2I_O = \frac{K}{4}(V_1 - V_2)^2 \tag{3.90}$$

The voltage squarer presented in Fig. 3.20 [12, 13] is based on the second mathematical principle and uses a symmetrical structure, M1–M2.

The output current expression has a linear dependence on the drain currents of M1, M2 and M3 transistors:

$$I_{OUT} = I_1 + I_2 - I_3 \tag{3.91}$$

Considering a biasing in saturation of all MOS devices from Fig. 3.20, the previous currents will have the following expressions:

$$I_1 = \frac{K}{2}(V_{12} - V - V_T)^2 \tag{3.92}$$

$$I_2 = \frac{K}{2}(V_{21} - V - V_T)^2 \tag{3.93}$$

$$I_3 = \frac{2K}{2}(-V - V_T)^2 \tag{3.94}$$

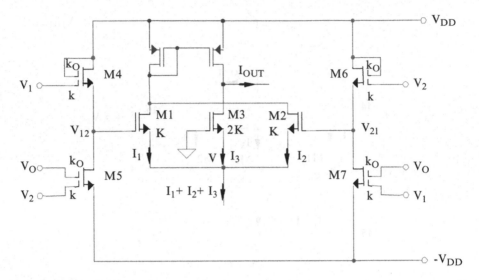

Fig. 3.21 Squaring circuit (13) based on PR 3.2

Because M4 and M5 transistors are identical and biased at the same drain current, their gate-source voltages will be equal, so $V_1 - V_{12} = V_2$, resulting:

$$V_{12} = V_1 - V_2 \tag{3.95}$$

Similarly:

$$V_{21} = V_2 - V_1 \tag{3.96}$$

From (3.91) – (3.96), the output current can be expressed as:

$$I_{OUT} = \frac{K}{2}[(V_1 - V_2) - (V + V_T)]^2 + \frac{K}{2}[(V_2 - V_1) - (V + V_T)]^2$$
$$- \frac{2K}{2}(V + V_T)^2 = \frac{K}{2}(V_1 - V_2)[(V_1 - V_2) - 2(V + V_T)]$$
$$+ \frac{K}{2}(V_2 - V_1)[(V_2 - V_1) - 2(V + V_T)] = K(V_1 - V_2)^2 \tag{3.97}$$

An alternate implementation [14] of the voltage squaring circuit uses FGMOS transistors for extending the range of the minimal supply voltage (Fig. 3.21). The ground potential is replaced, in this case, with $-V_{DD}$. All FGMOS transistors are identical and they have different weights of their inputs, k_O and k.
The drain currents of M4 and M5 transistors are equal, so:

$$\frac{K}{2}\left(\frac{k_O V_{DD} + k V_1}{k + k_O} - V_{12} - V_T\right)^2 = \frac{K}{2}\left(\frac{k_O V_O + k V_2}{k + k_O} + V_{DD} - V_T\right)^2 \tag{3.98}$$

resulting:

$$V_{12} = \frac{k(V_1 - V_2)}{k + k_O} - \frac{kV_{DD} + k_O V_O}{k + k_O} \tag{3.99}$$

Similarly:

$$V_{21} = -\frac{k(V_1 - V_2)}{k + k_O} - \frac{kV_{DD} + k_O V_O}{k + k_O} \tag{3.100}$$

equivalent with a voltage shifting with $(kV_{DD} + k_O V_O)/(k + k_O)$ of the differential $V_1 - V_2$ input voltage. In order to cancel out this shifting, the V_O biasing voltage, must be chosen to have the following expression:

$$V_O = -\frac{k}{k_O} V_{DD} \tag{3.101}$$

resulting, in this hypothesis:

$$V_{12} = \frac{k(V_1 - V_2)}{k + k_O} \tag{3.102}$$

and:

$$V_{21} = -\frac{k(V_1 - V_2)}{k + k_O} \tag{3.103}$$

The expression of the output current will be:

$$I_{OUT} = I_1 + I_2 - I_3 = \frac{K}{2}(V_{12} - V - V_T)^2$$
$$+ \frac{K}{2}(V_{21} - V - V_T)^2 - \frac{2K}{2}(-V - V_T)^2 \tag{3.104}$$

equivalent with:

$$I_{OUT} = \frac{K}{2}\left[\frac{k(V_1 - V_2)}{k + k_O} - V - V_T\right]^2 +$$
$$+ \frac{K}{2}\left[-\frac{k(V_1 - V_2)}{k + k_O} - V - V_T\right]^2 + \frac{2K}{2}(-V - V_T)^2 \tag{3.105}$$

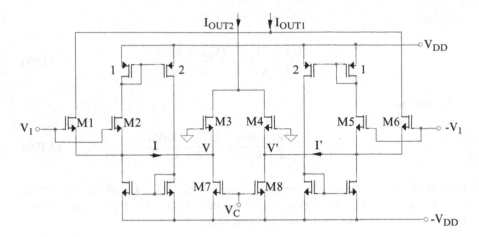

Fig. 3.22 Squaring circuit (14) based on PR 3.2

So:

$$I_{OUT} = \frac{K}{2} \frac{k(V_1 - V_2)}{k + k_O} \left[\frac{k(V_1 - V_2)}{k + k_O} - 2V - 2V_T \right]$$
$$- \frac{K}{2} \frac{k(V_1 - V_2)}{k + k_O} \left[-\frac{k(V_1 - V_2)}{k + k_O} - 2V - 2V_T \right]$$

(3.106)

resulting:

$$I_{OUT} = \frac{K k^2}{(k + k_O)^2} (V_1 - V_2)^2$$

(3.107)

The circuit presented in Fig. 3.22 [15] also represents a voltage squaring circuit based on PR 3.2.

The circuit connections and the current mirrors from the circuit impose zero values for I and I'' currents. Thus, because M3 and M7 transistors are identical and they are biased at the same drain current, their gate-source voltages will be equal, so $V = -V_C - V_{DD}$ and, similarly, $V' = -V_C - V_{DD}$. The output differential current of the voltage squaring circuit can be expressed as follows:

$$I_{OUT1} - I_{OUT2} = I_{D1} + I_{D6} - I_{D3} - I_{D4}$$

(3.108)

resulting:

$$I_{OUT1} - I_{OUT2} = \frac{K}{2}(V_1 - V - V_T)^2 + \frac{K}{2}(-V_1 - V' - V_T)^2$$
$$- \frac{2K}{2}(-V - V_T)^2 = \frac{K}{2}(V_1 + V_C + V_{DD} - V_T)^2$$
$$+ \frac{K}{2}(-V_1 + V_C + V_{DD} - V_T)^2 - \frac{2K}{2}(V_C + V_{DD} - V_T)^2$$

(3.109)

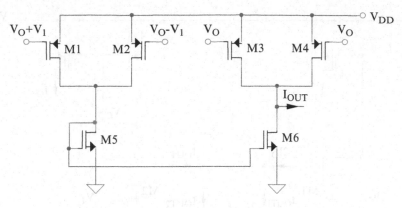

Fig. 3.23 Squaring circuit (15) based on PR 3.2

resulting:

$$I_{OUT1} - I_{OUT2} = \frac{K}{2}V_1(V_1 + 2V_C + 2V_{DD} - 2V_T)$$

$$-\frac{K}{2}V_1(-V_1 + 2V_C + 2V_{DD} - 2V_T) = KV_1^2 \qquad (3.110)$$

So, the output differential current will be proportional with the square of the input voltage.

The circuit presented in Fig. 3.23 [16] computes an output current proportional with the square of the input voltage.

The output current can be expressed as follows:

$$I_{OUT} = I_{D3} + I_{D4} - I_{D1} - I_{D2} = 2\frac{K}{2}(V_{DD} - V_O - V_T)^2$$

$$-\frac{K}{2}(V_{DD} - V_O - V_1 - V_T)^2 - \frac{K}{2}(V_{DD} - V_O + V_1 - V_T)^2 \qquad (3.111)$$

resulting:

$$I_{OUT} = \frac{K}{2}V_1(V_{DD} - 2V_O - 2V_T - V_1)$$

$$-\frac{K}{2}V_1(V_{DD} - 2V_O - 2V_T + V_1) = -KV_1^2 \qquad (3.112)$$

A voltage squaring circuit can be obtained using the circuit presented in Fig. 3.24 [17].

As M1 and M2 transistors are biased at I_O drain currents, V_A and V_B potentials can be expressed as follows:

$$V_A = V_2 - V_T - \sqrt{\frac{2I_O}{K}} \qquad (3.113)$$

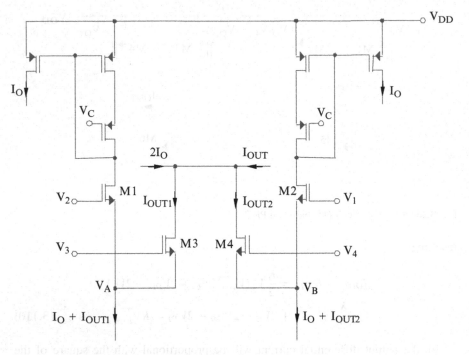

Fig. 3.24 Squaring circuit (16) based on PR 3.2

and:

$$V_B = V_1 - V_T - \sqrt{\frac{2I_O}{K}} \tag{3.114}$$

The expressions of I_{OUT1} and I_{OUT2} currents will be:

$$I_{OUT1} = \frac{K}{2}(V_3 - V_A - V_T)^2 \tag{3.115}$$

and:

$$I_{OUT2} = \frac{K}{2}(V_4 - V_B - V_T)^2 \tag{3.116}$$

Replacing (3.113) and (3.114) in (3.115) and (3.116), it results:

$$I_{OUT1} = \frac{K}{2}\left(V_3 - V_2 + \sqrt{\frac{2I_O}{K}}\right)^2 \tag{3.117}$$

and:

$$I_{OUT2} = \frac{K}{2}\left(V_4 - V_1 + \sqrt{\frac{2I_O}{K}}\right)^2 \tag{3.118}$$

The input potentials are chosen to have both common-mode and differential-mode components:

$$V_1 = V_C - V_{i1} \tag{3.119}$$

$$V_2 = V_C + V_{i1} \tag{3.120}$$

$$V_3 = V_C + V_{i2} \tag{3.121}$$

and:

$$V_4 = V_C - V_{i2} \tag{3.122}$$

resulting that the output current of the circuit presented in Fig. 3.24 will have the following expression:

$$I_{OUT} = I_{OUT1} + I_{OUT2} - 2I_O = \frac{K}{2}\left(V_{i2} - V_{i1} + \sqrt{\frac{2I_O}{K}}\right)^2$$

$$+ \frac{K}{2}\left(V_{i1} - V_{i2} + \sqrt{\frac{2I_O}{K}}\right)^2 - 2I_O = K(V_{i1} - V_{i2})^2 \tag{3.123}$$

A voltage squaring circuit using FGMOS transistors is presented in Fig. 3.25 [18].

Considering that FGMOS transistors from Fig. 3.25 have different inputs, as it is shown in the figure, the expressions of their drain currents will be:

$$I_{OUT1} = \frac{K}{2}\left(\frac{V_1 + V_{POL1}}{2} - V - V_T\right)^2 \tag{3.124}$$

$$I_{OUT2} = \frac{K}{2}\left(\frac{V_2 + V_{POL2}}{2} - V - V_T\right)^2 \tag{3.125}$$

and:

$$I_O = \frac{K}{2}\left(\frac{V_1 + V_2 + 2V_{POL3}}{4} - V - V_T\right)^2 \tag{3.126}$$

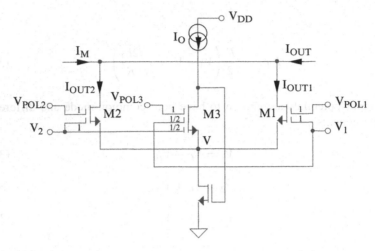

Fig. 3.25 Squaring circuit (17) based on PR 3.2

Replacing in (3.124) and (3.125) the expression of V potential from (3.126), it results:

$$I_{OUT1} = \frac{K}{2}\left(\frac{V_1 - V_2 + 2V_{POL1} - 2V_{POL3}}{4} + \sqrt{\frac{2I_O}{K}}\right)^2 \tag{3.127}$$

and:

$$I_{OUT2} = \frac{K}{2}\left(\frac{V_2 - V_1 + 2V_{POL2} - 2V_{POL3}}{4} + \sqrt{\frac{2I_O}{K}}\right)^2 \tag{3.128}$$

Designing a symmetrical structure ($V_{POL1} = V_{POL2}$), the output current will have the following expression:

$$I_{OUT} = I_{OUT1} + I_{OUT2} - I_M$$

$$= \frac{K}{2}\left[\left(\frac{V_1 - V_2}{4}\right) + \left(\frac{V_{POL1} - V_{POL3}}{2} + \sqrt{\frac{2I_O}{K}}\right)\right]^2$$

$$+ \frac{K}{2}\left[\left(-\frac{V_1 - V_2}{4}\right) + \left(\frac{V_{POL1} - V_{POL3}}{2} + \sqrt{\frac{2I_O}{K}}\right)\right]^2 - I_M \tag{3.129}$$

resulting:

$$I_{OUT} = \frac{K}{16}(V_1 - V_2)^2 + K\left(\frac{V_{POL1} - V_{POL3}}{2} + \sqrt{\frac{2I_O}{K}}\right)^2 - I_M \tag{3.130}$$

Fig. 3.26 Squaring circuit
(1) based on PR 3.3

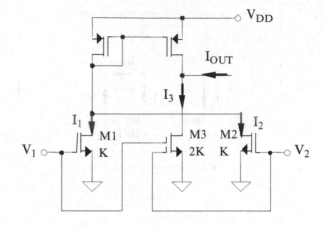

For obtaining an output current proportional with the square of the differential input voltage, I_M current must have the following expression:

$$I_M = K \left(\frac{V_{POL1} - V_{POL3}}{2} + \sqrt{\frac{2I_O}{K}} \right)^2 \tag{3.131}$$

In this case, it results:

$$I_{OUT} = \frac{K}{16} (V_1 - V_2)^2 \tag{3.132}$$

3.2.1.3 Squaring Circuits Based on the Third Mathematical Principle (PR 3.3)

An alternate implementation of a voltage squaring circuit using a FGMOS transistor is presented in Fig. 3.26 [47].

The output current expression has a linear dependence on the drain currents of M1, M2 and M3 transistors:

$$I_{OUT} = I_1 + I_2 - I_3 \tag{3.133}$$

Considering a biasing in saturation of all MOS devices from Fig. 3.26, I_{OUT} current will have the following dependence on the differential input voltage $V_1 - V_2$:

$$I_{OUT} = \frac{K}{2} (V_1 - V_T)^2 + \frac{K}{2} (V_2 - V_T)^2 - \frac{2K}{2} \left(\frac{V_1 + V_2}{2} - V_T \right)^2 \tag{3.134}$$

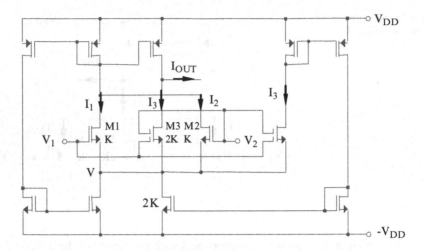

Fig. 3.27 Squaring circuit (2) based on PR 3.3

resulting:

$$I_{OUT} = \frac{K}{4}(V_1 - V_2)^2 \qquad (3.135)$$

Another possible implementation of a voltage squarer circuit using FGMOS transistors is based on the perfect symmetrical structure presented in Fig. 3.27 [19, 20].

The output current expression has a linear dependence on the drain currents of M1, M2 and M3 transistors:

$$I_{OUT} = I_1 + I_2 - I_3 \qquad (3.136)$$

For a biasing in saturation of all MOS devices from Fig. 3.27, the previous currents will have the following expressions:

$$I_1 = \frac{K}{2}(V_1 - V - V_T)^2 \qquad (3.137)$$

$$I_2 = \frac{K}{2}(V_2 - V - V_T)^2 \qquad (3.138)$$

$$I_3 = \frac{2K}{2}\left(\frac{V_1 + V_2}{2} - V - V_T\right)^2 \qquad (3.139)$$

From the previous relations, it results a quadratic dependence of the I_{OUT} output current on the differential input voltage, $V_1 - V_2$:

$$I_{OUT} = \frac{K}{4}(V_1 - V_2)^2 \qquad (3.140)$$

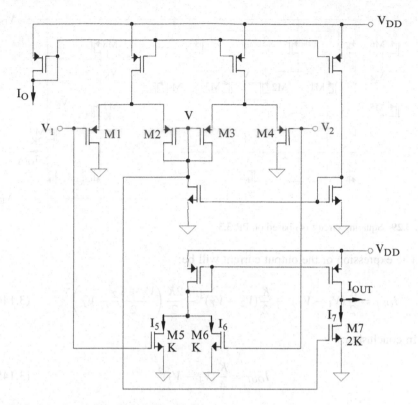

Fig. 3.28 Squaring circuit (3) based on PR 3.3

The realization of the voltage squaring circuit presented in Fig. 3.28 [21] uses the arithmetical mean of the input potentials, computed by M1–M4 transistors.

Because $I_{D1} + I_{D2} = I_{D2} + I_{D3} = I_{D3} + I_{D4} = I_O$, it results that $I_{D1} = I_{D3}$ and $I_{D2} = I_{D4}$. So, as M1–M4 transistors are identical, it is possible to conclude that $V_{SG1} = V_{SG3}$ and $V_{SG2} = V_{SG4}$. So:

$$V - V_1 = V_{SG1} - V_{SG2} = V_{SG3} - V_{SG4} = V_2 - V \tag{3.141}$$

resulting:

$$V = \frac{V_1 + V_2}{2} \tag{3.142}$$

The output current of the voltage squaring circuit will be linearly dependent on the drain currents of M5–M7 transistors:

$$I_{OUT} = I_5 + I_6 - I_7 \tag{3.143}$$

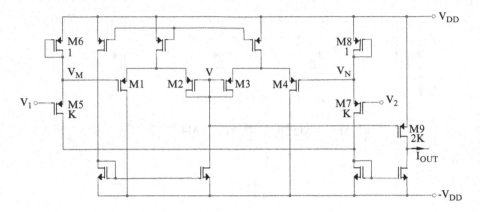

Fig. 3.29 Squaring circuit (4) based on PR 3.3

The expression of the output current will be:

$$I_{OUT} = \frac{K}{2}(V_1 - V_T)^2 + \frac{K}{2}(V_2 - V_T)^2 - \frac{2K}{2}\left(\frac{V_1 + V_2}{2} - V_T\right)^2 \qquad (3.144)$$

In conclusion:

$$I_{OUT} = \frac{K}{4}(V_1 - V_2)^2 \qquad (3.145)$$

The squaring circuit presented in Fig. 3.29 [22] is also based on the computation of the arithmetical mean of input potentials using four MOS transistors (M1–M4). So:

$$V = \frac{V_M + V_N}{2} \qquad (3.146)$$

The gate-source voltages of M5 and M6 transistors are equal, as they are identical and biased at the same drain current, so:

$$V_M = \frac{V_1 + V_{DD}}{2} \qquad (3.147)$$

and, similarly:

$$V_N = \frac{V_2 + V_{DD}}{2} \qquad (3.148)$$

resulting:

$$V = \frac{V_1 + V_2 + 2V_{DD}}{4} \qquad (3.149)$$

The output differential current will have the following expression:

$$I_{OUT} = I_{D9} - I_{D5} - I_{D7} = \frac{2K}{2}(V_{DD} - V - V_T)^2$$
$$- \frac{K}{2}(V_M - V_1 - V_T)^2 - \frac{K}{2}(V_N - V_2 - V_T)^2 \quad (3.150)$$

So:

$$I_{OUT} = \frac{2K}{2}\left(\frac{2V_{DD} - V_1 - V_2}{4} - V_T\right)^2$$
$$- \frac{K}{2}\left(\frac{V_{DD} - V_1}{2} - V_T\right)^2 - \frac{K}{2}\left(\frac{V_{DD} - V_2}{2} - V_T\right)^2 \quad (3.151)$$

equivalent with:

$$I_{OUT} = \frac{K}{2}\frac{V_1 - V_2}{4}\left(\frac{4V_{DD} - 3V_1 - V_2}{4} - 2V_T\right)$$
$$- \frac{K}{2}\frac{V_1 - V_2}{4}\left(\frac{4V_{DD} - 3V_2 - V_1}{4} - 2V_T\right) \quad (3.152)$$

It results that the output current of the circuit is proportional with the square of the differential input voltage:

$$I_{OUT} = -\frac{K}{16}(V_1 - V_2)^2 \quad (3.153)$$

3.2.1.4 Squaring Circuits Based on Different Mathematical Principles (PR 3. Da)

For the squaring circuit presented in Fig. 3.30 [23], the differential input voltage of the circuit can be expressed as follows:

$$V_1 - V_2 = V_{GS1} - V_{GS2} = \sqrt{\frac{2}{K}}(\sqrt{I_1} - \sqrt{I_2}) \quad (3.154)$$

It results:

$$I_1 + I_2 = 2\sqrt{I_1 I_2} + \frac{K}{2}(V_1 - V_2)^2 \quad (3.155)$$

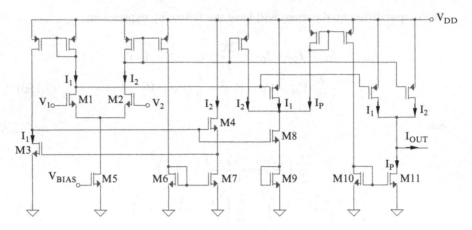

Fig. 3.30 Squaring circuit (1) based on PR 3.Da

The translinear loop containing M3, M4, M8 and M9 transistors has the following characteristic equation:

$$V_{GS3} + V_{GS4} = V_{GS8} + V_{GS9} \tag{3.156}$$

equivalent with:

$$\sqrt{I_1} + \sqrt{I_2} = \sqrt{I_1 + I_2 + I_P} \tag{3.157}$$

It was considered that $(W/L)_8 = (W/L)_9 = 4(W/L)_3 = 4(W/L)_4$. So:

$$I_P = 2\sqrt{I_1 I_2} \tag{3.158}$$

The output current of the circuit can be expressed as follows:

$$I_{OUT} = I_1 + I_2 - I_P = \frac{K}{2}(V_1 - V_2)^2 \tag{3.159}$$

The squaring circuit presented in Fig. 3.31 is realized using a parallel connection of two differential amplifiers, M1–M3 and M2–M4, their differential output currents being expressed as follows:

$$\begin{aligned}
I_1 - I_3 &= \frac{K}{2}(V_1 - V_X - V_T)^2 - \frac{K}{2}(V_2 - V_X - V_T)^2 \\
&= \frac{K}{2}(V_1 - V_2)(V_1 + V_2 - 2V_X - 2V_T)
\end{aligned} \tag{3.160}$$

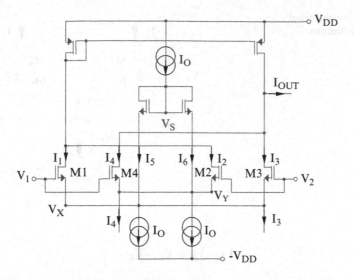

Fig. 3.31 Squaring circuit (2) based on PR 3.Da

and:

$$
\begin{aligned}
I_2 - I_4 &= \frac{K}{2}(V_2 - V_Y - V_T)^2 - \frac{K}{2}(V_1 - V_Y - V_T)^2 \\
&= \frac{K}{2}(V_2 - V_1)(V_1 + V_2 - 2V_Y - 2V_T)
\end{aligned}
\tag{3.161}
$$

resulting that the output current of the differential amplifier presented in Fig. 3.31 will have the following expression:

$$
\begin{aligned}
I_{OUT} &= (I_1 - I_3) + (I_2 - I_4) = \frac{K}{2}(V_1 - V_2)(2V_Y - 2V_X) \\
&= K(V_1 - V_2)(V_Y - V_X)
\end{aligned}
\tag{3.162}
$$

Between the currents from the circuit it exists the following linear relations:

$$I_1 + I_5 = I_O \tag{3.163}$$

$$I_2 + I_6 = I_O \tag{3.164}$$

and:

$$I_5 + I_6 = I_O \tag{3.165}$$

resulting $I_1 = I_6$ and $I_2 = I_5$. So:

$$V_1 - V_X = V_S - V_Y \tag{3.166}$$

Fig. 3.32 Asymmetrical
differential structure

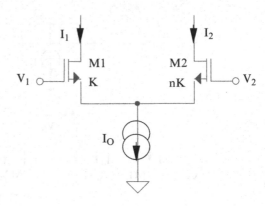

and:

$$V_2 - V_Y = V_S - V_X \tag{3.167}$$

equivalent with:

$$V_X - V_Y = V_1 - V_S \tag{3.168}$$

and:

$$V_X - V_Y = V_S - V_2 \tag{3.169}$$

So:

$$V_S = \frac{V_1 + V_2}{2} \tag{3.170}$$

and:

$$V_X - V_Y = V_1 - \frac{V_1 + V_2}{2} = \frac{V_1 - V_2}{2} \tag{3.171}$$

Replacing (3.171) in (3.162), the output current will be proportional with the square of the differential input voltage:

$$I_{OUT} = -\frac{K}{2}(V_1 - V_2)^2 \tag{3.172}$$

A method for designing a voltage squaring circuit is based on a differential amplifier (Fig. 3.32) [24, 25] having a controllable asymmetry between the geometries of two composing transistors. This difference between the aspect ratios of MOS transistors will introduce in the output currents of the differential amplifier a term proportional with the squaring of the differential input voltage.

Noting with $V = V_1 - V_2$ the differential input voltage, it can be expressed as follows:

$$V = V_{GS1} - V_{GS2} = \sqrt{\frac{2I_1}{K}} - \sqrt{\frac{2(I_O - I_1)}{nK}} \qquad (3.173)$$

resulting:

$$\frac{K}{2}V^2 = I_1 + \frac{I_O - I_1}{n} - 2\sqrt{\frac{I_1(I_O - I_1)}{n}} \qquad (3.174)$$

The expression of the I_1 unknown current can be obtained solving the following second-order equation, derived from (3.174):

$$I_1^2\left[\left(\frac{n-1}{n}\right)^2 + \frac{4}{n}\right] + I_1\left[2\frac{n-1}{n}\left(\frac{I_O}{n} - \frac{KV^2}{2}\right) - \frac{4I_O}{n}\right] + \left(\frac{I_O}{n} - \frac{KV^2}{2}\right)^2 = 0 \qquad (3.175)$$

So:

$$I_1 = \frac{I_O}{n+1} + \frac{n(n-1)}{2(n+1)^2}KV^2 + \frac{nV}{(n+1)^2}\sqrt{2KI_O(n+1) - K^2nV^2} \qquad (3.176)$$

and:

$$I_2 = I_O - I_1 = \frac{nI_O}{n+1} - \frac{n(n-1)}{2(n+1)^2}KV^2 - \frac{nV}{(n+1)^2}\sqrt{2KI_O(n+1) - K^2nV^2} \qquad (3.177)$$

The complete realization of a voltage squaring circuit uses a cross-coupling of two differential amplifiers having controllable asymmetries between their geometries, M1–M2 and M3–M4 (Fig. 3.33) [24, 25].
Using (3.176) and (3.177), it results:

$$I_{OUT}' = I_{D2} + I_{D4} = \frac{nI_O}{n+1} - \frac{n(n-1)}{2(n+1)^2}KV^2$$

$$-\frac{nV}{(n+1)^2}\sqrt{2KI_O(n+1) - K^2nV^2} + \frac{nI_O}{n+1} - \frac{n(n-1)}{2(n+1)^2}KV^2$$

$$+\frac{nV}{(n+1)^2}\sqrt{2KI_O(n+1) - K^2nV^2} = \frac{2nI_O}{n+1} - \frac{n(n-1)}{(n+1)^2}KV^2 \qquad (3.178)$$

So:

$$I_{OUT} = \frac{2nI_O}{n+1} - I_{OUT}' = \frac{n(n-1)}{(n+1)^2}KV^2 \qquad (3.179)$$

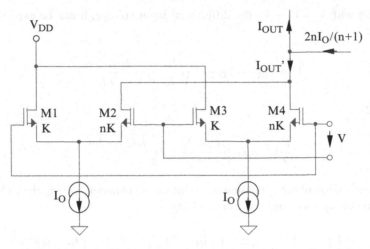

Fig. 3.33 Squaring circuit (3) based on PR 3.Da

3.2.2   Design of Current Squaring Circuits

The current squaring circuits are grouped in five classes, corresponding to the last five mathematical principles (PR 3.5 – PR 3.Db).

3.2.2.1   Squaring Circuits Based on the Fifth Mathematical Principle (PR 3.5)

A realization of a current squaring circuit using the fifth mathematical principle is presented in Fig. 3.34 [26]. The equation on the translinear loop containing M1–M4 transistors can be written as:

$$V_{GS1} + V_{GS2} = V_{GS3} + V_{GS4} \tag{3.180}$$

Because M1 and M2 transistors are biased at I_O drain current, M3 – at $I + I_{IN}$ current, while M4 is working at $I - I_{IN}$ drain current, for a biasing in saturation of all MOS transistors from the circuit, the previous relation becomes:

$$2\left(V_T + \sqrt{\frac{2I_O}{K}}\right) = \left(V_T + \sqrt{\frac{2(I + I_{IN})}{K}}\right) + \left(V_T + \sqrt{\frac{2(I - I_{IN})}{K}}\right) \tag{3.181}$$

So, $2\sqrt{I_O} = \sqrt{I + I_{IN}} + \sqrt{I - I_{IN}}$, resulting:

$$4I_O = 2I + 2\sqrt{I^2 - I_{IN}^2} \tag{3.182}$$

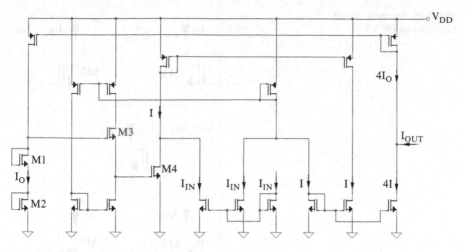

Fig. 3.34 Squaring circuit (1) based on PR 3.5

equivalent with:

$$4I_O^2 + I^2 - 4I_O I = I^2 - I_{IN}^2 \tag{3.183}$$

The I current can be expressed as follows:

$$I = I_O + \frac{I_{IN}^2}{4I_O} \tag{3.184}$$

The output current linearly depends on the currents from the circuit as follows:

$$I_{OUT} = 4I - 4I_O = \frac{I_{IN}^2}{I_O} \tag{3.185}$$

An alternative implementation of a current squaring circuit using the same principle is presented in Fig. 3.35.

The translinear loop containing M1–M4 transistors has the following characteristic equation:

$$V_{GS1} + V_{GS4} = V_{GS2} + V_{GS3} \tag{3.186}$$

resulting:

$$2\sqrt{I_O} = \sqrt{I_{O1}} + \sqrt{I_{IN} + I_{O1}} \tag{3.187}$$

Fig. 3.35 Squaring circuit
(2) based on PR 3.5

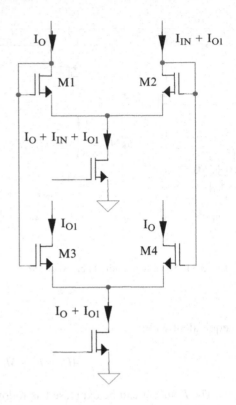

equivalent with the following dependence of I_{O1} current on the I_{IN} input current and on the reference current, I_O:

$$I_{O1} = I_O - \frac{I_{IN}}{2} + \frac{I_{IN}^2}{16I_O} \tag{3.188}$$

The entire implementation in CMOS technology of the previous current squaring circuit is presented in Fig. 3.36.

Designing a linear relation between the currents from the previous circuit:

$$I_{OUT} = 16I_{O_1} - 16I_O + 8I_{IN} \tag{3.189}$$

it results that the output current will be proportional with the square of the input current:

$$I_{OUT} = \frac{I_{IN}^2}{I_O} \tag{3.190}$$

The circuit presented in Fig. 3.37 [27, 28] implements the current squaring function. The core of the circuit is represented by the translinear loop realized using M1–M4 transistors. The characteristic equation of the loop is:

$$V_{GS1} + V_{GS2} = V_{GS3} + V_{GS4} \tag{3.191}$$

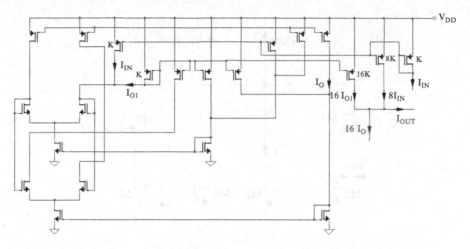

Fig. 3.36 Squaring circuit (3) based on PR 3.5

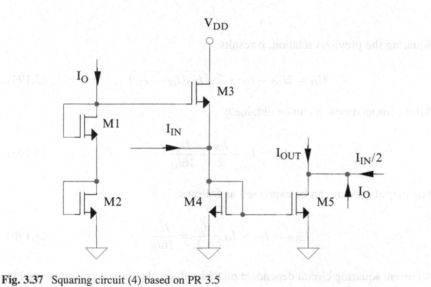

Fig. 3.37 Squaring circuit (4) based on PR 3.5

resulting:

$$2\left(V_T + \sqrt{\frac{2I_O}{K}}\right) = V_T + \sqrt{\frac{2I_{D5}}{K}} + V_T + \sqrt{\frac{2(I_{D5} - I_{IN})}{K}} \tag{3.192}$$

equivalent with:

$$2\sqrt{I_O} = \sqrt{I_{D5}} + \sqrt{I_{D5} - I_{IN}} \tag{3.193}$$

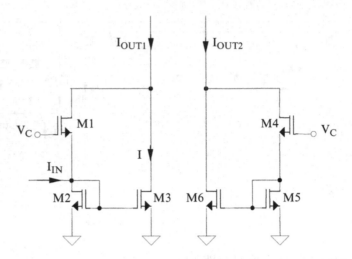

Fig. 3.38 Squaring circuit (5) based on PR 3.5

Squaring the previous relation, it results:

$$4I_O = 2I_{D5} - I_{IN} + 2\sqrt{I_{D5}(I_{D5} - I_{IN})} \tag{3.194}$$

After computations, it can be obtained:

$$I_{D5} = I_O + \frac{I_{IN}}{2} + \frac{I_{IN}^2}{16I_O} \tag{3.195}$$

The output current can be expressed as follows:

$$I_{OUT} = I_{D5} - I_O - \frac{I_{IN}}{2} = \frac{I_{IN}^2}{16I_O} \tag{3.196}$$

A current squaring circuit dependent on technological parameters is presented in Fig. 3.38 [29].

For the left part of the circuit, it can write that:

$$V_C = V_{GS1} + V_{GS2} \tag{3.197}$$

or:

$$V_C = \left(V_T + \sqrt{\frac{2I}{K}}\right) + \left(V_T + \sqrt{\frac{2(I - I_{IN})}{K}}\right) \tag{3.198}$$

resulting:

$$\sqrt{I} + \sqrt{I - I_{IN}} = \sqrt{\frac{K}{2}(V_C - 2V_T)} \qquad (3.199)$$

So:

$$2I - I_{IN} + 2\sqrt{I(I - I_{IN})} = A \qquad (3.200)$$

where:

$$A = \frac{K}{2}(V_C - 2V_T)^2 \qquad (3.201)$$

It results:

$$I = \frac{A}{4} + \frac{I_{IN}^2}{4A} + \frac{I_{IN}}{2} \qquad (3.202)$$

The I_{OUT1} current will have the following expression:

$$I_{OUT1} = 2I - I_{IN} = \frac{A}{2} + \frac{I_{IN}^2}{2A} \qquad (3.203)$$

The expression of I_{OUT2} current can be obtained from (3.177) for $I_{IN} = 0$, so:

$$I_{OUT2} = A/2 \qquad (3.204)$$

In conclusion, the differential output current can be expressed as follows:

$$I_{OUT1} - I_{OUT2} = \frac{I_{IN}^2}{2A} \qquad (3.205)$$

The disadvantage of this implementation of the current squarer consists in the dependence of the output current on technological parameters. In order to avoid this inconvenient, the V_C voltage can be generated using two gate-source voltages of MOS transistors, biased at a reference current I_O. In this particular case, V_C voltage will have the following expression:

$$V_C = 2\left(V_T + \sqrt{\frac{2I_O}{K}}\right) \qquad (3.206)$$

resulting the particular expression of A:

$$A = 4I_O \qquad (3.207)$$

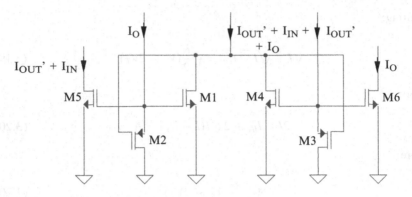

Fig. 3.39 Squaring circuit (6) based on PR 3.5

So:

$$I_{OUT} = \frac{I_{IN}^2}{8I_O} \tag{3.208}$$

The current squarer presented in Fig. 3.39 [30] uses a translinear loop for implementing the relation between the currents from the circuit. The characteristic equation of the loop is:

$$V_{SG2} + V_{GS4} = V_{GS1} + V_{SG3} \tag{3.209}$$

As M2 and M4 transistors are biased at I_O drain current, M1 is working at $I_{OUT}' + I_{IN}$ current, while M3 – at I_{OUT}' current, the previous relation becomes:

$$2\sqrt{I_O} = \sqrt{I_{OUT}'} + \sqrt{I_{OUT}' + I_{IN}} \tag{3.210}$$

resulting:

$$4I_O = 2I_{OUT}' + I_{IN} + 2\sqrt{I_{OUT}'(I_{OUT}' + I_{IN})} \tag{3.211}$$

So:

$$I_{OUT}' = I_O - \frac{I_{IN}}{2} + \frac{I_{IN}^2}{16I_O} \tag{3.212}$$

It is possible to implement the following linear relation between currents from the circuit, resulting:

$$I_{OUT} = I_{OUT}' - I_O + \frac{I_{IN}}{2} = \frac{I_{IN}^2}{16I_O} \tag{3.213}$$

Fig. 3.40 Squaring circuit
(7) based on PR 3.5

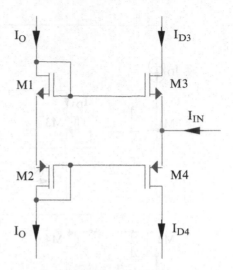

The circuit presented in Fig. 3.40 [31] represents a class AB amplifier.
The relation between the gate-source voltages from the translinear loop is:

$$V_{GS1} + V_{SG2} = V_{GS3} + V_{SG4} \qquad (3.214)$$

For a biasing in saturation of all MOS transistors, because M1 and M2 transistors
are working at I_O current, M3 transistor – at I_{D3} current, while M4 – at $I_{D3} + I_{IN}$, it
can be obtained:

$$2\sqrt{I_O} = \sqrt{I_{D3}} + \sqrt{I_{D3} + I_{IN}} \qquad (3.215)$$

Squaring the previous relation, it results:

$$4I_O = 2I_{D3} + I_{IN} + 2\sqrt{I_{D3}(I_{D3} + I_{IN})} \qquad (3.216)$$

So:

$$I_{D3} = I_O - \frac{I_{IN}}{2} + \frac{I_{IN}^2}{16I_O} \qquad (3.217)$$

The principle of operation of the circuit from Fig. 3.40 can be easily extended for
realizing the squaring function (Fig. 3.41) [31].
For this circuit, the expression of I_{D5} current can be obtained replacing I_{IN} with
$- I_{IN}$ in the expression (3.217) of I_{D3} current:

$$I_{D5} = I_O + \frac{I_{IN}}{2} + \frac{I_{IN}^2}{16I_O} \qquad (3.218)$$

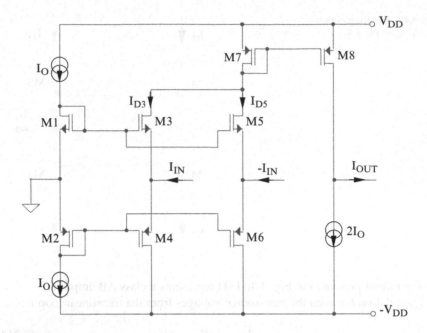

Fig. 3.41 Squaring circuit (8) based on PR 3.5

The output current of the squaring circuit is linearly dependent on the currents from the circuit:

$$I_{OUT} = I_{D3} + I_{D5} - 2I_O = \frac{I_{IN}^2}{8I_O} \tag{3.219}$$

The squaring circuit presented in Fig. 3.42 [32] uses a translinear loop implemented with M1–M4 transistors.

The characteristic equation of the translinear loop is:

$$V_{SG3} + V_{SG4} = V_{SG1} + V_{SG2} \tag{3.220}$$

Considering a biasing in saturation of all identical MOS transistors from Fig. 3.42, it results:

$$2\sqrt{I_O} = \sqrt{I_A} + \sqrt{I_B} = \sqrt{I_A} + \sqrt{I_{IN} + I_A} \tag{3.221}$$

Squaring the previous relation, the expression of I_A current will be:

$$I_A = I_O - \frac{I_{IN}}{2} + \frac{I_{IN}^2}{16I_O} \tag{3.222}$$

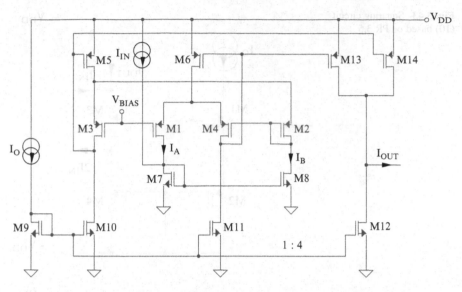

Fig. 3.42 Squaring circuit (9) based on PR 3.5

The output current of the circuit presented in Fig. 3.42 can be expressed as follows:

$$I_{OUT} = I_{D13} + I_{D14} - I_{D12} = (I_A + I_O) + (I_A + I_O + I_{IN}) - 4I_O \tag{3.223}$$

resulting:

$$I_{OUT} = \frac{I_{IN}^2}{8I_O} \tag{3.224}$$

The squaring circuit presented in Fig. 3.43 [33] uses a translinear loop implemented with M1–M4 transistors.

The characteristic equation of the translinear loop is:

$$V_{GS3} + V_{SG4} = V_{GS1} + V_{SG2} \tag{3.225}$$

For a biasing in saturation of all identical MOS transistors from Fig. 3.43, it results:

$$2\sqrt{I_O} = \sqrt{I_{OUT} + I_{IN}} + \sqrt{I_{OUT} - I_{IN}} \tag{3.226}$$

The output current of the circuit presented in Fig. 3.43 can be expressed as follows:

$$I_{OUT} = I_O + \frac{I_{IN}^2}{4I_O} \tag{3.227}$$

Fig. 3.43 Squaring circuit
(10) based on PR 3.5

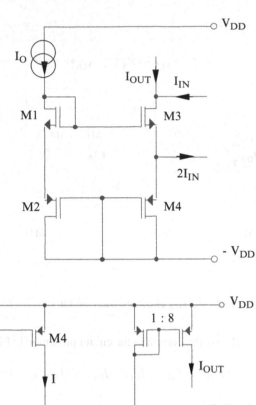

Fig. 3.44 Squaring circuit (11) based on PR 3.5

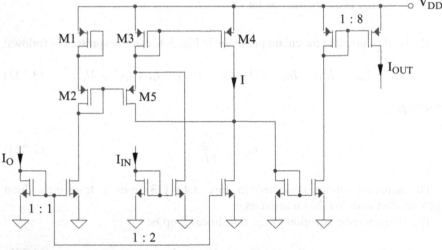

The squaring circuit presented in Fig. 3.44 [34] contains a translinear loop realized using M1, M2, M3 and M5 transistors, having the following characteristic equation:

$$V_{SG1} + V_{SG2} = V_{SG3} + V_{SG5} \qquad (3.228)$$

Considering a biasing in saturation of all MOS transistors from Fig. 3.44, it can be obtained:

$$2\sqrt{I_O} = \sqrt{I} + \sqrt{I - I_{IN}} \qquad (3.229)$$

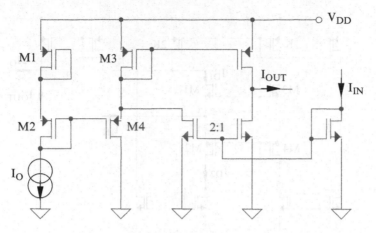

Fig. 3.45 Squaring circuit (12) based on PR 3.5

resulting:

$$I = I_O + \frac{I_{IN}}{2} + \frac{I_{IN}^2}{16I_O} \qquad (3.230)$$

The output current of the circuit presented in Fig. 3.44 will have the following expression:

$$I_{OUT} = [I + (I - I_{IN}) - 2I_O] = \frac{I_{IN}^2}{8I_O} \qquad (3.231)$$

Another squaring circuit based on the same mathematical principle is presented in Fig. 3.45 [35].

The characteristic equation of the translinear loop including M1–M4 transistors is:

$$V_{SG1} + V_{SG2} = V_{SG3} + V_{SG4} \qquad (3.232)$$

resulting:

$$2\sqrt{I_{D1}} = \sqrt{I_{D3}} + \sqrt{I_{D4}} \qquad (3.233)$$

equivalent with:

$$2\sqrt{I_O} = \sqrt{I_{OUT} + I_{IN}} + \sqrt{I_{OUT} - I_{IN}} \qquad (3.234)$$

The expression of the output current will be:

$$I_{OUT} = I_O + \frac{I_{IN}^2}{4I_O} \qquad (3.235)$$

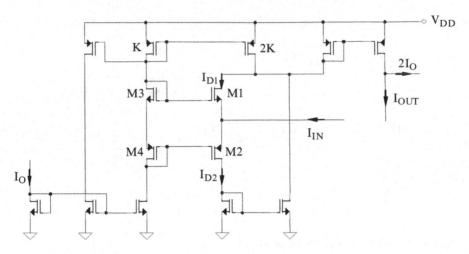

Fig. 3.46 Squaring circuit (13) based on PR 3.5

A current squaring circuit based on the fifth mathematical principle is presented in Fig. 3.46 [36].

The circuit implements the following relation between the gate-source voltages of MOS transistors:

$$V_{GS1} + V_{SG2} = V_{GS3} + V_{SG4} \tag{3.236}$$

Because all MOS transistors are biased in the saturation region, it is possible to write that:

$$2\sqrt{I_O} = \sqrt{I_{D1}} + \sqrt{I_{D1} + I_{IN}} \tag{3.237}$$

Thus:

$$I_{D1} = I_O - \frac{I_{IN}}{2} + \frac{I_{IN}^2}{16 I_O} \tag{3.238}$$

The expression of the output current can be written as:

$$I_{OUT} = 2 I_{D1} + I_{IN} - 2 I_O \tag{3.239}$$

resulting:

$$I_{OUT} = \frac{I_{IN}^2}{8 I_O} \tag{3.240}$$

So, the output current is proportional with the square of the input current.

Fig. 3.47 Squaring circuit
(1) based on PR 3.6

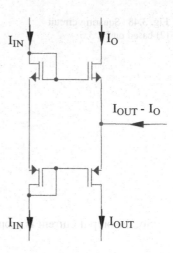

3.2.2.2 Squaring Circuits Based on the Sixth Mathematical Principle (PR 3.6)

A realization of a current squaring circuit based on a translinear loop with MOS transistors biased in weak inversion is presented in Fig. 3.47 [37].

For this circuit, it can write that:

$$2V_{GS}(I_{IN}) = V_{GS}(I_{OUT}) + V_{GS}(I_O) \tag{3.241}$$

where:

$$V_{GS}(I_{IN}) = V_T + nV_{th} \ln\left(\frac{I_{IN}}{\frac{W}{L}I_{DO}}\right) \tag{3.242}$$

$$V_{GS}(I_{OUT}) = V_T + nV_{th} \ln\left(\frac{I_{OUT}}{\frac{W}{L}I_{DO}}\right) \tag{3.243}$$

and:

$$V_{GS}(I_O) = V_T + nV_{th} \ln\left(\frac{I_O}{\frac{W}{L}I_{DO}}\right) \tag{3.244}$$

resulting:

$$2\ln(I_{IN}) = \ln(I_{OUT}) + \ln(I_O) \tag{3.245}$$

Fig. 3.48 Squaring circuit
(2) based on PR 3.6

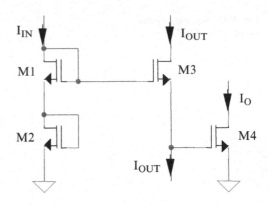

So, the output current is proportional with the squaring of the input current, I_{IN}:

$$I_{OUT} = \frac{I_{IN}^2}{I_O} \qquad (3.246)$$

A possible realization of a current squaring circuit (Fig. 3.48) [38, 39, 40, 43] uses also MOS transistors biased in weak inversion region. M1 and M2 transistors are working at the I_{IN} input current, M3 – at the I_{OUT} output current, while M4 transistor is biased at a reference current, I_O.

Because $V_{GS1} + V_{GS2} = V_{GS3} + V_{GS4}$ and using logarithmical dependencies of gate-source voltages on drain currents (similar with relations (3.216) – (3.218)), it results:

$$I_{OUT} = \frac{I_{IN}^2}{I_O} \qquad (3.247)$$

3.2.2.3   Squaring Circuits Based on the Seventh Mathematical Principle (PR 3.7)

A circuit that computes the squaring function starting from the arithmetical mean of input potentials, being based on the seventh mathematical principle is presented in Fig. 3.49 [21].

Transistors M5–M8 represent the arithmetical mean circuit, resulting:

$$V_3 = \frac{V_1 + V_2}{2} \qquad (3.248)$$

where V_1 and V_2 potentials are equal with the gate-source voltages of M1 and M4 transistors, respectively:

$$V_1 = V_{GS1} = V_T + \sqrt{\frac{2I_O}{K}} \qquad (3.249)$$

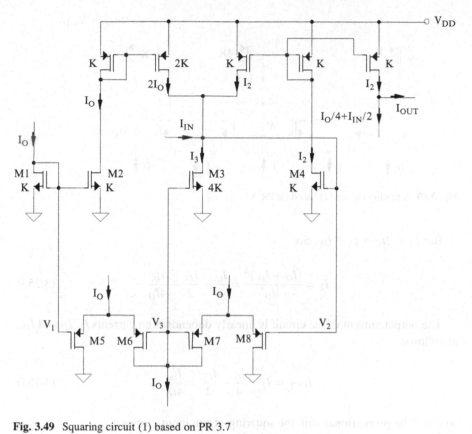

Fig. 3.49 Squaring circuit (1) based on PR 3.7

and:

$$V_2 = V_{GS4} = V_T + \sqrt{\frac{2I_2}{K}} \tag{3.250}$$

resulting:

$$V_3 = V_T + \frac{1}{2}\left(\sqrt{\frac{2I_O}{K}} + \sqrt{\frac{2I_2}{K}}\right) \tag{3.251}$$

Choosing the values of K parameters shown in Fig. 3.49, I_3 current will have the following expression:

$$I_3 = \frac{4K}{2}(V_3 - V_T)^2 = I_O + I_2 + 2\sqrt{I_O I_2} \tag{3.252}$$

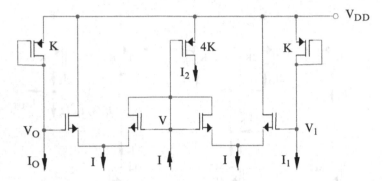

Fig. 3.50 Squaring circuit (2) based on PR 3.7

But $I_3 = 2I_O + I_2 + I_{IN}$. So:

$$I_2 = \frac{(I_O + I_{IN})^2}{4I_O} = \frac{I_O}{4} + \frac{I_{IN}}{2} + \frac{I_{IN}^2}{4I_O} \qquad (3.253)$$

The output current of the circuit is linearly dependent on currents I_2, I_O and I_{IN}, as follows:

$$I_{OUT} = I_2 - \frac{I_O}{4} - \frac{I_{IN}}{2} = \frac{I_{IN}^2}{4I_O} \qquad (3.254)$$

so it will be proportional with the squaring of the input current.

The circuit presented in Fig. 3.50 [41, 42] implements the squaring function, being based on the computation of the arithmetical mean of input potentials.

Because V potential represents the arithmetic mean of V_O and V_1 potentials (fixed by I_O and I_1 currents, respectively), the expression of I_2 current can be written as follows:

$$I_2 = \frac{4K}{2}(V_{DD} - V - V_T)^2 \qquad (3.255)$$

resulting:

$$I_2 = 2K\left(V_{DD} - \frac{V_O + V_1}{2} - V_T\right)^2 \qquad (3.256)$$

So:

$$I_2 = I_O + I_1 + 2\sqrt{I_O I_1} \qquad (3.257)$$

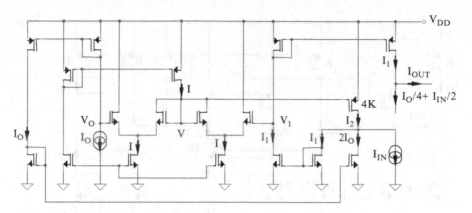

Fig. 3.51 Squaring circuit (3) based on PR 3.7

Using this square-root function, the desired squaring function can be easily obtained by subtracting I_{IN} and $2I_O$ currents from I_2 current expression. The full implementation in CMOS technology of the squaring function is presented in Fig. 3.51 [41, 42].

Using NMOS current mirrors, I_2 current is forced to be equal with:

$$I_2 = I_1 + I_{IN} + 2I_O \tag{3.258}$$

From the two previous relations it results:

$$I_1 = \frac{(I_O + I_{IN})^2}{4I_O} = \frac{I_O}{4} + \frac{I_{IN}}{2} + \frac{I_{IN}^2}{4I_O} \tag{3.259}$$

Because the output current has the following linear dependence on the circuit currents (implemented using simple current mirrors):

$$I_{OUT} = I_1 - \frac{I_O}{4} - \frac{I_{IN}}{2} \tag{3.260}$$

it results an output current proportional with the square of the input current:

$$I_{OUT} = \frac{I_{IN}^2}{4I_{OUT}} \tag{3.261}$$

The utilization of a FGMOS transistor usually simplifies the schematic of a current squaring circuit using MOS transistors biased in saturation region (Fig. 3.52) [43].

The expression of the drain current of the FGMOS transistor can be written as:

$$I_D = \frac{4K}{2} \left(\frac{1}{2} V_{GS1} + \frac{1}{2} V_{GS2} - V_T \right)^2 \tag{3.262}$$

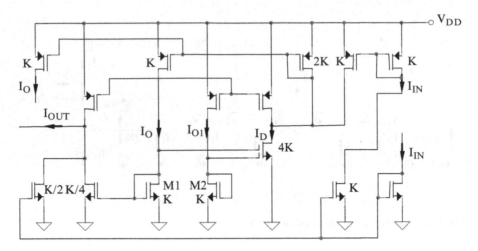

Fig. 3.52 Squaring circuit (4) based on PR 3.7

where V_{GS1} and V_{GS2} represent the gate-source voltages of M1 and M2 transistors, having the following expressions:

$$V_{GS1} = V_T + \sqrt{\frac{2I_O}{K}} \tag{3.263}$$

$$V_{GS2} = V_T + \sqrt{\frac{2I_{O1}}{K}} \tag{3.264}$$

From the previous three relations, it results the dependence of the FGMOS transistor drain current on I_O and I_{O1} currents:

$$I_D = I_O + I_{O1} + 2\sqrt{I_O I_{O1}} \tag{3.265}$$

Because of the PMOS multiple current mirrors, it can write that:

$$I_D = 2I_O + I_{O1} + I_{IN} \tag{3.266}$$

So:

$$I_{O1} = \frac{(I_O + I_{IN})^2}{4I_O} = \frac{I_O}{4} + \frac{I_{IN}}{2} + \frac{I_{IN}^2}{4I_O} \tag{3.267}$$

Thus, the output current expression will have the following expression:

$$I_{OUT} = I_{O1} - \frac{I_O}{4} - \frac{I_{IN}}{2} = \frac{I_{IN}^2}{4I_O} \tag{3.268}$$

Fig. 3.53 Squaring circuit
based on PR 3.8

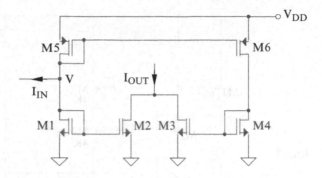

3.2.2.4 Squaring Circuits Based on the Eighth Mathematical Principle (PR 3.8)

A realization of a current squarer using the eight mathematical principle, whose performances are dependent on technological parameters is presented in Fig. 3.53 [44].

The input current can be expressed as a function of the drain currents of M1 and M5 transistors:

$$I_{IN} = I_{D5} - I_{D1} = \frac{K}{2}(V_{DD} - V - V_T)^2$$

$$-\frac{K}{2}(V - V_T)^2 = \frac{K}{2}(V_{DD} - 2V_T)(V_{DD} - 2V) \qquad (3.269)$$

From the previous relation, V potential can be expressed as follows:

$$V = \frac{1}{2}\left[V_{DD} - \frac{2I_{IN}}{K(V_{DD} - 2V_T)}\right] \qquad (3.270)$$

resulting:

$$I_{D1} = \frac{K}{2}(V - V_T)^2 = \frac{K}{2}\left[\frac{V_{DD} - 2V_T}{2} - \frac{I_{IN}}{K(V_{DD} - 2V_T)}\right]^2 \qquad (3.271)$$

and:

$$I_{D5} = \frac{K}{2}(V_{DD} - V - V_T)^2 = \frac{K}{2}\left[\frac{V_{DD} - 2V_T}{2} + \frac{I_{IN}}{K(V_{DD} - 2V_T)}\right]^2 \qquad (3.272)$$

The output current of the squarer is equal with:

$$I_{OUT} = I_{D2} + I_{D3} = I_{D1} + I_{D5} = \frac{K}{4}(V_{DD} - 2V_T)^2 + \frac{I_{IN}^2}{K(V_{DD} - 2V_T)^2} \qquad (3.273)$$

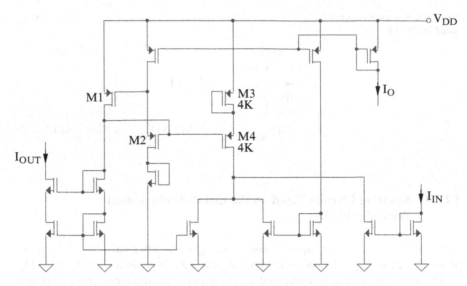

Fig. 3.54 Squaring circuit based on PR 3.9

Using the notation $A = V_{DD} - 2V_T$, it results the following expression of the output current:

$$I_{OUT} = \frac{KA^2}{4} + \frac{I_{IN}^2}{KA^2} \tag{3.274}$$

The operation of the circuit is affected by the technological errors introduced by the variations of V_T.

3.2.2.5 Squaring Circuits Based on the Ninth Mathematical Principle (PR 3.9)

Another current squaring circuit based on a square-rooting circuit is presented in Fig. 3.54 [45].

The characteristic equation of the translinear loop including M1–M4 transistors is:

$$V_{SG1} + V_{SG2} = V_{SG3} + V_{SG4} \tag{3.275}$$

resulting:

$$\sqrt{I_{D1}} + \sqrt{I_{D2}} = \frac{1}{2}\left(\sqrt{I_{D3}} + \sqrt{I_{D4}}\right) \tag{3.276}$$

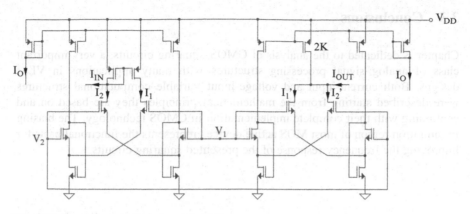

Fig. 3.55 Squaring circuit based on PR 3.Db

equivalent with:

$$\sqrt{I_{OUT}} + \sqrt{I_O} = \sqrt{I_{OUT} + I_O + I_{IN}} \qquad (3.277)$$

$$I_{IN} = 2\sqrt{I_O I_{OUT}} \qquad (3.278)$$

The expression of the output current will be:

$$I_{OUT} = \frac{I_{IN}^2}{4I_O} \qquad (3.279)$$

3.2.2.6 Squaring Circuits Based on Different Mathematical Principles (PR 3. Db)

The current squaring circuit presented in Fig. 3.55 [46] contains similar blocks, the left part of the circuit being characterized by the following relation:

$$I_{IN} = I_1 - I_2 = \sqrt{8KI_O(V_1 - V_2)} \qquad (3.280)$$

while the right part of the structure having a squaring dependence of the output current on the differential input voltage:

$$I_{OUT} = I_1' + I_2' - 2I_O = K(V_1 - V_2)^2 \qquad (3.281)$$

So:

$$I_{OUT} = \frac{I_{IN}^2}{8I_O} \qquad (3.282)$$

3.3 Conclusions

Chapter is dedicated to the analysis of CMOS squaring circuits, a very important class of analog signal processing structures with many applications in VLSI designs. Both current-input and voltage-input variable computational structures were described starting from the mathematical principles they are based on and continuing with their complete implementation in CMOS technology. The biasing in saturation region of most MOS active devices represents the functional basis for improving the frequency response of the presented squaring circuits.

References

1. Sato H, Hyogo A, Sekine K (2002) A V_t-zero equivalent MOSFET and its applications. In: IEEE international symposium on circuits and systems V-497–V-500, Arizona, USA
2. Filanovsky IM, Baltes H (1992) CMOS two-quadrant multiplier using transistor triode regime. IEEE J Solid-State Circuits 27:831–833
3. Popa C (2009) High accuracy CMOS multifunctional structure for analog signal processing. In: International semiconductor conference, pp 427–430, Sinaia, Romania
4. De La Cruz Blas CA, Feely O (2008) Limit cycle behavior in a class-AB second-order square root domain filter. In: IEEE international conference on electronics, circuits and systems, St. Julien's, pp 117–120, Malta
5. Zarabadi SR, Ismail M, Chung-Chih H (1998) High performance analog VLSI computational circuits. IEEE J Solid-State Circuits 33:644–649
6. Zele RH, Allstot DJ, Fiez TS (1991) Fully-differential CMOS current-mode circuits and applications. In: IEEE international symposium on circuits and systems, Westin Stamford, pp 1817–1820, Raffles City, Singapore
7. Demosthenous A, Panovic M (2005) Low-voltage MOS linear transconductor/squarer and four-quadrant multiplier for analog VLSI. IEEE Trans Circuits Syst I, Reg Pap 52:1721–1731
8. Lee BW, Sheu BJ (1990) A high slew-rate CMOS amplifier for analog signal processing. IEEE J Solid-State Circuits 25:885–889
9. Kumar JV, Rao KR (2002) A low-voltage low power square-root domain filter. In: Asia-Pacific conference on circuits and systems, pp 375–378, Bali, Indonesia
10. Klumperink E, van der Zwan E, Seevinck E (1989) CMOS variable transconductance circuit with constant bandwidth. Electron Lett 25:675–676
11. El Mourabit A, Sbaa MH, Alaoui-Ismaili Z, Lahjomri F (2007) A CMOS transconductor with high linear range. In: IEEE international conference on electronics, circuits and systems, pp 1131–1134, Marrakech, Morocco
12. Popa C (2006) Improved linearity active resistor using equivalent FGMOS devices. In: International conference on microelectronics, 396–399, Nis, Serbia and Montenegro
13. Popa C (2006) Improved linearity active resistor with increased frequency response for VLSI applications. IEEE international conference on automation, quality and testing, robotics, Cluj-Napoca, pp 114–116, Romania
14. Vlassis S, Siskos S (2001) Differential-voltage attenuator based on floating-gate MOS transistors and its applications. IEEE Trans Circuits Syst I, Fundam Theory Appl 48:1372–1378
15. Shen-Iuan L, Cheng-Chieh C (1996) A CMOS square-law vector summation circuit. IEEE Trans Circuits Syst II, Analog Digit Signal Process 43:520–523

16. Giustolisi G, Palmisano G, Palumbo G (1997) 1.5 V power supply CMOS voltage squarer. Electron Lett 33:1134–1136
17. Kimura K (1994) Analysis of "An MOS four-quadrant analog multiplier using simple two-input squaring circuits with source followers". IEEE Trans Circuits Syst I, Fundam Theory Appl 41:72–75
18. El Mourabit A, Lu GN, Pittet P (2005) Wide-linear-range subthreshold OTA for low-power, low-voltage and low-frequency applications. IEEE Trans Circuits Syst I, Reg Pap 52:1481–1488
19. Popa C (2010) Improved linearity CMOS active resistor based on complementary computational circuits. In: IEEE international conference on electronics, circuits, and systems, Athens, 455–458, Greece
20. Popa C (2004) A new FGMOS active resistor with improved linearity and frequency response. In: International semiconductor conference, 2:295–298, Sinaia, Romania
21. Manolescu AM, Popa C (2009) Low-voltage low-power improved linearity CMOS active resistor circuits. Springer J Analog Integr Circuits Signal Process 62:373–387
22. Popa C (2008) Programmable CMOS active resistor using computational circuits. In: International semiconductor conference, Sinaia, pp 389–392, Romania
23. Jong-Kug S, Charlot JJ (1999) Design and applications of precise analog computational circuits. Midwest Symposium on Circuits and Systems, Las Cruces, pp 275–278
24. Xiang-Luan Jia WH, Shi-Cai Q (1995) A new CMOS analog multiplier with improved input linearity. In: IEEE region 10 international conference on microelectronics and VLSI, pp 135–136, Hong Kong
25. Jong-Kug S, Charlot JJ (2000) A CMOS inverse trigonometric function circuit. In: IEEE midwest symposium on circuits and systems, pp 474–477, Michigan, USA
26. Popa C (2004) A digital-selected current-mode function generator for analog signal processing applications. In: International semiconductor conference, 2: 495–498, Sinaia, Romania
27. Quoc-Hoang D, Hoang-Nam D, Trung-Kien N, Sang-Gug L (2004) All CMOS current-mode exponential function generator. In: International conference on advanced communication technology, Korea, pp 528–531
28. Landolt O, Vittoz E, Heim P (1992) CMOS selfbiased Euclidean distance computing circuit with high dynamic range. Electron Lett 28:352–354
29. Cheng-Chieh C, Shen-Iuan L (2000) Current-mode full-wave rectifier and vector summation circuit. Electron Lett 36:1599–1600
30. Singh S, Radhakrishna Rao K (2006) Low voltage analogue multiplier. In: IEEE Asia pacific conference on circuits and systems, pp 1772–1775, Singapore
31. Boonchu B, Surakampontom W (2003) A CMOS current-mode squarer/rectifier circuit. In: International symposium on circuits and systems, pp I-405–I-408, Bangkok, Thailand
32. De La Blas CA, Lopez A (2008) A novel two quadrant MOS translinear squarer-divider cell. In: IEEE international conference on electronics, circuits and systems, St. Julien's, pp 5–8, Malta
33. Naderi A, Khoei A, Hadidi K (2007) High speed, low power four-quadrant CMOS current-mode multiplier. In: IEEE international conference on electronics, circuits and systems, Marrakech, pp 1308–1311, Morocco
34. Chuen-Yau C, Ju-Ying T, Bin-Da L(1998) Current-mode circuit to realize fuzzy classifier with maximum membership value decision. In: IEEE international symposium on circuits and systems, Monterey, 3:243–246, USA
35. Naderi A et al (2009) Four-quadrant CMOS analog multiplier based on new current squarer circuit with high-speed. In: IEEE international conference on "computer as a tool", St.-Petersburg, pp 282–287, Russia
36. Popa C (2009) A new CMOS current-mode classifier circuit for statistics applications. In: International conference on neural networks, pp 17–20, Prague, Czech Republic
37. Popa C (2006) CMOS quadratic circuits with applications in VLSI designs. In: International conference on signals and electronic systems, pp 627–630, Lodz, Poland

38. Popa C (2008) Low-power high precision integrated nanostructure with superior-order curvature-corrected logarithmic core. In: International conference on IC design and technology, pp xii–xvii, Grenoble, France
39. Popa C (2009) Logarithmical curvature-corrected voltage reference with improved temperature behavior. J Circuits, Syst Comput 18:519–534
40. Popa C (2009) Logarithmic compensated voltage reference. In: Spanish conference on electron devices, Santiago de Compostela, pp 215–218, Spain
41. Popa C (2007) Improved accuracy function generator circuit for analog signal processing. In: International conference on "computer as a tool", Warsaw, pp 231–236, Poland
42. Sawigun C, Serdijn WA (2009) Ultra-low-power, class-AB, CMOS four-quadrant current multiplier. Electron Lett 45:483–484
43. Popa C (2004) FGMOST-based temperature-independent Euclidean distance circuit. In: International conference on optimization of electric and electronic equipment, pp 29–32, Brasov, Romania
44. Kumngern M, Chanwutitum J, Dejhan K (2008) Simple CMOS current-mode exponential function generator circuit. In: International conference on electrical engineering/electronics, computer, telecommunications and information technology, Krabi, pp 709–712, Thailand
45. Kircay A, Keserlioglu MS, Cam U (2009) A new current-mode square-root-domain notch filter. In: european conference on circuit theory and design, Antalya, pp 229–232, Turkey
46. De La Cruz-Blas CA, Lopez-Martin AJ, Carlosena A (2005) 1.5-V square-root domain second-order filter with on-chip tuning. IEEE Trans Circuits Syst I, Reg Pap 52:1996–2006
47. Vlassis S, Fikos G, Siskos S (2001) A floating gate CMOS Euclidean distance calculator and its application to hand-written digit recognition. In: International conference on image processing, pp 350–353, Thessaloniki, Greece
48. Popa C (2005) CMOS logarithmic curvature-corrected voltage reference using a multiple differential structure. In: International symposium on signals, circuits and systems, pp 413–416, Iasi, Romania
49. Popa C (2003) DTMOST low-voltage low-power voltage references with superior-order curvature-corrections. In: European conference on circuits theory and design, pp 38–41, Krakow, Poland
50. Hidayat R, Dejhan K, Moungnoul P, Miyanaga Y (2008) OTA-based high frequency CMOS multiplier and squaring circuit. In: International symposium on intelligent signal processing and communications systems, pp 1–4, Bangkok, Thailand
51. Machowski W, Kuta S, Jasielski J, Kolodziejski W (2010) Quarter-square analog four-quadrant multiplier based on CMOS invertes and using low voltage high speed control circuits. In: International conference on mixed design of integrated circuits and systems, Warsaw, pp 333–336, Poland
52. Raikos G, Vlassis S (2009) Low-voltage CMOS voltage squarer. In: IEEE international on electronics, circuits, and systems, pp 159–162, Medina, Tunisia
53. Muralidharan R, Chip-Hong C (2009) Fixed and variable multi-modulus squarer architectures for triple moduli base of RNS. In: IEEE international conference on circuits and systems, Taipei, pp 441–444, Taiwan
54. Garofalo V et al (2010) A novel truncated squarer with linear compensation function. In: IEEE international symposium on circuits and systems, Paris, pp 4157–4160, France
55. Kumbun J, Lawanwisut S, Siripruchyanun M (2009) A temperature-insensitive simple current-mode squarer employing only multiple-output CCTAs. In: IEEE region 10 conference, Singapore, pp 1–4

Chapter 4
Square-Root Circuits

4.1 Mathematical Analysis for Synthesis of Square-Root Circuits

An important class of VLSI computational structures is represented by the square-root circuits. Usually implemented using a translinear loop, they exploit the squaring characteristic of MOS transistors biased in saturation region. The square-root circuits find a lot of applications in analog signal processing, such as square-root domain filters, Euclidean distance circuits, vector summation structures or real time image processing circuits. The presented design techniques are based on five different elementary mathematical principles, each of them being illustrated by concrete implementations in CMOS technology of their functional relations.

4.1.1 First Mathematical Principle (PR 4.1)

The first mathematical principle used for implementing square-rooting circuits [1–17] is based on the following relation:

$$I_{OUT} = \left(\sqrt{I_1} + \sqrt{I_2}\right)^2 - I_1 - I_2 = 2\sqrt{I_1 I_2} \qquad (4.1)$$

4.1.2 Second Mathematical Principle (PR 4.2)

The mathematical relation that models this principle is

$$\sqrt{I_1 + I_2 + aI_{OUT}} = \sqrt{I_1} + \sqrt{I_2} \Rightarrow I_{OUT} = \frac{2}{a}\sqrt{I_1 I_2} \qquad (4.2)$$

C.R. Popa, *Synthesis of Computational Structures for Analog Signal Processing*, 249
DOI 10.1007/978-1-4614-0403-3_4, © Springer Science+Business Media, LLC 2011

4.1.3   Third Mathematical Principle (PR 4.3)

The third mathematical principle is used for obtaining a differential output current proportional with the difference between the square-root of the input currents:

$$I_{OUT1} - I_{OUT2} = a(\sqrt{I_1} - \sqrt{I_2}) \qquad (4.3)$$

4.1.4   Fourth Mathematical Principle (PR 4.4)

The fourth mathematical principle used for implementing square-rooting circuits is based on the following relation:

$$2\sqrt{I_1} = \sqrt{I_1 + aI_2 - \frac{I_{OUT}}{2}} + \sqrt{I_1 + aI_2 + \frac{I_{OUT}}{2}} \Rightarrow I_{OUT} = 4\sqrt{aI_1I_2} \qquad (4.4)$$

4.1.5   Different Mathematical Principle (PR 4.D)

A class of square-rooting circuits is based on different mathematical principles.

4.2   Analysis and Design of Square-Root Circuits

4.2.1   Square-Root Circuits Based on the First Mathematical Principle (PR 4.1)

The circuit presented in Fig. 4.1 [1] implements the square-rooting function using the arithmetical mean of input potentials.

The I current can be expressed as follows:

$$I = \frac{4K}{2}(V - V_T)^2 \qquad (4.5)$$

As M1–M4 transistors compute the arithmetical mean of V_1 and V_2 potentials, it results:

$$I = 2K\left(\frac{V_1 + V_2}{2} - V_T\right)^2 \qquad (4.6)$$

These potentials are imposed by the input currents I_1 and I_2:

$$V_1 = V_{GS5} = V_T + \sqrt{\frac{2I_1}{K}} \qquad (4.7)$$

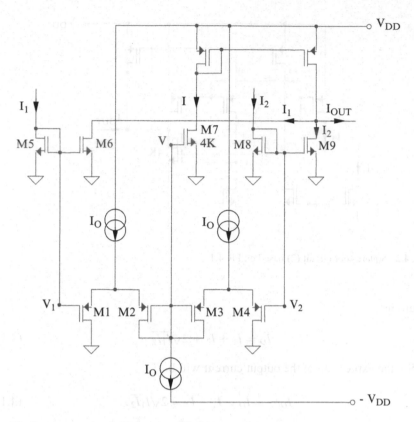

Fig. 4.1 Square-root circuit (1) based on PR 4.1

and

$$V_2 = V_{GS8} = V_T + \sqrt{\frac{2I_2}{K}} \qquad (4.8)$$

Replacing (4.7) and (4.8) in (4.6), the expression of I current will be

$$I = I_1 + I_2 + 2\sqrt{I_1 I_2} \qquad (4.9)$$

The output current of the circuit will be proportional with the square-root of the product between their input currents:

$$I_{OUT} = I - I_1 - I_2 = 2\sqrt{I_1 I_2} \qquad (4.10)$$

Another possible implementation of the square-root circuit is based on a similar structure. The square-root circuit using MOS transistors working in saturation and a FGMOS transistor for reducing the circuit complexity is presented in Fig. 4.2 [2, 3]. The expression of FGMOST drain current from Fig. 4.2 is

$$I_D = \frac{4K}{2} \left[\frac{V_{GS}(I_1) + V_{GS}(I_2)}{2} - V_T \right]^2 \qquad (4.11)$$

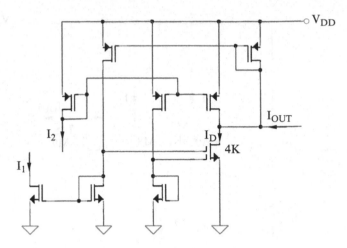

Fig. 4.2 Square-root circuit (2) based on PR 4.1

resulting:

$$I_D = I_1 + I_2 + 2\sqrt{I_1 I_2} \tag{4.12}$$

So, the expression of the output current will be

$$I_{OUT} = I_D - I_1 - I_2 = 2\sqrt{I_1 I_2} \tag{4.13}$$

4.2.2 Square-Root Circuits Based on the Second Mathematical Principle (PR 4.2)

The square-root circuit presented in Fig. 4.3 [4] implements the required function using a translinear loop realized using M1–M4 transistors.

The characteristic equation of the translinear loop is

$$V_{SG3} + V_{SG4} = V_{SG1} + V_{SG2} \tag{4.14}$$

Considering a biasing in saturation of all identical MOS transistors from Fig. 4.3, it results:

$$\sqrt{I_3} + \sqrt{I_4} = \sqrt{I_1} + \sqrt{I_2} \tag{4.15}$$

Fig. 4.3 Square-root circuit
(1) based on PR 4.2

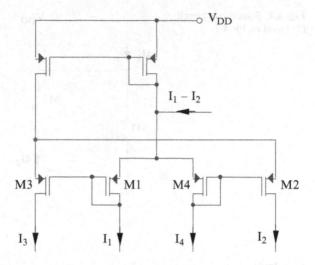

Imposing by external current mirrors the following relation between the currents from the circuit:

$$I_3 = I_4 = \frac{I_1 + I_2 + 2I_{OUT}}{4} \tag{4.16}$$

relation (4.15) becomes:

$$\sqrt{I_1 + I_2 + 2I_{OUT}} = \sqrt{I_1} + \sqrt{I_2} \tag{4.17}$$

So

$$I_{OUT} = \sqrt{I_1 I_2} \tag{4.18}$$

The circuit presented in Fig. 4.4 [5] implements the square-rooting function using a translinear loop containing M1–M4 transistors.

The characteristic equation of the translinear loop is

$$V_{GS1} + V_{GS2} = V_{GS3} + V_{GS4} \tag{4.19}$$

So

$$\sqrt{4I_1} + \sqrt{4I_2} = 2\sqrt{I_1 + I_2 + I_{OUT}} \tag{4.20}$$

The output current will be proportional with the square-root of the product between the input currents:

$$I_{OUT} = 2\sqrt{I_1 I_2} \tag{4.21}$$

Fig. 4.4 Square-root circuit
(2) based on PR 4.2

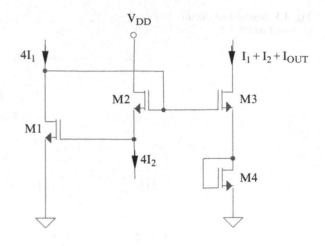

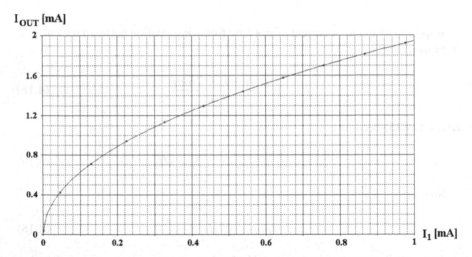

Fig. 4.5 The $I_{OUT}(I_1)$ characteristic for the square-root circuit from Fig. 4.4 for $I_2 = 1\ mA$

The operation of the circuit presented in Fig. 4.4 is simulated for $I_2 = 1\ mA$ and a variation range of I_1 current between 0 and $1\ mA$. The $I_{OUT}(I_1)$ characteristic is presented in Fig. 4.5.

A comparison between the simulated and the theoretical estimated results is shown in Table 4.1.

For extremely low values of input currents, some of the circuit transistors could operate in weak inversion region. In this case, the circuit doesn't implement the required square-root dependence (4.21). However, the circuit has a relatively extended range of the input currents, between hundreds nanoamperes and few miliamperes, the error of achieving the square-root function being relatively small. The frequency of operation of the square-root circuit presented in Fig. 4.4

Table 4.1 Comparison between the simulated and the theoretical estimated results for the square-root circuit presented in Fig. 4.4

I_1 (mA)	$I_{OUT\ sim}$ (mA)	$I_{OUT\ theor}$ (mA)
0	0	0
0.1	0.629	0.632
0.2	0.886	0.894
0.3	1.082	1.095
0.4	1.246	1.265
0.5	1.390	1.414
0.6	1.519	1.549
0.7	1.637	1.673
0.8	1.746	1.788
0.9	1.848	1.897
1	1.944	2

strongly depends on its biasing currents. For a proper operation of the structure (all MOS active devices biased in saturation region), it is expected to obtain an excellent frequency response.

The square-root circuit presented in Fig. 4.6 [6] contains a translinear loop using M4, M7, M10 and M11 transistors, having the following characteristic equation:

$$V_{GS4} + V_{GS7} = V_{GS10} + V_{GS11} \tag{4.22}$$

It results

$$\sqrt{I} = \sqrt{I_1} + \sqrt{I_2} \tag{4.23}$$

So

$$I = I_1 + I_2 + 2\sqrt{I_1 I_2} \tag{4.24}$$

The output current will have the following expression:

$$I_{OUT} = I - I_1 - I_2 = 2\sqrt{I_1 I_2} \tag{4.25}$$

Another current square-rooting circuit is presented in Fig. 4.7 [7].
The characteristic equation of the translinear loop including M1–M4 transistors is

$$V_{SG1} + V_{SG2} = V_{SG3} + V_{SG4} \tag{4.26}$$

It can be obtained:

$$\sqrt{I_{D1}} + \sqrt{I_{D2}} = \frac{1}{2}\left(\sqrt{I_{D3}} + \sqrt{I_{D4}}\right) \tag{4.27}$$

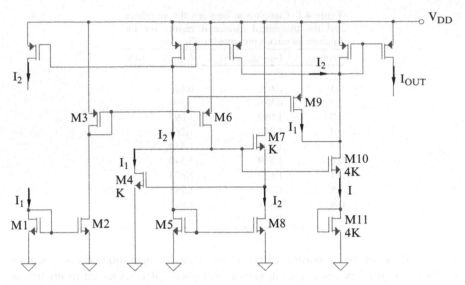

Fig. 4.6 Square-root circuit (3) based on PR 4.2

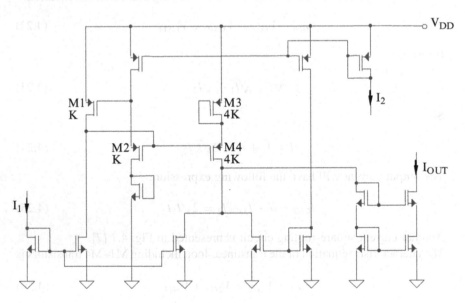

Fig. 4.7 Square-root circuit (4) based on PR 4.2

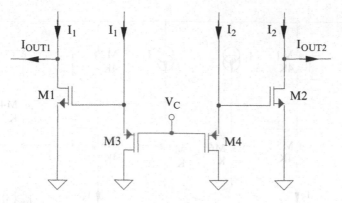

Fig. 4.8 Square-root circuit (1) based on PR 4.3

equivalent with:

$$\sqrt{I_1} + \sqrt{I_2} = \sqrt{I_1 + I_2 + I_{OUT}} \tag{4.28}$$

The expression of the output current will be

$$I_{OUT} = 2\sqrt{I_1 I_2} \tag{4.29}$$

4.2.3 Square-Root Circuits Based on the Third Mathematical Principle (PR 4.3)

A circuit that implements the current squaring function is presented in Fig. 4.8 [8]. The differential output current of the circuit can be expressed as follows:

$$I_{OUT1} - I_{OUT2} = (I_1 - I_{D1}) - (I_2 - I_{D2}) \tag{4.30}$$

The V_C constant potential is equal with the difference between two gate-source voltages. Supposing a biasing in saturation of all identical MOS transistors, it results:

$$V_C = V_{GS1} - V_{SG3} = \sqrt{\frac{2I_{D1}}{K}} - \sqrt{\frac{2I_1}{K}} \tag{4.31}$$

So

$$\sqrt{I_{D1}} = \sqrt{I_1} + \sqrt{\frac{K}{2}} V_C \tag{4.32}$$

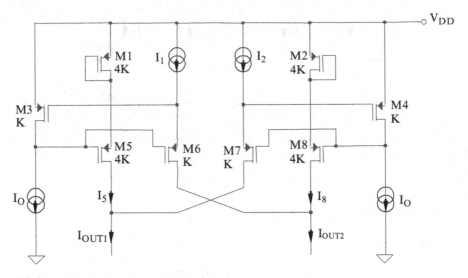

Fig. 4.9 Square-root circuit (2) based on PR 4.3

It can be obtained:

$$I_{D1} = I_1 + \frac{K}{2}V_C^2 + \sqrt{2KI_1}\,V_C \tag{4.33}$$

and, similarly:

$$I_{D2} = I_2 + \frac{K}{2}V_C^2 + \sqrt{2KI_2}\,V_C \tag{4.34}$$

From (4.30), (4.33) and (4.34), it results a square-root dependence of the differential output current on input currents:

$$I_{OUT1} - I_{OUT2} = \sqrt{2K}V_C\left(\sqrt{I_2} - \sqrt{I_1}\right) \tag{4.35}$$

Another implementation of a differential squaring circuit is presented in Fig. 4.9 [8].
The M1, M3, M5 and M6 transistors form a translinear loop, the characteristic equation being:

$$V_{SG3} + V_{SG6} = V_{SG1} + V_{SG5} \tag{4.36}$$

equivalent with:

$$2\sqrt{\frac{2I_5}{4K}} = \sqrt{\frac{2I_O}{K}} + \sqrt{\frac{2I_1}{K}} \tag{4.37}$$

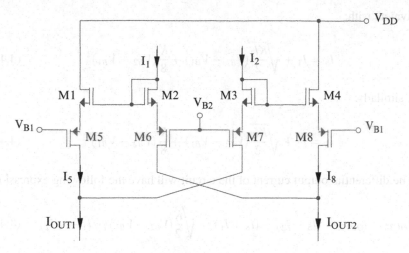

Fig. 4.10 Square-root circuit (3) based on PR 4.3

resulting:

$$I_5 = I_O + I_1 + 2\sqrt{I_O I_1} \qquad (4.38)$$

and, similarly:

$$I_8 = I_O + I_2 + 2\sqrt{I_O I_2} \qquad (4.39)$$

The differential output current of the circuit can be expressed as follows:

$$I_{OUT1} - I_{OUT2} = (I_5 + I_2) - (I_8 + I_1) = 2\sqrt{I_O}\left(\sqrt{I_1} - \sqrt{I_2}\right) \qquad (4.40)$$

An alternate realization of a differential squaring circuit is shown in Fig. 4.10 [8]. The $V_{B2} - V_{B1}$ differential voltage can be expressed as function on the gate-source voltages, as follows:

$$V_{B2} - V_{B1} = 2V_{GS}(I_5) - 2V_{GS}(I_1) \qquad (4.41)$$

resulting:

$$V_{B2} - V_{B1} = 2\sqrt{\frac{2}{K}}\left(\sqrt{I_5} - \sqrt{I_1}\right) \qquad (4.42)$$

So

$$\sqrt{I_5} = \sqrt{I_1} + \sqrt{\frac{K}{8}}(V_{B2} - V_{B1}) \qquad (4.43)$$

equivalent with:

$$I_5 = I_1 + \sqrt{\frac{KI_1}{2}}(V_{B2} - V_{B1}) + \frac{K}{8}(V_{B2} - V_{B1})^2 \qquad (4.44)$$

and, similarly:

$$I_8 = I_2 + \sqrt{\frac{KI_2}{2}}(V_{B2} - V_{B1}) + \frac{K}{8}(V_{B2} - V_{B1})^2 \qquad (4.45)$$

The differential output current of the circuit will have the following expression:

$$I_{OUT1} - I_{OUT2} = (I_5 + I_2) - (I_8 + I_1) = \sqrt{\frac{K}{2}}(V_{B2} - V_{B1})\left(\sqrt{I_1} - \sqrt{I_2}\right) \qquad (4.46)$$

4.2.4   Square-Root Circuits Based on the Fourth Mathematical Principle (PR 4.4)

A square-root circuit using a translinear loop containing four MOS transistors biased in saturation region is presented in Fig. 4.11 [9, 10]. The circuit operation is derived from the fourth mathematical principle with $a = 1$. The characteristic equation of the loop is

$$2V_{GS}(I_1) = V_{GS}(I_D) + V_{GS}(I_D + I_{OUT}) \qquad (4.47)$$

resulting:

$$2\sqrt{I_1} = \sqrt{I_D} + \sqrt{I_D + I_{OUT}} \qquad (4.48)$$

After some computations, it results:

$$I_D = I_1 - \frac{I_{OUT}}{2} + \frac{I_{OUT}^2}{16I_1} \qquad (4.49)$$

Because

$$I_D + \frac{I_{OUT}}{2} = I_1 + I_2 \qquad (4.50)$$

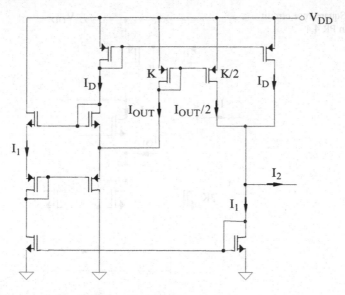

Fig. 4.11 Square-root circuit (1) based on PR 4.4

it can be obtained:

$$I_{OUT} = 4\sqrt{I_1 I_2} \tag{4.51}$$

Another application of the fourth mathematical principle ($a = 1/2$) is presented in Fig. 4.12 [11].

The characteristic equation of the translinear loop from Fig. 4.12 is

$$2V_{GS}(I_1) = V_{GS}(I_D) + V_{GS}(I_D + I_{OUT}) \tag{4.52}$$

resulting:

$$2\sqrt{I_1} = \sqrt{I_D} + \sqrt{I_D + I_{OUT}} \tag{4.53}$$

So

$$I_D = I_1 - \frac{I_{OUT}}{2} + \frac{I_{OUT}^2}{16 I_1} \tag{4.54}$$

Because

$$I_D + (I_D + I_{OUT}) = 2I_1 + I_2 \tag{4.55}$$

Fig. 4.12 Square-root circuit
(2) based on PR 4.4

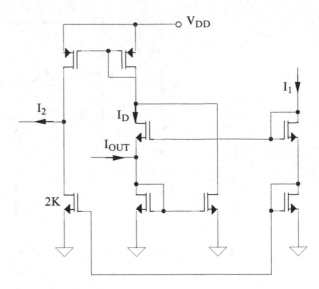

It results, after some computations, that the output current of the circuit from Fig. 4.12 can be expressed as follows:

$$I_{OUT} = 2\sqrt{2I_1I_2} \tag{4.56}$$

4.2.5   Square-Root Circuits Based on Different Mathematical Principles (PR 4.D)

A method for obtaining the square-rooting function using the squaring circuit shown in Fig. 3.3 from Chap. 3 is presented in Fig. 4.13 [12]. The V_1 and V_2 potentials are obtained using four gate-drain connected MOS transistors, biased at I_1 and I_2 input currents.

The sum of the output currents of the squaring circuit is

$$I_{OUT1} + I_{OUT2} = 2I_O + K(V_1 - V_2)^2 \tag{4.57}$$

The $V_1 - V_2$ differential input voltage can be expressed as a function of the gate-source voltages, as follows:

$$V_1 - V_2 = 2V_{GS}(I_1) - 2V_{GS}(I_2) \tag{4.58}$$

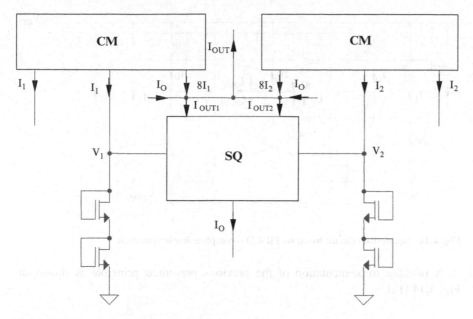

Fig. 4.13 Square-root circuit based on PR 4.D – principle circuit

Replacing (4.58) in (4.57) and using the square-root dependence of the drain current on the gate-source voltage for a MOS transistor biased in saturation, it results:

$$I_{OUT1} + I_{OUT2} = 2I_O + K\left(2\sqrt{\frac{2I_1}{K}} - 2\sqrt{\frac{2I_2}{K}}\right)^2 \tag{4.59}$$

equivalent with:

$$I_{OUT1} + I_{OUT2} = 2I_O + 8I_1 + 8I_2 - 16\sqrt{I_1 I_2} \tag{4.60}$$

The output current can be expressed using a linear relation between the currents from the circuit:

$$I_{OUT} = 8I_1 + 8I_2 + 2I_O - (I_{OUT1} + I_{OUT2}) \tag{4.61}$$

resulting a square-root dependence of the I_{OUT} output current on I_1 and I_2 input currents:

$$I_{OUT} = 16\sqrt{I_1 I_2} \tag{4.62}$$

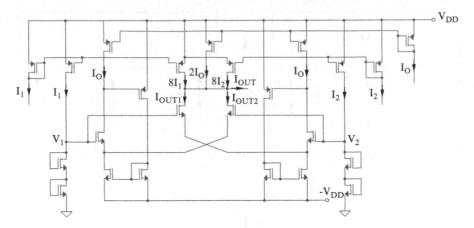

Fig. 4.14 Square-root circuit based on PR 4.D – complete implementation

A possible implementation of the previous presented principle is shown in Fig. 4.14 [12].

4.3 Conclusions

Chapter presents a large number of square-root circuits designed for analog signal processing. In order to improve the circuits' frequency operation, usually MOS transistors biased in saturation have been used, the current-mode operation of most presented computational structures being responsible for an additional improvement of the frequency behavior.

References

1. Psychalinos C, Vlassis S (2002) A systematic design procedure for square-root-domain circuits based on the signal flow graph approach. IEEE Transactions on Circuits and Systems I. Fundam Theory Appl 49:1702–1712
2. Manolescu AM, Popa C (2009) Low-voltage low-power improved linearity CMOS active resistor circuits. Springer J Analog Integr Circuits Signal Process 62:373–387
3. Popa C (2008) Optimal superior-order curvature-corrected voltage reference based on the weight difference of gate-source voltages. Springer J Analog Integr Circuits Signal Process 1:1–6
4. De La Blas CA, Lopez A (2008) A novel two quadrant MOS translinear squarer-divider cell. In: IEEE International conference on electronics, circuits and systems, St. Julien's, pp 5–8, Malta
5. De La Cruz-Blas CA, Lopez-Martin A, Carlosena A (2003) 1.5-V MOS translinear loops with improved dynamic range and their applications to current-mode signal processing. actions on and. IEEE Trans Circuits Syst II: Analog Digit Signal Process 50:918–927

6. Jong-Kug S, Charlot JJ (1999) Design and applications of precise analog computational circuits. In: Midwest symposium on circuits and systems 275–278, Michigan, USA
7. Kircay A, Keserlioglu MS, Cam U (2009) A new current-mode square-root-domain notch filter. In: European conference on circuit theory and design, Antalya, pp 229–232, Turkey
8. Ngamkham W, Kiatwarin N et al (2008) A linearized source-couple pair transconductor using a low-voltage square root circuit. In: International conference on electrical engineering/electronics, computer, telecommunications and information technology, Krabi, pp 701–704
9. Popa C (2005) A new current-mode CMOS Euclidean distance circuit with improved accuracy and frequency response. In: IEEE conference on microelectronics, electronics and electronic technologies, pp 99–102, Opatija, Croatia
10. Popa C (2010) Superior-order curvature-corrected voltage reference using a current generator. In: Lecture notes in computer science, Springer, pp 12–21
11. Popa C (2010) Low-power low-voltage superior-order curvature corrected voltage reference. Inter J Electron 97:613–622
12. Popa C (2009) High accuracy CMOS multifunctional structure for analog signal processing. In: International semiconductor conference, Sinaia, pp 427–430, Romania
13. Popa C (2005) A new improved linearity active resistor using complementary functions. In: International semiconductor conference, pp 391–394
14. Popa C (2006) Digitally-selected optimal curvature-corrected voltage reference using FGMOS devices. In: IEEE conference on microelectronics, electronics and electronic technologies, pp 90–93, Opatija, Croatia
15. Popa C (2002) Autoprogrammable superior-order curvature-correction CMOS thermal system. In: International semiconductor conference, pp 369–372, Sinaia, Romania
16. Popa C (2005) Power-efficient superior-order curvature corrected voltage reference using CMOS computational circuits. In: International symposium on signals, circuits and systems, pp 23–26, Iasi, Romania
17. Kircay A, Keserlioglu MS (2009) Novel current-mode second-order square-root-domain highpass and allpass filter. In: International conference on electrical and electronics engineering, pp II-242–II-246, Bursa, Turkey

Chapter 5
Exponential Circuits

5.1 Mathematical Analysis for Synthesis of Exponential Circuits

Exponential circuits represent important building blocks for VLSI signal processing structures, telecommunication applications, medical equipments, hearing aid or disk drives. The exponential function can be obtained in bipolar technology from the exponential characteristic of the bipolar transistor. The nonzero value of the base current and the temperature dependence of the bipolar transistor parameters introduce relatively large errors in the computation of the exponential function. In CMOS technology, the exponential law is available only for the weak inversion operation of the MOS transistor. The great disadvantage of the circuits using MOS transistors in weak inversion is represented by their poor circuit frequency response, caused by the much smaller drain currents available for charging and discharging the parasite capacitances of the MOS transistors. Thus, circuits realized in CMOS technology that require a good frequency response can be designed using exclusively MOS transistors biased in saturation region. Because it exists a relative limited number of mathematical principles that are used for implementing the exponential circuits, the first part of the chapter is dedicated to the analysis of the mathematical relations representing the functional core of the designed circuits. In the second part of the chapter, starting from these elementary principles, there are analyzed and designed concrete exponential circuits, grouped following the mathematical principles they are based on.

The mathematical analysis that represents the basis for designing exponential circuits [1–14] is structured in nine mathematical principles.

5.1.1 First Mathematical Principle (PR 5.1)

The classical approximation of the exponential function $f(x) = \exp(x)$ is given by their limited Taylor series expansion:

C.R. Popa, *Synthesis of Computational Structures for Analog Signal Processing*, DOI 10.1007/978-1-4614-0403-3_5, © Springer Science+Business Media, LLC 2011

$$g(x) = 1 + x + \frac{x^2}{2!} + \frac{x^3}{3!} + \frac{x^4}{4!} + \cdots \tag{5.1}$$

For a nth order approximation, all the first $n + 1$ polynomial terms from the Taylor series expansions of $f(x)$ and $g(x)$ must be equal, the approximation error being mainly caused by the difference between the $(n + 1)$th order terms from these expansions.

5.1.2   Second Mathematical Principle (PR 5.2)

An approximation of the $f(x) = \exp(2ax)$ function, a and k being constants, is given by the following function:

$$g(x) = \frac{1 + ax + k\frac{(ax)^2}{2}}{1 - ax + k\frac{(ax)^2}{2}} \tag{5.2}$$

The third-order limited Taylor series expansion of $f(x)$ function has the following expression:

$$f(x) \cong 1 + 2ax + 2a^2x^2 + \frac{4a^3}{3}x^3 + \cdots \tag{5.3}$$

In order to evaluate the third-order limited Taylor series expansion of $g(x)$ function, its superior-order derivates must be determined. The first-order derivate of $g(x)$ function is

$$g'(x) = \frac{2a - 3a^3Kx^2}{\left[1 - ax + k\frac{(ax)^2}{2}\right]^2} \tag{5.4}$$

while the second-order derivate can be expressed as follows:

$$g''(x) = \frac{-10Ka^3x + 3K^2a^5x^3 + 4a^2}{\left[1 - ax + k\frac{(ax)^2}{2}\right]^3} \tag{5.5}$$

The third-order derivate of $g(x)$ function is

$$g'''(x) = \frac{-10Ka^3 - 32Ka^4x + 34K^2a^5x^2 - \frac{9}{2}K^3a^7x^4 + 12a^3}{\left[1 - ax + k\frac{(ax)^2}{2}\right]^4} \tag{5.6}$$

Using the general Taylor series expansion, the function $g(x)$ can be third-order approximated by the following polynomial function:

$$g(x) \cong 1 + 2ax + 2a^2x^2 + 2a^3x^3 + \cdots \tag{5.7}$$

Comparing the limited Taylor series expansions of $f(x)$ and $g(x)$ functions expressed by (5.3) and (5.7), the errors of the $f(x) \cong g(x)$ approximation are mainly given by the third-order term from these expansions:

$$\varepsilon_{f(x)}^{g(x)}(x) \cong \frac{2a^3 - \frac{4a^3}{3}}{\exp(2ax)}x^3 = \frac{2a^3x^3}{3\exp(2ax)} \tag{5.8}$$

5.1.3 Third Mathematical Principle (PR 5.3)

In order to obtain a second-order approximation of the following exponential function:

$$f(x) = \exp(2mx) \tag{5.9}$$

m being a constant, the $g(x)$ function can be used, having the important advantage of allowing a simple implementation in CMOS technology, using multipliers and squaring circuits:

$$g(x) = \left(\frac{1+x}{1-x}\right)^m \tag{5.10}$$

The third-order limited Taylor series expansion of $f(x)$ function has the following expression:

$$f(x) \cong 1 + 2mx + 2m^2x^2 + \frac{4m^3}{3}x^3 + \cdots \tag{5.11}$$

In order to evaluate the third-order limited Taylor series expansion of $g(x)$ function, its superior-order derivates must be determined. The first-order derivate of $g(x)$ function is

$$g'(x) = \frac{2m(1+x)^{m-1}}{(1-x)^{m+1}} \tag{5.12}$$

while the second-order derivate can be expressed as follows:

$$g''(x) = \frac{4m(m+x)(1+x)^{m-2}}{(1-x)^{m+2}} \tag{5.13}$$

The third-order derivate of $g(x)$ function is

$$g'''(x) = \frac{4m(2m^2 + 6mx + 3x^2 + 1)(1+x)^{m-3}}{(1-x)^{m+3}} \tag{5.14}$$

Using the general Taylor series expansion, the $g(x)$ function can be third-order approximated by the following polynomial function:

$$g(x) \cong 1 + 2mx + 2m^2x^2 + \frac{2m}{3}(2m^2 + 1)x^3 + \cdots \tag{5.15}$$

Comparing the limited Taylor series expansions of $f(x)$ and $g(x)$ functions expressed by (5.11) and (5.15), the errors of the $f(x) \cong g(x)$ approximation are mainly given by the third-order term from these expansions:

$$\varepsilon_{f(x)}^{g(x)}(x) \cong \frac{\frac{2m}{3}(2m^2 + 1) - \frac{4m^3}{3}}{\exp(2mx)}x^3 = \frac{2mx^3}{3\exp(2mx)} \tag{5.16}$$

5.1.4 Fourth Mathematical Principle (PR 5.4)

The $f(x) = \exp(x)$ exponential function can be second-order approximated using the following general function:

$$g(x) = \frac{1 + ax}{1 + bx} \tag{5.17}$$

a and b being constants having values that can be determined from the condition that $g(x)$ function must represent the second-order approximation of $f(x)$ function.

The superior-order derivates of $g(x)$ function can be expressed as follows:

$$g'(x) = \frac{a - b}{(1 + bx)^2} \tag{5.18}$$

$$g''(x) = \frac{2b(b - a)}{(1 + bx)^3} \tag{5.19}$$

and

$$g'''(x) = \frac{6b^2(a - b)}{(1 + bx)^4} \tag{5.20}$$

The conditions that must be fulfilled for obtaining a second-order approximation of $f(x)$ using $g(x)$ are

$$g(0) = 1 \tag{5.21}$$

$$g'(x)|_{x=0} = 1 \tag{5.22}$$

$$g''(x)|_{x=0} = 1 \tag{5.23}$$

resulting:

$$a - b = 1 \tag{5.24}$$

and

$$2b(b - a) = 1 \tag{5.25}$$

equivalent with:

$$a = \frac{1}{2} \tag{5.26}$$

and

$$b = -\frac{1}{2} \tag{5.27}$$

The value of the third-order derivate of $g(x)$ for $x = 0$ will be

$$g'''(x)|_{x=0} = \frac{3}{2} \tag{5.28}$$

So, the expressions of $g(x)$ function and of their Taylor series expansion will be

$$g(x) = \frac{1 + \frac{x}{2}}{1 - \frac{x}{2}} \tag{5.29}$$

and

$$g(x) \cong 1 + x + \frac{x^2}{2} + \frac{x^3}{4} + \cdots \tag{5.30}$$

Comparing the limited Taylor series expansions of $f(x)$ and $g(x)$ functions, the errors of the $f(x) \cong g(x)$ approximation are mainly given by the third-order term from these expansions:

$$\varepsilon_{f(x)}^{g(x)}(x) \cong \frac{\frac{x^3}{4} - \frac{x^3}{6}}{\exp(x)} = \frac{x^3}{12 \exp(x)} \tag{5.31}$$

5.1.5 Fifth Mathematical Principle (PR 5.5)

A third-order approximation of the $f(x) = \exp(x)$ exponential function can be achieved using the following general function:

$$g(x) = \frac{1 + ax}{1 + bx} + cx \tag{5.32}$$

a, b and c being constants having values that can be determined from the condition that $g(x)$ function must represent the third-order approximation of $f(x)$ function.

The superior-order derivates of $g(x)$ function can be expressed as follows:

$$g'(x) = \frac{a-b}{(1+bx)^2} + c \tag{5.33}$$

$$g''(x) = \frac{2b(b-a)}{(1+bx)^3} \tag{5.34}$$

$$g'''(x) = \frac{6b^2(a-b)}{(1+bx)^4} \tag{5.35}$$

and

$$g''''(x) = \frac{24b^3(b-a)}{(1+bx)^5} \tag{5.36}$$

The conditions that must be fulfilled for obtaining a third-order approximation of $f(x)$ using $g(x)$ are

$$g(0) = 1 \tag{5.37}$$

$$g'(x)|_{x=0} = 1 \tag{5.38}$$

$$g''(x)|_{x=0} = 1 \tag{5.39}$$

$$g'''(x)|_{x=0} = 1 \tag{5.40}$$

resulting:

$$a - b + c = 1 \tag{5.41}$$

$$2b(b-a) = 1 \tag{5.42}$$

and

$$6b^2(a-b) = 1 \tag{5.43}$$

It results:

$$a = \frac{7}{6} \tag{5.44}$$

$$b = -\frac{1}{3} \tag{5.45}$$

and

$$c = -\frac{1}{2} \tag{5.46}$$

The value of the fourth-order derivate of $g(x)$ for $x = 0$ will be

$$g''''(x)|_{x=0} = \frac{4}{3} \tag{5.47}$$

So, the expressions of $g(x)$ function and of their Taylor series expansion will be

$$g(x) = \frac{1 + \frac{7x}{6}}{1 - \frac{x}{3}} - \frac{x}{2} \tag{5.48}$$

and

$$g(x) \cong 1 + x + \frac{x^2}{2} + \frac{x^3}{6} + \frac{x^4}{18} + \cdots \tag{5.49}$$

Comparing the limited Taylor series expansions of $f(x)$ and $g(x)$ functions, the errors of the $f(x) \cong g(x)$ approximation are mainly given by the fourth-order term from these expansions:

$$\varepsilon_{f(x)}^{g(x)}(x) \cong \frac{\frac{x^4}{18} - \frac{x^4}{24}}{\exp(x)} = \frac{x^4}{72 \exp(x)} \tag{5.50}$$

5.1.6 Sixth Mathematical Principle (PR 5.6)

A third-order approximation of the $f(x) = \exp(x)$ exponential function can be obtained using the following general function:

$$g(x) = \frac{9}{2} \frac{x}{3 - x} - \frac{x}{2} + 1 \tag{5.51}$$

The superior-order derivates of $g(x)$ function can be expressed as follows:

$$g'(x) = \frac{27}{2} \frac{1}{(3 - x)^2} - \frac{1}{2} \tag{5.52}$$

$$g''(x) = \frac{27}{(3 - x)^3} \tag{5.53}$$

$$g'''(x) = \frac{81}{(3 - x)^4} \tag{5.54}$$

$$g''''(x) = \frac{324}{(3-x)^5} \tag{5.55}$$

So, the expression of the Taylor series expansion of $f(x)$ function will be

$$g(x) = 1 + x + \frac{x^2}{2} + \frac{x^3}{6} + \frac{x^4}{18} + \cdots \tag{5.56}$$

Comparing the limited Taylor series expansions of $f(x)$ and $g(x)$ functions, the errors of the $f(x) \cong g(x)$ approximation are mainly given by the fourth-order term from these expansions:

$$\varepsilon_{f(x)}^{g(x)}(x) \cong \frac{\frac{x^4}{18} - \frac{x^4}{24}}{\exp(x)} = \frac{x^4}{72 \exp(x)} \tag{5.57}$$

5.1.7   Seventh Mathematical Principle (PR 5.7)

A fourth-order approximation of the $f(x) = \exp(x)$ exponential function can be achieved using the following function:

$$g(x) = \frac{1 + ax + bx^2}{1 + cx + dx^2} \tag{5.58}$$

the values of a, b, c and d constants being determined from the condition that $g(x)$ function must represent the fourth-order approximation of the $f(x) = \exp(x)$ exponential function. The conditions that must be fulfilled for obtaining a fourth-order approximation of $f(x)$ using $g(x)$ are

$$g(0) = 1 \tag{5.59}$$

$$g'(x)|_{x=0} = 1 \tag{5.60}$$

$$g''(x)|_{x=0} = 1 \tag{5.61}$$

$$g'''(x)|_{x=0} = 1 \tag{5.62}$$

$$g''''(x)|_{x=0} = 1 \tag{5.63}$$

In order to obtain the explicit Taylor series expansion for $g(x)$, its superior-order derivates must be computed. The first-order derivate is

$$g'(x) = \frac{a - c + 2x(b - d) + x^2(bc - ad)}{(1 + cx + dx^2)^2} \tag{5.64}$$

The (5.60) condition imposes the following relation between a and c constants:

$$a - c = 1 \tag{5.65}$$

Replacing (5.65) in (5.64), it results an equivalent expression for $g'(x)$:

$$g'(x) = \frac{1 + 2x(b-d) + x^2(bc - cd - d)}{(1 + cx + dx^2)^2} \tag{5.66}$$

The second-order derivate will have the following expression:

$$g''(x) = \frac{2(b-d) - 2c - 6dx + x^2(6d^2 - 6bd) + x^3(2d^2 - 2bcd + 2cd^2)}{(1 + cx + dx^2)^3} \tag{5.67}$$

The (5.61) condition imposes the following relation between b, c and d constants:

$$2(b - d) - 2c = 1 \tag{5.68}$$

Replacing (5.68) in (5.67), it results an equivalent expression for $g''(x)$:

$$g''(x) = \frac{1 - 6dx - 3dx^2(2c + 1) + x^3(2d^2 - 2c^2d - cd)}{(1 + cx + dx^2)^3} \tag{5.69}$$

The third-order derivate of $g(x)$ can be expressed as follows:

$$g'''(x) = \frac{-6d - 3c - 12xd + 36d^2x^2 + x^3(24cd^2 + 12d^2) + x^4(-6d^3 + 6c^2d^2 + 3cd^2)}{(1 + cx + dx^2)^4}$$
$$\tag{5.70}$$

The (5.62) condition imposes the following relation between c and d constants:

$$-6d - 3c = 1 \tag{5.71}$$

resulting:

$$g'''(x) = \frac{1 - 12xd + \alpha x^2 + \beta x^3 + \gamma x^4}{(1 + cx + dx^2)^4} \tag{5.72}$$

The following notations have been used:

$$\alpha = 36d^2 \tag{5.73}$$

$$\beta = 24cd^2 + 12d^2 \tag{5.74}$$

and

$$\gamma = -6d^3 + 6c^2d^2 + 3cd^2 \tag{5.75}$$

The fourth-order derivate of $g(x)$ will have the following expression:

$$g''''(x) = \frac{-12d + 2\alpha x + 3\beta x^2 + 4\gamma x^3}{(1 + cx + dx^2)^4}$$

$$-\frac{4(c + 2dx)(1 - 12xd + \alpha x^2 + \beta x^3 + \gamma x^4)}{(1 + cx + dx^2)^5} \tag{5.76}$$

From (5.63) and (5.76) it results:

$$-12d - 4c = 1 \tag{5.77}$$

Using (5.71) and (5.77), the values for c and d constants will be

$$c = -\frac{1}{2} \tag{5.78}$$

and

$$d = \frac{1}{12} \tag{5.79}$$

From (5.65) it results:

$$a = \frac{1}{2} \tag{5.80}$$

and, using (5.68), the value of b is

$$b = \frac{1}{12} \tag{5.81}$$

The particular expression of $g(x)$ that is able to fourth-order approximate the $f(x)$ exponential function will be

$$g(x) = \frac{1 + \frac{x}{2} + \frac{x^2}{12}}{1 - \frac{x}{2} + \frac{x^2}{12}} \tag{5.82}$$

5.1.8 Eighth Mathematical Principle (PR 5.8)

Another fourth-order approximation of the $f(x) = \exp(2x)$ exponential function is represented by the following function:

$$g(x) = \frac{1 + x + \frac{x^2}{3}}{1 - x + \frac{x^2}{3}} \tag{5.83}$$

The superior-order derivates of $g(x)$ function can be expressed as follows:

$$g'(x) = \frac{2 - \frac{2x^2}{3}}{\left(1 - x + \frac{x^2}{3}\right)^2} \tag{5.84}$$

$$g''(x) = \frac{4 - 4x + \frac{4x^3}{9}}{\left(1 - x + \frac{x^2}{3}\right)^3} \tag{5.85}$$

$$g'''(x) = \frac{8 - 16x + 8x^2 - \frac{4x^4}{9}}{\left(1 - x + \frac{x^2}{3}\right)^4} \tag{5.86}$$

$$g''''(x) = \frac{16 - \frac{160x}{3} + \frac{160x^2}{3} - \frac{64x^3}{3} + \frac{16x^5}{27}}{\left(1 - x + \frac{x^2}{3}\right)^5} \tag{5.87}$$

The value of the fifth-order derivate of $g(x)$ for $x = 0$ is

$$g'''''(x)\big|_{x=0} = \frac{80}{3} \tag{5.88}$$

resulting the following fourth-order Taylor series expansion for $g(x)$:

$$g(x) = 1 + 2x + \frac{(2x)^2}{2!} + \frac{(2x)^3}{3!} + \frac{(2x)^4}{4!} + \frac{2x^5}{9} + \cdots \tag{5.89}$$

while the fourth-order Taylor series expansion of $f(x)$ is

$$f(x) = 1 + 2x + \frac{(2x)^2}{2!} + \frac{(2x)^3}{3!} + \frac{(2x)^4}{4!} + \frac{(2x)^5}{5!} + \cdots \tag{5.90}$$

Comparing the limited Taylor series expansions of $f(x)$ and $g(x)$ functions, the errors of the $f(x) \cong g(x)$ approximation are mainly given by the fifth-order term from these expansions:

$$\varepsilon_{f(x)}^{g(x)}(x) \cong \frac{\frac{4x^5}{15} - \frac{2x^5}{9}}{\exp(2x)} = \frac{2x^5}{45 \exp(2x)} \tag{5.91}$$

5.1.9   Ninth Mathematical Principle (PR 5.9)

A fourth-order approximation of the $f(x) = \exp(x)$ exponential function can be obtained using the following general function:

$$g(x) = \frac{1 + ax}{1 + bx} + cx + dx^2 \tag{5.92}$$

a, b, c and d being constants with values that can be determined from the condition that $g(x)$ function must represent the fourth-order approximation of $f(x)$ function. The superior-order derivates of $g(x)$ function can be expressed as follows:

$$g'(x) = \frac{a - b}{(1 + bx)^2} + c + 2dx \tag{5.93}$$

$$g''(x) = \frac{2b(b - a)}{(1 + bx)^3} + 2d \tag{5.94}$$

$$g'''(x) = \frac{6b^2(a - b)}{(1 + bx)^4} \tag{5.95}$$

$$g''''(x) = \frac{24b^3(b - a)}{(1 + bx)^5} \tag{5.96}$$

and

$$g'''''(x) = \frac{120b^4(a - b)}{(1 + bx)^6} \tag{5.97}$$

The conditions that must be fulfilled for obtaining a fourth-order approximation of $f(x)$ using $g(x)$ are

$$g(0) = 1 \tag{5.98}$$

$$g'(x)|_{x=0} = 1 \tag{5.99}$$

$$g''(x)|_{x=0} = 1 \tag{5.100}$$

$$g'''(x)|_{x=0} = 1 \tag{5.101}$$

and

$$g''''(x)|_{x=0} = 1 \tag{5.102}$$

resulting:

$$a - b + c = 1 \tag{5.103}$$

$$2b(b - a) + 2d = 1 \tag{5.104}$$

$$6b^2(a - b) = 1 \tag{5.105}$$

and

$$24b^3(b - a) = 1 \tag{5.106}$$

From (5.105) and (5.106), it results:

$$a = \frac{29}{12} \tag{5.107}$$

$$b = -\frac{1}{4} \tag{5.108}$$

So, using (5.107) and (5.108), the values of c and d must be:

$$c = -\frac{5}{3} \tag{5.109}$$

and

$$d = -\frac{1}{6} \tag{5.110}$$

The value of the fifth-order derivate of $g(x)$ for $x = 0$ is

$$g'''''(x)|_{x=0} = \frac{5}{4} \tag{5.111}$$

Thus, the expressions of $g(x)$ function and of their Taylor series expansion will be

$$g(x) = \frac{1 + \frac{29x}{12}}{1 - \frac{x}{4}} - \frac{5x}{3} - \frac{x^2}{6} \tag{5.112}$$

and

$$g(x) \cong 1 + x + \frac{x^2}{2} + \frac{x^3}{6} + \frac{x^4}{24} + \frac{x^5}{96} + \cdots \tag{5.113}$$

Comparing the limited Taylor series expansions of $f(x)$ and $g(x)$ functions, the errors of the $f(x) \cong g(x)$ approximation are mainly given by the fifth-order term from these expansions:

$$\varepsilon_{f(x)}^{g(x)}(x) \cong \frac{\dfrac{x^5}{96} - \dfrac{x^5}{120}}{\exp(x)} = \frac{x^5}{480 \, \exp(x)} \tag{5.114}$$

5.2   Analysis and Design of Exponential Circuits

5.2.1   *Exponential Circuits Based on the First Mathematical Principle (PR 5.1)*

Using the second-order Taylor series expansion for approximating the exponential function having as variable $x = I_{IN}/I_{OUT}$, where I_{IN} is the input current and I_O is the reference current, it results:

$$I_O \exp\left(\frac{I_{IN}}{I_O}\right) \cong I_O \left[1 + \frac{I_{IN}}{I_O} + \frac{1}{2}\left(\frac{I_{IN}}{I_O}\right)^2 + \cdots \right] \tag{5.115}$$

A possible realization of an exponential circuit, based on this mathematical principle, is presented in Fig. 5.1 [1].

Considering a biasing in saturation for all MOS transistors from the circuit, the drain current of the FGMOS transistor will have the following expression:

$$I_2 = \frac{4K}{2} \left[\frac{1}{2}\left(2V_T + \sqrt{\frac{2I_1}{K}} + \sqrt{\frac{2I_O}{K}} \right) - V_T \right]^2 \tag{5.116}$$

equivalent with:

$$I_2 = I_1 + I_O + 2\sqrt{I_1 I_O} \tag{5.117}$$

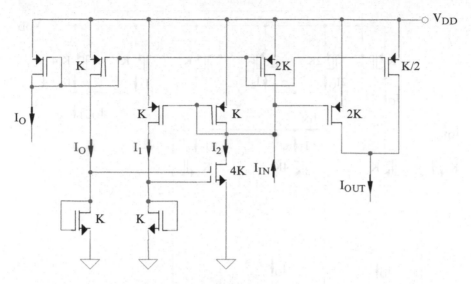

Fig. 5.1 Exponential circuit (1) based on PR 5.1

Because $I_2 = I_1 + 2I_O + I_{IN}$ it results that:

$$I_1 = \frac{(I_O + I_{IN})^2}{4I_O} = \frac{I_O}{4} + \frac{I_{IN}}{2} + \frac{I_{IN}^2}{4I_O} \tag{5.118}$$

The output current of the circuit presented in Fig. 5.1 is

$$I_{OUT} = 2I_1 + \frac{I_O}{2} = I_O + I_{IN} + \frac{I_{IN}^2}{2I_O} \tag{5.119}$$

Comparing (5.115) with (5.119), the approximate value of I_{OUT} current will be

$$I_{OUT} \cong I_O \, \exp\!\left(\frac{I_{IN}}{I_O}\right) \tag{5.120}$$

An exponential circuit based on the approximation of the exponential function using the second-order limited Taylor series is presented in Fig. 5.2 [2].

Because M1–M4 transistors implement an arithmetical mean circuit, V_3 potential will be equal with the arithmetical mean of V_1 and V_2 potentials:

$$V_3 = \frac{V_1 + V_2}{2} = V_T + \frac{1}{2}\left(\sqrt{\frac{2I_O}{K}} + \sqrt{\frac{2I_2}{K}}\right) \tag{5.121}$$

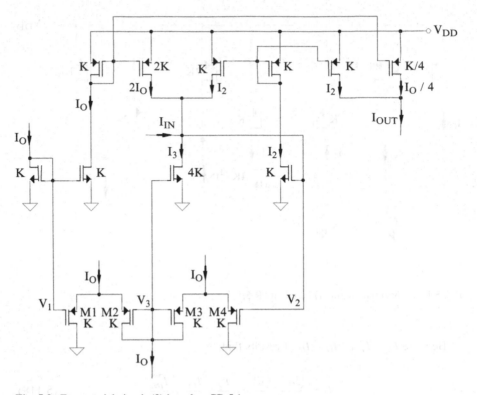

Fig. 5.2 Exponential circuit (2) based on PR 5.1

The expression of I_3 current is

$$I_3 = 2K(V_3 - V_T)^2 = I_O + I_2 + 2\sqrt{I_O I_2} \qquad (5.122)$$

But $I_3 = 2I_O + I_2 + I_{IN}$. It results:

$$I_2 = \frac{(I_O + I_{IN})^2}{4I_O} \qquad (5.123)$$

and

$$I_{OUT} = I_2 + \frac{I_O}{4} = \frac{I_O}{2} + \frac{I_{IN}}{2} + \frac{I_{IN}^2}{4I_O}$$
$$= \frac{I_O}{2}\left[1 + \left(\frac{I_{IN}}{I_O}\right) + \frac{1}{2}\left(\frac{I_{IN}}{I_O}\right)^2\right] \cong \frac{I_O}{2}\exp\left(\frac{I_{IN}}{I_O}\right) \qquad (5.124)$$

The circuit presented in Fig. 5.3 [3] implements the exponential function using the first mathematical principle (PR 5.1).

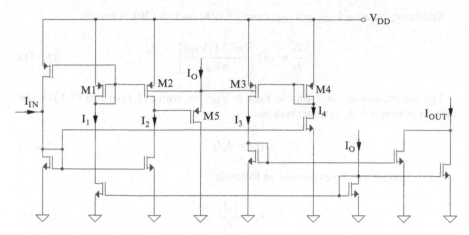

Fig. 5.3 Exponential circuit (3) based on PR 5.1

The M1–M5 transistors from Fig. 5.3 represent a multiplier circuit. All transistors are biased in weak inversion region, the dependence of their drain currents on gate-source voltages being expressed by the following relation:

$$I_D = I_{DO} \exp\left[\frac{V_{SG} + (n-1)V_{SB}}{nV_{th}}\right] \tag{5.125}$$

I_{DO} and n being model parameters, while V_{th} represents the thermal voltage. For M1–M4 transistors, this relation can be particularized as follows:

$$I_1 = I_{DO} \exp\left(\frac{V_{SG1}}{nV_{th}}\right) \tag{5.126}$$

$$I_2 = I_{DO} \exp\left[\frac{V_{SG2} + (n-1)V_{SB2}}{nV_{th}}\right] \tag{5.127}$$

$$I_3 = I_{DO} \exp\left[\frac{V_{SG3} + (n-1)V_{SB3}}{nV_{th}}\right] \tag{5.128}$$

and

$$I_4 = I_{DO} \exp\left(\frac{V_{SG4}}{nV_{th}}\right) \tag{5.129}$$

Because $V_{SG1} = V_{SG2}$, from (5.126) and (5.127), it can be written that:

$$\frac{I_2}{I_1} = \exp\left[\frac{(n-1)V_{SB2}}{nV_{th}}\right] \tag{5.130}$$

Similarly, because $V_{SG3} = V_{SG4}$, from (5.128) and (5.129), it results:

$$\frac{I_3}{I_4} = \exp\left[\frac{(n-1)V_{SB3}}{nV_{th}}\right] \tag{5.131}$$

The circuit connections impose $V_{SB3} = V_{SB2}$ so, using (5.130) and (5.131), the relation between I_1-I_4 currents will be

$$I_1 I_3 = I_2 I_4 \tag{5.132}$$

So, I_3 current can be expressed as follows:

$$I_3 = \frac{I_2 I_4}{I_1} \tag{5.133}$$

Because $I_1 = I_O$ and $I_2 = I_4 = I_{IN} + I_O$, it results:

$$I_3 = \frac{(I_O + I_{IN})^2}{I_O} = I_O + 2I_{IN} + \frac{I_{IN}^2}{I_O} \tag{5.134}$$

The output current of the circuit presented in Fig. 5.3 will have the following expression:

$$I_{OUT} = I_O + I_3 = 2I_O\left[1 + \left(\frac{I_{IN}}{I_O}\right) + \frac{1}{2}\left(\frac{I_{IN}}{I_O}\right)^2\right] \tag{5.135}$$

Using the notation $x = I_{IN}/I_O$ and (5.1) relation that models the first mathematical principle, it results that I_{OUT} current represents the second-order approximation of the exponential function:

$$I_{OUT} \cong I_O \exp\left(\frac{I_{IN}}{I_O}\right) \tag{5.136}$$

For applications requiring a better accuracy that can be obtained using the second-order approximation of the exponential function, a method is to use the third-order approximation of the exponential function, expressed as

$$\exp(x) \cong 1 + x + \frac{x^2}{2} + \frac{x^3}{6} \tag{5.137}$$

for $x = I_{IN}/I_O$. It results:

$$I_O \exp\left(\frac{I_{IN}}{I_O}\right) \cong I_O\left[1 + \frac{I_{IN}}{I_O} + \frac{1}{2}\left(\frac{I_{IN}}{I_O}\right)^2 + \frac{1}{6}\left(\frac{I_{IN}}{I_O}\right)^3 + \cdots\right] \tag{5.138}$$

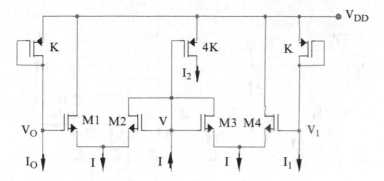

Fig. 5.4 Exponential circuit (4) based on PR 5.1 – circuit core

The core of the exponential circuit with third-order approximation is represented by a current squaring circuit (having a possible implementation presented in Fig. 5.4).

Because M1–M4 transistors implement an arithmetical mean circuit, the V potential will represent the algebraic mean of V_O and V potentials (fixed by I_O and I_1 currents, respectively). So, I_2 current expression can be written as

$$I_2 = \frac{4K}{2}(V_{DD} - V - V_T)^2 \qquad (5.139)$$

or

$$I_2 = 2K\left(V_{DD} - \frac{V_O + V_1}{2} - V_T\right)^2 \qquad (5.140)$$

resulting:

$$I_2 = I_O + I_1 + 2\sqrt{I_O I_1} \qquad (5.141)$$

Using this square-root function, the required squaring function can be obtained subtracting I_{IN} and $2I_O$ currents from I_2 current expression. The full implementation in CMOS technology of the squaring circuit is presented in Fig. 5.5 [4].

Using NMOS current mirrors, I_2 current is forced to be equal with:

$$I_2 = I_1 + I_{IN} + 2I_O \qquad (5.142)$$

From the two previous relations it results:

$$I_1 = \frac{(I_O + I_{IN})^2}{4I_O} = \frac{I_O}{4} + \frac{I_{IN}}{2} + \frac{I_{IN}^2}{4I_O} \qquad (5.143)$$

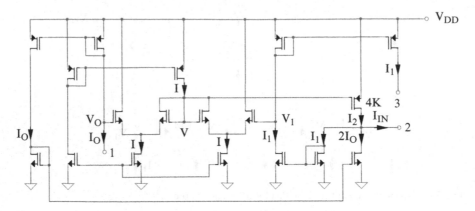

Fig. 5.5 Exponential circuit (4) based on PR 5.1 – complete implementation

In order to obtain the third-order approximation (5.138) of the exponential function, equivalent with the computation of the I_{IN}^3/I_O^2 term, two identical circuits from Fig. 5.5 must be used. For the second circuit, I_O current has to be replaced by I_{IN} current and I_{IN} current – by I_{IN}^2/I_O, respectively. Because I_{IN} current is smaller than I_O current, it is important to use a reasonable value of the I_{IN}/I_O ratio, in such a way that I_{IN}^2/I_O current to determine a biasing in saturation of all MOS transistors from Fig. 5.5. For extremely low values of I_{IN} current, some transistors could operate in weak inversion region, this fact fundamentally changing the function implemented by the circuit. In conclusion, the minimal value of I_{IN} current is imposed by the condition of operating at the limit of saturation region of all MOS transistors from Fig. 5.5. The I_{IN}^2/I_O term can be easily obtained by subtracting $I_O/4$ and $I_{IN}/2$ from I_1 current expression:

$$\frac{I_{IN}^2}{I_O} = 4I_1 - I_O - 2I_{IN} \tag{5.144}$$

The block diagram of the exponential circuit with third-order approximation is presented in Fig. 5.6 [4].

The $I_1{}'$ current can be expressed as follows:

$$I_1{}' = \frac{I_{IN}}{4} + \frac{I_{IN}^2}{2I_O} + \frac{I_{IN}^3}{4I_O^2} \tag{5.145}$$

The output current of the circuit presented in Fig. 5.6 will be

$$I_{OUT} = \frac{2}{3}I_1 + \frac{2}{3}I_1{}' + \frac{5}{6}I_O + \frac{I_{IN}}{2} \tag{5.146}$$

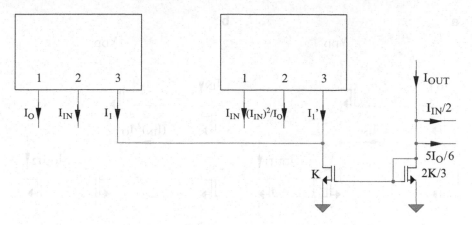

Fig. 5.6 Exponential circuit (5) based on PR 5.1 – block diagram

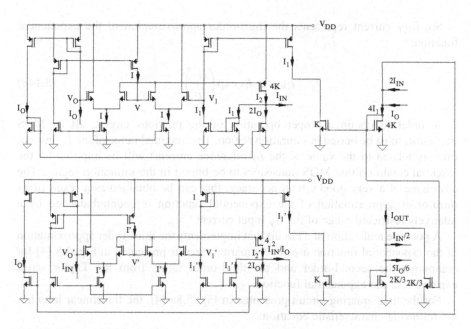

Fig. 5.7 Exponential circuit (5) based on PR 5.1 – complete implementation

The full implementation of the exponential circuit with third-order approximation is presented in Fig. 5.7 [4], resulting:

$$I_{OUT} = I_{IN} + I_O + \frac{I_{IN}^2}{2I_O} + \frac{I_{IN}^3}{6I_O^2} \qquad (5.147)$$

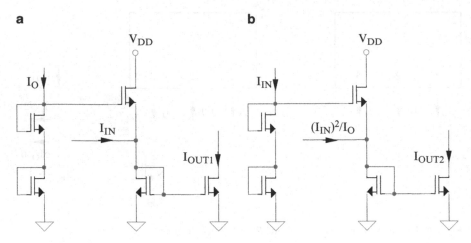

Fig. 5.8 Squaring circuits (1) (**a**) circuit for computing the second-order term (**b**) circuit for computing the third-order term

So, I_{OUT} current represents the third-order approximation of the exponential function:

$$I_{OUT} \cong I_O \, \exp\left(\frac{I_{IN}}{I_O}\right) \qquad (5.148)$$

In order to obtain a proper operation of the previous circuit, all its MOS transistors must be biased in saturation region. The maximal range of the I_{IN} current (mainly related to the value of the I_O reference current) will be imposed by the essential condition that MOS transistors to be biased in the saturation region. The advantage of a very good circuit accuracy, that can be obtained as a result of its third-order approximation of the exponential function is counterbalanced by a relatively restricted range of the I_{IN} input current.

A possible realization of a circuit that implements the third-order approximation of the exponential function uses two squaring circuits presented in Fig. 5.8 [4] for computing the second-order and the third-order terms from the Taylor series expansion of the exponential function.

For the first squaring circuit presented in Fig. 5.8a [4], the translinear loop has the following characteristic equation:

$$2V_{GS}(I_O) = V_{GS}(I_{OUT1}) + V_{GS}(I_{OUT1} - I_{IN}) \qquad (5.149)$$

Considering a biasing in saturation of all MOS transistors from Fig. 5.8a, it results:

$$2\sqrt{I_O} = \sqrt{I_{OUT1}} + \sqrt{I_{OUT1} - I_{IN}} \qquad (5.150)$$

After computations, I_{OUT1} current will have the following expression:

$$I_{OUT1} = I_O + \frac{I_{IN}}{2} + \frac{I_{IN}^2}{16I_O} \tag{5.151}$$

Similarly, the output current of the current squaring circuit presented in Fig. 5.8b can be expressed as follows:

$$I_{OUT2} = I_{IN} + \frac{I_{IN}^2}{2I_O} + \frac{I_{IN}^3}{16I_O^2} \tag{5.152}$$

As the ratio of I_{IN} and I_{OUT} currents could have relatively small values, the designer must pay attention to check if all MOS transistors from Fig. 5.8b are biased in saturation region (the I_{IN}^2/I_O current can decrease under the limit that generates the transition in weak inversion of some MOS transistors). This small limiting of the inferior limit of I_{IN} current is compensated by the extremely small approximation error of the circuit (fourth-order one).

Using the third-order limited Taylor expansion, the approximate expression of the exponential function will be

$$I_O \exp\left(\frac{I_{IN}}{I_O}\right) \cong I_O\left[1 + \frac{I_{IN}}{I_O} + \frac{1}{2}\left(\frac{I_{IN}}{I_O}\right)^2 + \frac{1}{6}\left(\frac{I_{IN}}{I_O}\right)^3\right] \tag{5.153}$$

Expressing the second-order and the third-order terms from (5.153) and (5.154) and replacing in (5.155), it results a linear dependence of the exponential function approximation as a function of the circuit currents:

$$I_O \exp\left(\frac{I_{IN}}{I_O}\right) \cong I_{OUT} = I_O + \frac{40}{3}(I_O - I_{OUT1}) + 5I_{IN} + \frac{8}{3}I_{OUT2} \tag{5.154}$$

The complete implementation of the exponential circuit with third-order approximation, based on (5.154) relation, is presented in Fig. 5.9 [4].

Another possible implementation [5] of an exponential circuit with third-order approximation uses the following squaring structure as circuit core:

Because $V_{GS1} + V_{SG2} = V_{GS3} + V_{SG5}$, it results:

$$2\sqrt{I_O} = \sqrt{I_{D3}} + \sqrt{I_{D5}} \tag{5.155}$$

where $I_{D5} = I_{D3} + I_{IN}$. So

$$I_{D3} = I_O - \frac{I_{IN}}{2} + \frac{I_{IN}^2}{16I_O} \tag{5.156}$$

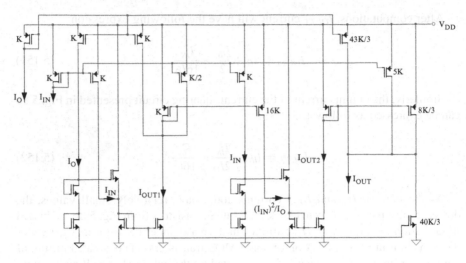

Fig. 5.9 Exponential circuit (6) based on PR 5.1

and

$$I_{D5} = I_O + \frac{I_{IN}}{2} + \frac{I_{IN}^2}{16I_O} \tag{5.157}$$

The quadratic term required for implementing the limited Taylor series expansion of the exponential function will have the following linear expression:

$$\frac{I_{IN}^2}{I_O} = 8(I_{D3} + I_{D5}) - 16I_O \tag{5.158}$$

In order to obtain the third-order term from the limited Taylor series expansion of the exponential function, the same squaring circuit presented in Fig. 5.10 can be used, having different biasing currents (Fig. 5.11) [5].

In this case, the $I_{D3}{}'$ and $I_{D5}{}'$ currents will have the following expressions:

$$I_{D3}{}' = I_{IN} - \frac{I_{IN}^2}{2I_O} + \frac{I_{IN}^3}{16I_O^2} \tag{5.159}$$

and

$$I_{D5}{}' = I_{IN} + \frac{I_{IN}^2}{2I_O} + \frac{I_{IN}^3}{16I_O^2} \tag{5.160}$$

resulting the following expression of the third-order term:

$$\frac{I_{IN}^3}{I_O^2} = 8(I_{D3}{}' + I_{D5}{}') - 16I_{IN} \tag{5.161}$$

Fig. 5.10 Exponential circuit
(7) based on PR 5.1 – circuit
core (1)

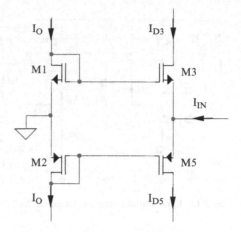

Fig. 5.11 Exponential circuit
(7) based on PR 5.1 – circuit
core (2)

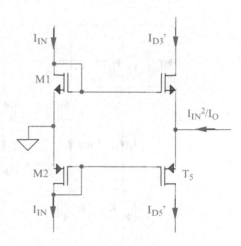

Considering (5.160) and (5.163), the third-order approximation of the exponential function will be

$$I_O \exp\left(\frac{I_{IN}}{I_O}\right) \cong I_{OUT} = 4(I_{D3} + I_{D5}) + \frac{4}{3}(I_{D3}' + I_{D5}') - \frac{5}{3}I_{IN} - 7I_O \quad (5.162)$$

so a linear expression of the circuit currents, which can be implemented using multiple PMOS and NMOS current mirrors. It results the following implementation of the third-order approximated exponential circuit (Fig. 5.12) [5].

For applications that require a better accuracy that can be obtained using the second or third-order approximation of the exponential function presented above, it is possible to use the n-th order approximation of the exponential function using the Taylor series, where n is given by the maximal value of the accepted approximation error. The result will be a tradeoff between the approximation error and the circuit complexity.

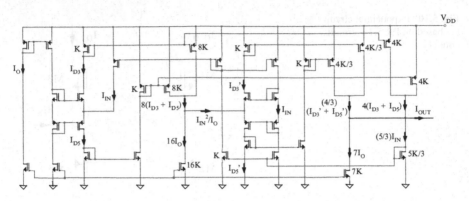

Fig. 5.12 Exponential circuit (7) based on PR 5.1 – complete implementation

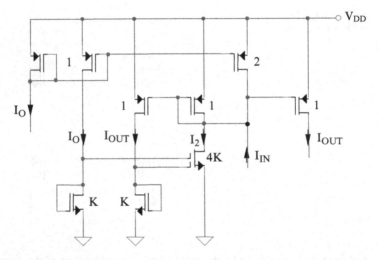

Fig. 5.13 Squaring circuit (2)

In order to implement the exponential function with nth order approximation, n identical squaring circuits must be used, having as possible implementation the structure presented in Fig. 5.13 [1].

The input and output currents for these n circuits and their connection are presented in Fig. 5.14 [6].

The "Σ" block implements a linear function for I_{OUT}, using b_O, b_{IN}, b_1, b_2, ..., b_{n-1} constants:

$$I_{OUT} = b_O I_O + b_{IN} I_{IN} + \sum_{k=1}^{n-1} b_k I_{OUT}^{(k)} \qquad (5.163)$$

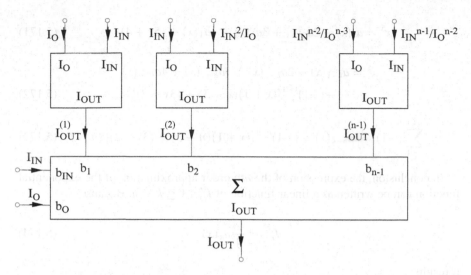

Fig. 5.14 Exponential circuit (8) based on PR 5.1 – block diagram

where b_k are constants coefficients which must be determined. Using the notation $x = I_{IN}/I_O$, the n output currents of these circuits can be expressed as follows:

$$I_{OUT}^{(1)} = \frac{I_O}{4}\left(1 + 2x + x^2\right) = \frac{I_O}{4} a_1(x) \qquad (5.164)$$

$$I_{OUT}^{(2)} = \frac{I_O}{4}\left(x + 2x^2 + x^3\right) = \frac{I_O}{4} a_2(x) \qquad (5.165)$$

$$I_{OUT}^{(3)} = \frac{I_O}{4}\left(x^2 + 2x^3 + x^4\right) = \frac{I_O}{4} a_3(x) \qquad (5.166)$$

$$I_{OUT}^{(n-1)} = \frac{I_O}{4}\left(x^{n-2} + 2x^{n-1} + x^n\right) = \frac{I_O}{4} a_{n-1}(x) \qquad (5.167)$$

In order to obtain the nth order approximation of the exponential function using the previous expressions of the output currents, it is necessary to obtain the expression of x^2, x^3, ..., x^n as linear functions of $a_1(x)$, $a_2(x)$, ..., $a_{n-1}(x)$, equivalent with the necessity of obtaining a linear dependence of I_{OUT} current as a function of $I_{OUT}^{(1)}, I_{OUT}^{(2)}, \ldots, I_{OUT}^{(n-1)}$ currents:

After some algebraic computations, it results the following expressions of x^2, x^3, ..., x^n:

$$x^2 = a_1(x) - (2x + 1) \qquad (5.168)$$

$$x^3 = a_2(x) - 2a_1(x) + (3x + 2) \qquad (5.169)$$

$$x^4 = a_3(x) - 2a_2(x) + 3a_1(x) - (4x + 3) \qquad (5.170)$$

$$x^5 = a_4(x) - 2a_3(x) + 3a_2(x) - 4a_1(x) + (5x + 4) \tag{5.171}$$

$$x^k = a_{k-1}(x) - 2a_{k-2}(x) + 3a_{k-3}(x) + 4a_{k-4}(x)$$
$$+ \cdots + (-1)^{k-1}[(x+1)(k-3) + (3x+2)] \tag{5.172}$$

$$x^n = \sum_{k=1}^{n-1} (-1)^{k-1} k a_{n-k}(x) + (-1)^{n-1}[(x+1)(n-3) + (3x+2)] \, (n > 2) \tag{5.173}$$

In conclusion, the expression of the nth order approximation of the exponential function can be written as a linear function of $I_O^{(k)}$, $1 \le k \le n$. Because

$$I_O^{(k)} = \frac{I_O}{4} a_k(x) \tag{5.174}$$

it results:

$$\exp(x) = 1 + x + \frac{\frac{4}{I_O} I_{OUT}^{(1)} - (2x+1)}{2!} + \frac{\frac{4}{I_O}\left[I_{OUT}^{(2)} - 2I_{OUT}^{(1)}\right] + (3x+2)}{3!}$$
$$+ \frac{\frac{4}{I_O}\left[I_{OUT}^{(3)} - 2I_{OUT}^{(2)} + 3I_{OUT}^{(1)}\right] - (4x+3)}{4!} + \cdots \tag{5.175}$$

$$\exp(x) = 1 + x\left(1 - \frac{2}{2!} + \frac{3}{3!} - \frac{4}{4!} + \frac{5}{5!} - \cdots\right) + \frac{4I_{OUT}^{(1)}}{I_O}\left(\frac{1}{2!} - \frac{2}{3!} + \frac{3}{4!} - \frac{4}{5!} + \cdots\right)$$
$$+ \frac{4I_{OUT}^{(2)}}{I_O}\left(\frac{1}{3!} - \frac{2}{4!} + \frac{3}{5!} - \cdots\right) + \frac{4I_{OUT}^{(3)}}{I_O}\left(\frac{1}{4!} - \frac{2}{5!} + \cdots\right) + \cdots \tag{5.176}$$

Replacing $x = I_{IN}/I_O$ and identifying the constants b_k as follows:

$$b_O = 0 \tag{5.177}$$

$$b_{IN} = 1 - \frac{2}{2!} + \frac{3}{3!} - \frac{8}{4!} + \frac{13}{5!} - \cdots \tag{5.178}$$

$$b_1 = 4\left(\frac{1}{2!} - \frac{2}{3!} + \frac{3}{4!} - \frac{4}{5!} + \cdots\right) \tag{5.179}$$

$$b_2 = 4\left(\frac{1}{3!} - \frac{2}{4!} + \frac{3}{5!} - \cdots\right) \tag{5.180}$$

$$b_3 = 4\left(\frac{1}{4!} - \frac{2}{5!} + \cdots\right) \tag{5.181}$$

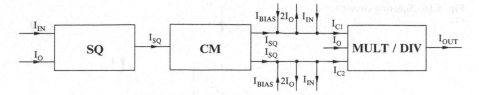

Fig. 5.15 Exponential circuit based on PR 5.2 – block diagram

it results:

$$I_O \exp\left(\frac{I_{IN}}{I_O}\right) \cong I_O + b_{IN}I_{IN} + b_1 I_{OUT}^{(1)} + b_2 I_{OUT}^{(2)} + b_3 I_{OUT}^{(3)} + \cdots \qquad (5.182)$$

Thus, the nth order approximation of the exponential function using a limited Taylor series expansion can be obtained if in Fig. 5.14 the b_O, b_{IN}, $b_1 - b_{n-1}$ constant coefficients are set to have values corresponding to (5.177)–(5.181) relations.

5.2.2   Exponential Circuits Based on the Second Mathematical Principle (PR 5.2)

The block diagram of the exponential circuit based on the second mathematical principle is presented in Fig. 5.15.

The composing blocks are: SQ – a current squaring circuit, having the implementation presented in Fig. 5.16 [5, 10], CM – a current mirror with two outputs and a transfer factor equal with 1 and MULT/DIV – a multiplier/divider circuit presented in Fig. 5.17 [9]. The operation of this circuit is detailed analyzed in Chap. 2 (Fig. 2.71).

The characteristic equation of the translinear loop for the squaring circuit presented in Fig. 5.16 is

$$2V_{GS}(I_O) = V_{GS}(I) + V_{GS}(I + I_{IN}) \qquad (5.183)$$

Considering that all MOS transistors from Fig. 5.16 are biased in saturation, it results:

$$2\sqrt{I_O} = \sqrt{I} + \sqrt{I + I_{IN}} \qquad (5.184)$$

So

$$I = I_O - \frac{I_{IN}}{2} + \frac{I_{IN}^2}{16I_O} \qquad (5.185)$$

Fig. 5.16 Squaring circuit
(3)

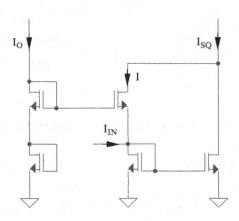

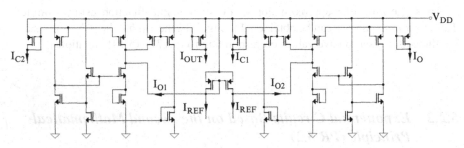

Fig. 5.17 MULT/DIV circuit implementation

The output current of the squaring circuit will contain a constant term and a term proportional with the squaring of the input current:

$$I_{SQ} = 2I + I_{IN} = 2I_O + \frac{I_{IN}^2}{8I_O} \tag{5.186}$$

The output current of the current multiplier circuit presented in Fig. 5.17 is

$$I_{OUT} = I_O \frac{I_{C1}}{I_{C2}} \tag{5.187}$$

The connections from Fig. 5.15 impose that:

$$I_{C1} = I_{SQ} + I_{BIAS} - 2I_O + I_{IN} \tag{5.188}$$

and

$$I_{C2} = I_{SQ} + I_{BIAS} - 2I_O - I_{IN} \tag{5.189}$$

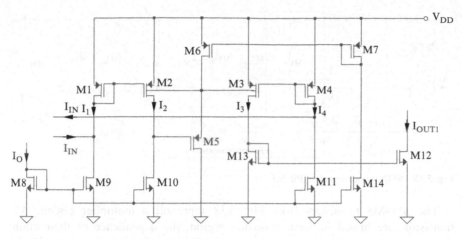

Fig. 5.18 Exponential circuit based on PR 5.3

Considering (5.186)–(5.189), it results:

$$I_{OUT} = I_O \frac{1 + \dfrac{I_{IN}}{I_{BIAS}} + \dfrac{I_{IN}^2}{8I_O I_{BIAS}}}{1 - \dfrac{I_{IN}}{I_{BIAS}} + \dfrac{I_{IN}^2}{8I_O I_{BIAS}}} \tag{5.190}$$

With the notations $a = 1/I_{BIAS}$, $k = I_{BIAS}/4I_O$ and $x = I_{IN}$, the output current expression will be

$$I_{OUT} = I_O \frac{1 + ax + k\frac{(ax)^2}{2}}{1 - ax + k\frac{(ax)^2}{2}} \tag{5.191}$$

So, using the second mathematical principle expressed by relation (5.2), it results that I_{OUT} current represents the second-order approximation of the exponential function:

$$I_{OUT} \cong I_O \exp(2ax) \tag{5.192}$$

5.2.3 Exponential Circuits Based on the Third Mathematical Principle (PR 5.3)

The circuit presented in Fig. 5.18 [10] implements the exponential function using the third mathematical principle (PR 5.3).

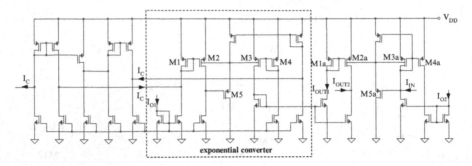

Fig. 5.19 VGA circuit based on PR 5.3

The M1–M5 transistors from Fig. 5.18 represent a multiplier circuit. All transistors are biased in weak inversion region, the dependence of their drain currents on gate-source voltages being expressed by the following relation:

$$I_D = I_{DO} \exp\left[\frac{V_{SG} + (n-1)V_{SB}}{nV_{th}}\right] \tag{5.193}$$

I_{DO} and n being model parameters, while V_{th} represents the thermal voltage. The relation between I_1–I_4 currents is

$$I_1 I_3 = I_2 I_4 \tag{5.194}$$

or, equivalent:

$$(I_O - I_{IN})I_{OUT1} = I_O(I_O + I_{IN}) \tag{5.195}$$

resulting:

$$I_{OUT1} = I_O \frac{1 + \dfrac{I_{IN}}{I_O}}{1 - \dfrac{I_{IN}}{I_O}} \tag{5.196}$$

Using as variable $x = I_{IN}/I_O$ and the approximation of the exponential function given by the fourth mathematical principle for $m = 1$, it results:

$$I_{OUT1} \cong I_O \exp\left(\frac{2I_{IN}}{I_O}\right) \tag{5.197}$$

Based on the same principle, it is possible to design a VGA circuit (Fig. 5.19) [10].

The M1a–M5a transistors biased in weak inversion implement another multiplier circuit, having a similar operation with the circuit realized using M1–M5 transistors, so

$$I_{1a}I_{3a} = I_{2a}I_{4a} \tag{5.198}$$

resulting:

$$I_{OUT1}(I_{O2} - I_{IN}) = (I_{OUT1} - I_{OUT2})I_{O2} \tag{5.199}$$

Thus

$$I_{OUT2} = \frac{I_{OUT1}I_{IN}}{I_{O2}} \tag{5.200}$$

The M1–M5 multiplier implements the following approximate relation:

$$I_{OUT1} \cong I_{O1} \exp\left(\frac{2I_C}{I_{O1}}\right) \tag{5.201}$$

From (5.200) and (5.201), it results:

$$I_{OUT2}(I_{IN}) = I_{IN}\frac{I_{O1}}{I_{O2}} \exp\left(\frac{2I_C}{I_{O1}}\right) = G_I I_{IN} \tag{5.202}$$

The value of the current gain can be exponentially controlled using the control current, I_C:

$$G_I = \frac{I_{O1}}{I_{O2}} \exp\left(\frac{2I_C}{I_{O1}}\right) \tag{5.203}$$

5.2.4 Exponential Circuits Based on the Fourth Mathematical Principle (PR 5.4)

The block diagram of the exponential circuit based on the fourth mathematical principle is presented in Fig. 5.20. The MULT/DIV circuit has the implementation presented in Fig. 5.17.

The expression of I_{OUT} current is

$$I_{OUT} = I_O \frac{I_{C1}}{I_{C2}} \tag{5.204}$$

Fig. 5.20 Exponential circuit
based on PR 5.4 – block
diagram

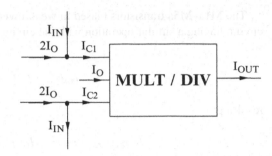

resulting:

$$I_{OUT} = I_O \frac{2I_O + I_{IN}}{2I_O - I_{IN}} = I_O \frac{1 + \frac{1}{2}\left(\frac{I_{IN}}{I_O}\right)}{1 - \frac{1}{2}\left(\frac{I_{IN}}{I_O}\right)} \tag{5.205}$$

Using the notation $x = I_{IN}/I_O$ and (5.29) relation that models the fourth mathematical principle, it results that I_{OUT} current represents the second-order approximation of the exponential function:

$$I_{OUT} \cong I_O \, \exp\left(\frac{I_{IN}}{I_O}\right) \tag{5.206}$$

5.2.5 Exponential Circuits Based on the Fifth Mathematical Principle (PR 5.5)

The block diagram of the exponential circuit based on the fifth mathematical principle is presented in Fig. 5.21. The MULT/DIV circuit has the implementation presented in Fig. 5.17.

The expression of I_{OUT}' current is

$$I_{OUT}' = I_O \frac{I_{C1}}{I_{C2}} \tag{5.207}$$

resulting:

$$I_{OUT}' = I_O \frac{6I_O + 7I_{IN}}{3I_O - I_{IN}} = 2I_O \frac{1 + \frac{7}{6}\left(\frac{I_{IN}}{I_O}\right)}{1 - \frac{1}{3}\left(\frac{I_{IN}}{I_O}\right)} \tag{5.208}$$

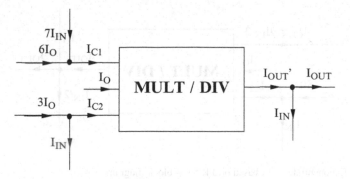

Fig. 5.21 Exponential circuit based on PR 5.5 – block diagram

The output current of the circuit having the block diagram presented in Fig. 5.21 will have the following expression:

$$I_{OUT} = I_{OUT}' - I_{IN} = 2I_O \left[\frac{1 + \frac{7}{6}\left(\frac{I_{IN}}{I_O}\right)}{1 - \frac{1}{3}\left(\frac{I_{IN}}{I_O}\right)} - \frac{1}{2}\left(\frac{I_{IN}}{I_O}\right) \right] \tag{5.209}$$

Using the notation $x = I_{IN}/I_O$ and (5.48) relation that models the fifth mathematical principle, it results that I_{OUT} current represents the third-order approximation of the exponential function:

$$I_{OUT} \cong I_O \exp\left(\frac{I_{IN}}{I_O}\right) \tag{5.210}$$

5.2.6 Exponential Circuits Based on the Sixth Mathematical Principle (PR 5.6)

The block diagram of the exponential circuit based on the sixth mathematical principle is presented in Fig. 5.22. The MULT/DIV circuit has the implementation presented in Fig. 5.17.

The expression of I_{OUT}' current is

$$I_{OUT}' = I_O \frac{I_{C1}}{I_{C2}} \tag{5.211}$$

So

$$I_{OUT}' = I_O \frac{9I_{IN}/2}{3I_O - I_{IN}} = I_O \frac{\frac{9}{2}\left(\frac{I_{IN}}{I_O}\right)}{3 - \left(\frac{I_{IN}}{I_O}\right)} \tag{5.212}$$

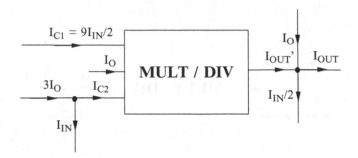

Fig. 5.22 Exponential circuit based on PR 5.6 – block diagram

The output current of the circuit having the block diagram presented in Fig. 5.22 will have the following expression:

$$I_{OUT} = I_{OUT}' + I_O - \frac{I_{IN}}{2} = I_O \left[\frac{\frac{9}{2}\left(\frac{I_{IN}}{I_O}\right)}{3 - \left(\frac{I_{IN}}{I_O}\right)} - \frac{1}{2}\left(\frac{I_{IN}}{I_O}\right) + 1 \right] \qquad (5.213)$$

Using the notation $x = I_{IN}/I_O$ and (5.51) relation that models the sixth mathematical principle, it results that I_{OUT} current represents the third-order approximation of the exponential function:

$$I_{OUT} \cong I_O \, \exp\left(\frac{I_{IN}}{I_O}\right) \qquad (5.214)$$

5.2.7   Exponential Circuits Based on the Seventh Mathematical Principle (PR 5.7)

The block diagram of the exponential circuit based on the seventh mathematical principle is presented in Fig. 5.23. The SQ and MULT/DIV circuits have the implementations presented in Figs. 5.16 and 5.17.

The output current of the circuit presented in Fig. 5.23 has the following relation:

$$I_{OUT} = I_O \frac{I_{C1}}{I_{C2}} \qquad (5.215)$$

Because

$$I_{C1} = \frac{4}{3}I_{SQ} + I_{IN} - \frac{2}{3}I_O \qquad (5.216)$$

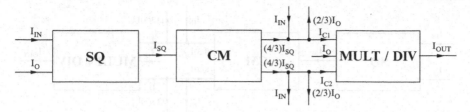

Fig. 5.23 Exponential circuit based on PR 5.7 – block diagram

and

$$I_{C2} = \frac{4}{3}I_{SQ} - I_{IN} - \frac{2}{3}I_O \qquad (5.217)$$

It results, using the (5.186) expression of I_{SQ}:

$$I_{OUT} = I_O \frac{2I_O + I_{IN} + \frac{I_{IN}^2}{6I_O}}{2I_O - I_{IN} + \frac{I_{IN}^2}{6I_O}} \qquad (5.218)$$

or, equivalent:

$$I_{OUT} = I_O \frac{1 + \frac{1}{2}\left(\frac{I_{IN}}{I_O}\right) + \frac{1}{12}\left(\frac{I_{IN}}{I_O}\right)^2}{1 - \frac{1}{2}\left(\frac{I_{IN}}{I_O}\right) + \frac{1}{12}\left(\frac{I_{IN}}{I_O}\right)^2} \qquad (5.219)$$

resulting, using (5.82) relation, that I_{OUT} current represents the fourth-order approximation of the exponential function:

$$I_{OUT} \cong I_O \exp\left(\frac{I_{IN}}{I_O}\right) \qquad (5.220)$$

5.2.8 Exponential Circuits Based on the Eighth Mathematical Principle (PR 5.8)

The block diagram of the exponential circuit based on the eighth mathematical principle is presented in Fig. 5.24. The SQ and MULT/DIV circuits have the implementations presented in Figs. 5.16 and 5.17.

The output current expression is

$$I_{OUT} = I_O \frac{I_{C1}}{I_{C2}} \qquad (5.221)$$

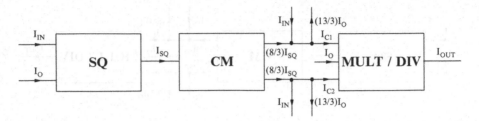

Fig. 5.24 Exponential circuit based on PR 5.8 – block diagram

Because

$$I_{C1} = \frac{8}{3}I_{SQ} + I_{IN} - \frac{13}{3}I_O \qquad (5.222)$$

and

$$I_{C2} = \frac{8}{3}I_{SQ} - I_{IN} - \frac{13}{3}I_O \qquad (5.223)$$

it results:

$$I_{OUT} = I_O \frac{I_O + I_{IN} + \frac{I_{IN}^2}{3I_O}}{I_O - I_{IN} + \frac{I_{IN}^2}{3I_O}} \qquad (5.224)$$

or, equivalent

$$I_{OUT} = I_O \frac{1 + \left(\frac{I_{IN}}{I_O}\right) + \frac{1}{3}\left(\frac{I_{IN}}{I_O}\right)^2}{1 - \left(\frac{I_{IN}}{I_O}\right) + \frac{1}{3}\left(\frac{I_{IN}}{I_O}\right)^2} \qquad (5.225)$$

resulting, using (5.83) relation, that I_{OUT} current represents the fourth-order approximation of the exponential function:

$$I_{OUT} \cong I_O \, \exp\left(\frac{I_{IN}}{I_O}\right) \qquad (5.226)$$

5.2.9 Exponential Circuits Based on the Ninth Mathematical Principle (PR 5.9)

The block diagram of the exponential circuit based on the ninth mathematical principle is presented in Fig. 5.25. The MULT/DIV circuit has the implementations presented in Fig. 5.17.

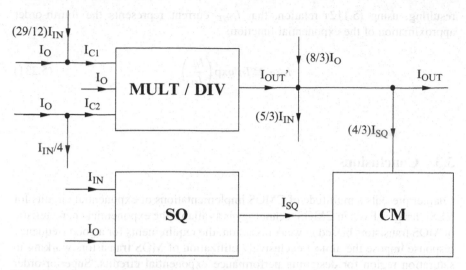

Fig. 5.25 Exponential circuit based on PR 5.9 – block diagram

The expression of I_{OUT}' current is

$$I_{OUT}' = I_O \frac{I_{C1}}{I_{C2}} = I_O \frac{I_O + \frac{29 I_{IN}}{12}}{I_O - \frac{I_{IN}}{4}} = I_O \frac{1 + \frac{29}{12}\left(\frac{I_{IN}}{I_O}\right)}{1 - \frac{1}{4}\left(\frac{I_{IN}}{I_O}\right)} \tag{5.227}$$

The output current of the current squaring circuit can be expressed as follows:

$$I_{SQ} = 2I_O + \frac{I_{IN}^2}{8I_O} \tag{5.228}$$

So, the expression of the output current of the exponential circuit presented in Fig. 5.25 will be

$$I_{OUT} = I_{OUT}' - \frac{4}{3}I_{SQ} + \frac{8}{3}I_O - \frac{5}{3}I_{IN} \tag{5.229}$$

equivalent with:

$$I_{OUT} = I_O \frac{1 + \frac{29}{12}\left(\frac{I_{IN}}{I_O}\right)}{1 - \frac{1}{4}\left(\frac{I_{IN}}{I_O}\right)} - \frac{5}{3}I_{IN} - \frac{I_{IN}^2}{6I_O}$$

$$= I_O \left[\frac{1 + \frac{29}{12}\left(\frac{I_{IN}}{I_O}\right)}{1 - \frac{1}{4}\left(\frac{I_{IN}}{I_O}\right)} - \frac{5}{3}\left(\frac{I_{IN}}{I_O}\right) - \frac{1}{6}\left(\frac{I_{IN}}{I_O}\right)^2 \right] \tag{5.230}$$

resulting, using (5.112) relation, that I_{OUT} current represents the fourth-order approximation of the exponential function:

$$I_{OUT} \cong I_O \, \exp\left(\frac{I_{IN}}{I_O}\right) \tag{5.231}$$

5.3 Conclusions

Chapter presents a multitude of CMOS implementations of exponential circuits for VLSI designs. Even in CMOS technology is available the exponential characteristic of MOS transistors biased in weak inversion, the requirements for a good frequency response impose the almost exclusively utilization of MOS transistors working in saturation region for designing performance exponential circuits. Superior-order approximation functions based on limited Taylor series expansions have been used in order to accurately approximate the exponential function.

References

1. Popa C (2004) FGMOST-based temperature-independent Euclidean distance circuit. In: International conference on optimization of electric and electronic equipment, pp 29–32, Brasov, Romania
2. Manolescu AM, Popa C (2009) Low-voltage low-power improved linearity CMOS active resistor circuits. Springer J Analog Integr Circuits Signal Process 62:373–387
3. Kao CH, Lin WP, Hsieh CS (2005) Low-voltage low-power current mode exponential circuit. In: IEE proceedings on circuits, devices and systems pp 633–635
4. Popa C (2003) CMOS current-mode pseudo-exponential circuits with superior-order approximation. In: IEEE-EURASIP workshop on nonlinear signal and image processing, Trieste, Italy (only CD)
5. Popa C (2004) CMOS current-mode high-precision exponential circuit with improved frequency response. In: International conference on automation, quality & testing, robotics pp 279–284, Cluj, Romania
6. Popa C (2005) A new current-mode pseudo-exponential circuit with an n-th order approximation. In: The international symposium on system theory, automation, robotics, computers, informatics, electronics and instrumentation pp 404–407, Craiova, Romania
7. Popa C (2010) Improved linearity CMOS active resistor based on complementary computational circuits. In: IEEE international conference on electronics, circuits, and systems, Athens, pp 455–458, Greece
8. Landolt O, Vittoz E, Heim P (1992) CMOS selfbiased Euclidean distance computing circuit with high dynamic range. Electron Lett 28:352–354
9. Lopez-Martin AJ, Carlosena A (1999) Geometric-mean based current-mode CMOS multiplier/divider. In: International symposium on circuits and systems, Orlando, pp 342–345
10. Kao CH, Tseng CC, Hsieh CS (2005) Low-voltage exponential function converter. In: IEE proceedings on circuits, devices and systems pp 485–487

11. Jia L, Fengyi H, Xinrong H, Xusheng T (2010) A 1GHz, 68dB CMOS variable gain amplifier with an exponential-function circuit. In: International symposium on signals systems and eElectronics, Nanjing, pp 1–4
12. Ethier S, Sawan M (2010) Exponential current pulse generation for efficient very high-impedance multisite stimulation. In: IEEE transactions on biomedical circuits and systems pp 1–9
13. Hedayati H, Bakkaloglu B (2009) A 3GHz wideband $\sum\Delta$ fractional-N synthesizer with voltage-mode exponential CP-PFD. In: Radio frequency integrated circuits symposium, Boston, pp 325–328, USA
14. Moro-Frias D, Sanz-Pascual MT, De La Cruz-Bias CA (2010) Linear-in-dB variable gain amplifier with PWL exponential gain control. In: IEEE international symposium on circuits and systems, Paris, pp 2824–2827, France

Chapter 6
Euclidean Distance Circuits

6.1 Mathematical Analysis for Synthesis of Euclidean Distance Circuits

The Euclidean distance function is very important in instrumentation circuits, communication, neural networks, display systems or classification algorithms, useful for vector quantization or nearest neighbor classification. In order to obtain a good frequency response, the Euclidean distance circuits are implemented using exclusively MOS transistors working in saturation. Depending on their input variable, the Euclidean distance circuits can be classified in computational structures having current-input or voltage-input vectors.

6.1.1 Euclidean Distance of Voltage Input Vectors

The Euclidean distance between two n-dimensional vectors:

$$V_a = (V_{a1}, V_{a2}, ..., V_{an}) \tag{6.1}$$

and

$$V_b = (V_{b1}, V_{b2}, ..., V_{bn}) \tag{6.2}$$

is defined [1–10] as:

$$V_{OUT} = \|V_a - V_b\| = \sqrt{\sum_{k=1}^{n}(V_{ak} - V_{bk})^2}, \tag{6.3}$$

being a direct measure of similarity between the V_a and V_b vectors.

The structure of an Euclidian distance circuit with n inputs, designed for voltage input vectors is represented in Fig. 6.1 [1, 2], consisting in n current squarer circuits and a square-rooting circuit.

C.R. Popa, *Synthesis of Computational Structures for Analog Signal Processing*, DOI 10.1007/978-1-4614-0403-3_6, © Springer Science+Business Media, LLC 2011

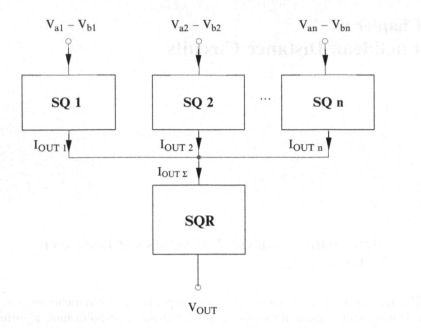

Fig. 6.1 Euclidian distance circuit with n inputs for voltage vectors

6.1.2   Euclidean Distance of Current Input Vectors

An equivalent function based on current-mode operation can be written as:

$$I_{OUT} = \|I_a - I_b\| = \sqrt{\sum_{k=1}^{n} (I_{ak} - I_{bk})^2}, \qquad (6.4)$$

where:

$$I_a = (I_{a1}, I_{a2}, ..., I_{an}) \qquad (6.5)$$

and:

$$V_b = (V_{b1}, V_{b2}, ..., V_{bn}) \qquad (6.6)$$

represent the input current-mode n-dimensional vectors.

The structure of an Euclidian distance circuit with n inputs designed for current vectors is represented in Fig. 6.2 [1, 2], consisting in n current squarer circuits and a square-root circuit.

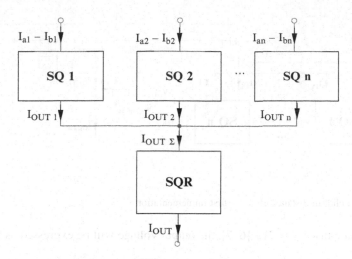

Fig. 6.2 Euclidian distance circuit with n inputs for current vectors

6.2 Analysis and Design of Euclidean Distance Circuits

6.2.1 Euclidean Distance Circuits for Voltage Input Vectors

A possible realization of an Euclidean distance circuit is presented in Fig. 6.3.
The expression of I_{OUT} current is:

$$I_{OUT} = I_{OUT1} - I_{OUT2} = \sum_{k=1}^{n} I_{kX} - \sum_{k=1}^{n} I_{kY} = \sum_{k=1}^{n} (I_{kX} - I_{kY}) \tag{6.7}$$

Considering that the differential output current of each voltage squaring circuit is equal with $K \Delta V^2 / 4$, ΔV being its differential input voltage, the expression of I_{OUT} current for the entire structure presented in Fig. 6.3 will be:

$$I_{OUT} = \frac{K}{4} \sum_{k=1}^{n} (V_{kX} - V_{kY})^2, \tag{6.8}$$

I_{OUT} being also the drain current of the FGMOS transistor, it results:

$$I_{OUT} = \frac{2K}{2} \left(\frac{V_B + V_{OUT}}{2} - V_T \right)^2. \tag{6.9}$$

So:

$$V_{OUT} = 2V_T - V_B + 2\sqrt{\frac{2I_{OUT}}{2K}}. \tag{6.10}$$

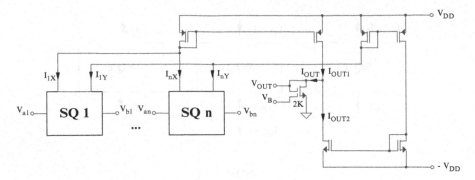

Fig. 6.3 Euclidean distance circuit – first implementation

Imposing that $V_B = 2V_T$ [6, 7], the output voltage will be expressed as follows:

$$V_{OUT} = 2\sqrt{\frac{I_{OUT}}{K}}$$ (6.11)

Replacing (6.8) in (6.11), it can be obtained:

$$V_{OUT} = \sqrt{\sum_{k=1}^{n} (V_{kX} - V_{kY})^2}.$$ (6.12)

6.2.2 Euclidean Distance Circuits for Current Input Vectors

A possible realization of an Euclidean distance circuit is presented in Fig. 6.4 [8] and it contains n current squaring circuits and a current square-rooting circuit in order to implement the required function.

The first squaring circuit is realized using the translinear loop implemented with M1–M4 transistors, having the following characteristic equation:

$$V_{GS1} + V_{SG2} = V_{GS3} + V_{SG4}.$$ (6.13)

The relation between the currents from the circuit are:

$$I_{IN1} + I_{IN1}' = I_{OUT1} - I_{IN1}'.$$ (6.14)

So:

$$I_{IN1}' = \frac{I_{OUT1} - I_{IN1}}{2}.$$ (6.15)

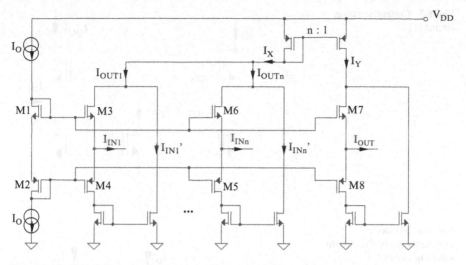

Fig. 6.4 Euclidean distance circuit – second implementation

Considering a biasing in saturation of all MOS transistors from Fig. 6.4, it results:

$$2\sqrt{I_O} = \sqrt{\frac{I_{OUT1} + I_{IN1}}{2}} + \sqrt{\frac{I_{OUT1} - I_{IN1}}{2}}. \tag{6.16}$$

The I_{OUT1} current will be expressed as follows:

$$I_{OUT1} = 2I_O + \frac{I_{IN1}^2}{8I_O}. \tag{6.17}$$

Similarly, analyzing the translinear loop implemented using M1, M2, M5 and M6 transistors, it results:

$$I_{OUTn} = 2I_O + \frac{I_{INn}^2}{8I_O}. \tag{6.18}$$

The expression of I_X current is:

$$I_X = \sum_{k=1}^{n} I_{OUTk} = 2nI_O + \frac{1}{8I_O} \sum_{k=1}^{n} I_{INk}^2, \tag{6.19}$$

while I_Y current has the following expression:

$$I_Y = \frac{I_X}{n} = 2I_O + \frac{1}{8nI_O} \sum_{k=1}^{n} I_{INk}^2. \tag{6.20}$$

Fig. 6.5 Current squaring circuit (1)

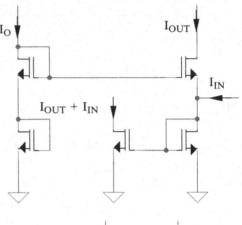

Fig. 6.6 Symbolic representation of the current squaring circuit (1)

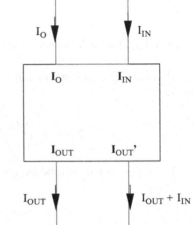

In order to implement the square-rooting function, the translinear loop containing M1, M2, M7 and M8 transistors is used, resulting:

$$I_Y = 2I_O + \frac{I_{OUT}^2}{8I_O}.$$ (6.21)

From (6.20) and (6.21), the output current can be expressed as follows:

$$I_{OUT} = \sqrt{\frac{1}{n}\sum_{k=1}^{n} I_{INk}^2}.$$ (6.22)

Another possible implementation of a current squaring circuit for an Euclidean distance circuit is based on the utilization of two groups of cascaded MOS transistors, the devices from the first group being biased at the same drain current, while each MOS transistor from the second group works at different drain currents (Fig. 6.5) [4].

The symbolic representation of the current squarer is presented in Fig. 6.6 [4].

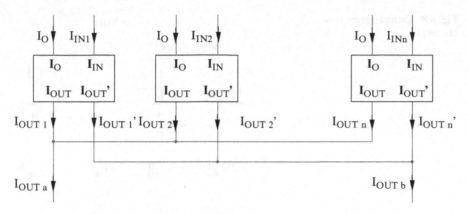

Fig. 6.7 Euclidean distance circuit – third implementation (block diagram)

Considering a biasing in saturation of all MOS transistors from Fig. 6.5, it can be written that:

$$2\sqrt{I_O} = \sqrt{I_{OUT}} + \sqrt{I_{OUT} + I_{IN}}.$$ (6.23)

After some algebraic computations, it results the expression of the output current, I_{OUT}:

$$I_{OUT} = I_O - \frac{I_{IN}}{2} + \frac{I_{IN}^2}{16I_O}$$ (6.24)

and

$$I'_{OUT} = I_{OUT} + I_{IN} = I_O + \frac{I_{IN}}{2} + \frac{I_{IN}^2}{16I_O}.$$ (6.25)

Because the Euclidean distance circuit must have n inputs, it is obviously the necessity of using n identical circuits from Fig. 6.5, having the same I_O reference current, but different input currents, I_{IN1}, I_{IN2}, ..., I_{INn}. The output currents of these squaring circuits will be noted with I_{OUT1}, I_{OUT2}, ..., I_{OUTn} and I_{OUT1}', ..., I_{OUTn}', respectively and they will be summed in order to obtain the output currents, I_{OUTa} and I_{OUTb} (Fig. 6.7 [4]).

The expression of I_{OUTa} current can be written as:

$$I_{OUTa} = \sum_{k=1}^{n} I_{OUTk},$$ (6.26)

equivalent with:

$$I_{OUTa} = nI_O - \frac{1}{2}\sum_{k=1}^{n} I_{INk} + \frac{1}{16I_O}\sum_{k=1}^{n} I_{INk}^2,$$ (6.27)

Fig. 6.8 Current square-root circuit (1)

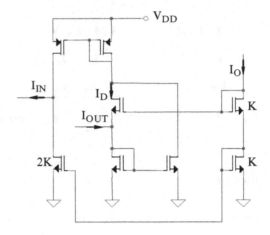

while I_{OUTb} current expression is:

$$I_{OUTb} = nI_O + \frac{1}{2} \sum_{k=1}^{n} I_{INk} + \frac{1}{16I_O} \sum_{k=1}^{n} I_{INk}^2. \tag{6.28}$$

The total output current of the squarer circuit is the sum of I_{OUTa} and I_{OUTb} currents:

$$I_{OUT\Sigma} = I_{OUTa} + I_{OUTb} = 2nI_O + \frac{1}{8I_O} \sum_{k=1}^{n} I_{INk}^2. \tag{6.29}$$

The implementation of the square-rooting circuit is presented in Fig. 6.8 [4]. Similarly with the current squarer, it can be written that:

$$2\sqrt{I_O} = \sqrt{I_D} + \sqrt{I_D + I_{OUT}}. \tag{6.30}$$

It results:

$$I_D = I_O - \frac{I_{OUT}}{2} + \frac{I_{OUT}^2}{16I_O} \tag{6.31}$$

and

$$2I_D + I_{OUT} - 2I_O = I_{IN} \tag{6.32}$$

So, the I_{OUT} output current will be proportional with the square-root of the input current, I_{IN}:

$$I_{OUT} = \sqrt{8I_O I_{IN}}. \tag{6.33}$$

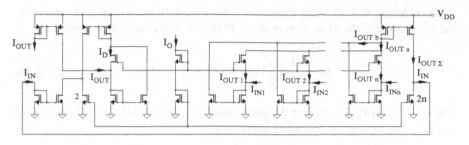

Fig. 6.9 Euclidean distance circuit – third implementation (complete circuit)

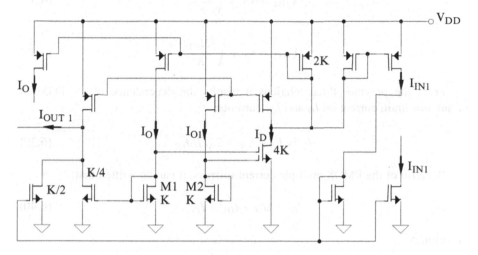

Fig. 6.10 Current squaring circuit (2)

In order to obtain the Euclidean distance of the input currents, n current squarers and a square-root circuit are used in Fig. 6.9 [4].

The I_{IN} input current, is imposed to be equal with:

$$I_{IN} = I_{OUT\Sigma} - 2nI_O = \frac{1}{8I_O} \sum_{k=1}^{n} I_{INk}^2. \tag{6.34}$$

From (6.33) and (6.34), the output current of the entire circuit could be expressed as:

$$I_{OUT} = \sqrt{\sum_{k=1}^{n} I_{INk}^2}. \tag{6.35}$$

The third implementation of the Euclidean distance circuit uses the current squarer presented in Fig. 6.10 [3].

Considering that all MOS transistors from Fig. 6.10 are working in saturation, the expression of the drain current of the FGMOS transistor can be written as:

$$I_D = \frac{4K}{2}\left(\frac{1}{2}V_{GS1} + \frac{1}{2}V_{GS2} - V_T\right)^2, \tag{6.36}$$

where V_{GS1} and V_{GS2} represent the gate-source voltages of M1 and M2 transistors, respectively, having the following expressions:

$$V_{GS1} = V_T + \sqrt{\frac{2I_O}{K}}, \tag{6.37}$$

$$V_{GS2} = V_T + \sqrt{\frac{2I_{O1}}{K}}. \tag{6.38}$$

From the previous three relations it results the dependence of the FGMOS transistor drain current on I_O and I_{O1} currents:

$$I_D = I_O + I_{O1} + 2\sqrt{I_O I_{O1}}. \tag{6.39}$$

Because of the PMOS multiple current mirrors, it can be written that:

$$I_D = 2I_O + I_{O1} + I_{IN1}, \tag{6.40}$$

resulting:

$$I_{O1} = \frac{(I_O + I_{IN1})^2}{4I_O} = \frac{I_O}{4} + \frac{I_{IN1}}{2} + \frac{I_{IN1}^2}{4I_O}. \tag{6.41}$$

Thus, the output current expression is:

$$I_{OUT1} = I_{O1} - \frac{I_O}{4} - \frac{I_{IN1}}{2} = \frac{I_{IN1}^2}{4I_O}. \tag{6.42}$$

A possible implementation of the square-rooting circuit is based on a structure that is similar with the circuit used for obtaining the squaring function. The square-rooting circuit using MOS transistors working in saturation and a FGMOS transistor for reducing the circuit complexity is presented in Fig. 6.11 [9].

The expression of the drain current for the FGMOS transistor from Fig. 6.11 is:

$$I_D = I_O + I_{OUT\Sigma} + \sqrt{4I_O I_{OUT\Sigma}} \tag{6.43}$$

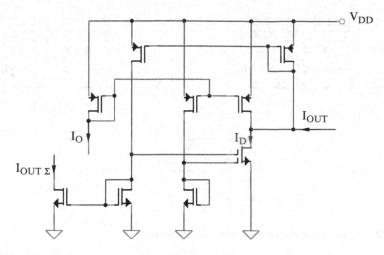

Fig. 6.11 Current square-root circuit (2)

where I_{OUT_Σ} represents the sum of output currents obtained from the n squaring circuits:

$$I_{OUT_\Sigma} = \sum_{k=1}^{n} I_{OUTk} = \frac{1}{4I_O} \sum_{k=1}^{n} I_{INk}^2. \tag{6.44}$$

Because of the PMOS current mirrors from Fig. 6.11, it exists the following linear relation between the currents from the circuit:

$$I_D = I_O + I_{OUT_\Sigma} + I_{OUT}. \tag{6.45}$$

So, the expression of the output current for the entire Euclidean distance circuit will be:

$$I_{OUT} = \sqrt{4I_O I_{OUT_\Sigma}} = \sqrt{\sum_{k=1}^{n} I_{INk}^2}. \tag{6.46}$$

The complete implementation of the Euclidean distance circuit is presented in Fig. 6.12 [3].

A CMOS implementation of a current squaring circuit using MOS transistors biased in weak inversion region is presented in Fig. 6.13.

The translinear loop from Fig. 6.13 has the following characteristic equation:

$$2V_{GS}(I_{IN}) = V_{GS}(I_{OUT}) + V_{GS}(I_O), \tag{6.47}$$

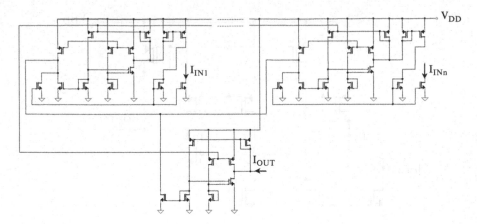

Fig. 6.12 Euclidean distance circuit – fourth implementation

Fig. 6.13 Current squaring
circuit (3)

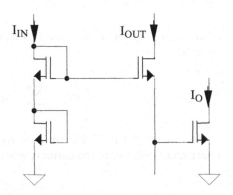

where:

$$V_{GS}(I_{IN}) = V_T + nV_{th} \ln\left(\frac{I_{IN}}{\frac{W}{L}I_{DO}}\right), \tag{6.48}$$

$$V_{GS}(I_{OUT}) = V_T + nV_{th} \ln\left(\frac{I_{OUT}}{\frac{W}{L}I_{DO}}\right) \tag{6.49}$$

and:

$$V_{GS}(I_O) = V_T + nV_{th} \ln\left(\frac{I_O}{\frac{W}{L}I_{DO}}\right). \tag{6.50}$$

Fig. 6.14 Current square-root circuit (3)

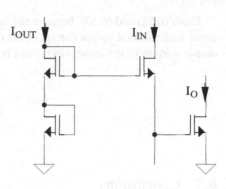

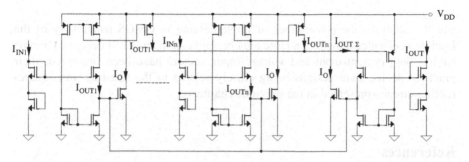

Fig. 6.15 Euclidean distance circuit – fifth implementation

Considering that all MOS active devices are identical, the output current expression of the CMOS current squarer will have the following expression:

$$I_{OUT} = \frac{I_{IN}^2}{I_O} \tag{6.51}$$

A simple realization of the square-rooting circuit (Fig. 6.14) is derived from the current squarer presented in Fig. 6.7.

Similarly with the current squarer, it results:

$$I_{OUT} = \sqrt{I_O I_{IN}}. \tag{6.52}$$

In order to obtain the Euclidean distance of the input currents, n current squarers and a square-rooting circuit must be used, the complete implementation of the Euclidean distance circuit being presented in Fig. 6.15.

The total output current, $I_{OUT\Sigma}$, will be:

$$I_{OUT\Sigma} = \sum_{k=1}^{n} I_{OUT k} = \frac{1}{I_O} \sum_{k=1}^{n} I_{IN k}^2. \tag{6.53}$$

From (6.52) and (6.53), because the input current of the square rooting circuit is equal with the total output current of the current squaring circuits ($I_{IN} = I_{OUT\Sigma}$), the output current of the entire circuit can be expressed as:

$$I_{OUT} = \sqrt{\sum_{k=1}^{n} I_{IN\,k}^2}.$$ (6.54)

6.3 Conclusions

Chapter analyzes the possibilities of implementing in CMOS technology of the Euclidean distance, covering a large area of applications in VLSI designs. Circuits having both current-input and voltage-input vectors have been presented, their practical device-level designs being closely related to the squaring and square-rooting circuits presented in the previous chapters.

References

1. Popa C (2009) A new FGMOST Euclidean distance computational circuits based on algebraic mean of the input potentials. In: Lecture notes in computer science, Springer, pp 459–466
2. Vlassis S, Fikos G, Siskos S (2001) A floating gate CMOS Euclidean distance calculator and its application to hand-written digit recognition. In: International conference on image processing, pp 350–353
3. Popa C (2004) FGMOST-based temperature-independent Euclidean distance circuit. In: International conference on optimization of electric and electronic equipment, pp 29–32
4. Popa C (2005) Current-mode Euclidean distance circuit independent on technological parameters. In: International semiconductor conference, pp 459–462
5. Vlassis S, Yiamalis T, Siskos S (1999) Analogue computational circuits based on floating-gate transistors. In: International conference on electronics, circuits and systems, 5–8 Sept 1999, pp 129–132
6. Popa C (2003) Low-voltage accurate CMOS threshold voltage extractors. In: IEEE-EURASIP workshop on nonlinear signal and image processing, Trieste, Italy (only CD)
7. Popa C (2004) CMOS current-mode Euclidean distance circuit using floating-gate MOS transistors. In: International conference on microelectronics, 16–19 May 2004, pp 585–588
8. Netbut C, Kumngern M, Prommee P, Dejhan K (2006) A versatile vector summation circuit. In: International symposium on communications and information technologies, Bangkok, pp 1093–1096
9. Manolescu AM, Popa C (2009) Low-voltage low-power improved linearity CMOS active resistor circuits. Springer J Analog Integr Circuits Signal Process 62:373–387
10. Hyo-Jin A, Chang-Seok C, Hanho L (2009) High-speed low-complexity folded degree-computationless modified Euclidean algorithm architecture for RS decoders. In: International symposium on integrated circuits, Singapore, 14–16 Dec 2009, pp 582–585

Chapter 7
Active Resistor Circuits

7.1 Mathematical Analysis for Synthesis of Active Resistor Circuits

The diversity of mathematical principles that represent the basis of designing active resistor circuits [1–25] is relatively restricted, existing about six fundamental theoretical methods for implementing this class of circuits. In the process of designing this circuits, V_1 and V_2 represent the input potentials, V_O and I_O are a constant voltage and a constant current, respectively, while R_{ECH} notation is used for the equivalent resistance between the input terminals.

7.1.1 First Mathematical Principle (PR 7.1)

The active resistor circuits designed based on the first mathematical principle uses a linear differential amplifier as constitutive core and two input–output connections, in order to implement a linear current–voltage characteristic between two input pins.

7.1.2 Second Mathematical Principle (PR 7.2)

The linearization techniques based on the second mathematical principle uses a proper voltage biasing of a classical differential amplifier, that modifies the transfer characteristic of this circuit, in order to have a linear behavior for the resulting active resistor structure.

C.R. Popa, *Synthesis of Computational Structures for Analog Signal Processing*, DOI 10.1007/978-1-4614-0403-3_7, © Springer Science+Business Media, LLC 2011

7.1.3   Third Mathematical Principle (PR 7.3)

The method for designing a linear active resistor, based on the third mathematical principle, consists in the passing between two pins of a current obtained at the output of a differential amplifier, considering these pins as differential circuit inputs.

7.1.4   Fourth Mathematical Principle (PR 7.4)

The linearization technique of the active resistors designed using the fourth mathematical principle is based on the utilization of two blocks implementing complementary functions (usually, squaring and square-rooting functions).

7.1.5   Fifth Mathematical Principle (PR 7.5)

The active resistors designed using the fifth mathematical principle are based on the "mirroring" of the Ohm law from two control terminals to the external pins of active resistor circuit, the equivalent resistance being equal with the ratio of a reference voltage, V_O and a reference current, I_O.

7.1.6   Different Mathematical Principles (PR 7.D)

Active resistors using this mathematical principle present different possible implementations, their common point being the exclusively utilization of MOS transistors biased in saturation.

7.2   Analysis and Design of Active Resistor Circuits

7.2.1   Active Resistor Circuits Based on the First Mathematical Principle (PR 7.1)

Active resistor circuits designed based on the first mathematical principle use a linear differential amplifier as constitutive core in order to implement a linear current–voltage characteristic between two input pins. Two input–output connections allow to compute the differential output current, $I_1 - I_2$, and to pass

Fig. 7.1 Active resistor with
positive equivalent resistance
based on PR 7.1 – block
diagram

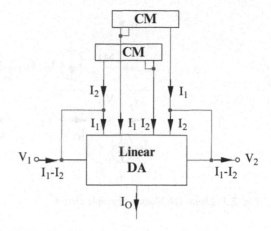

Fig. 7.2 Active resistor with
negative equivalent resistance
based on PR 7.1 – block
diagram

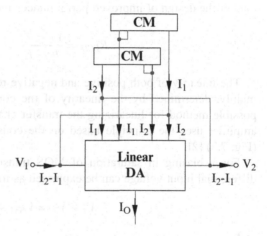

it through the input pins of the entire structure (Fig. 7.1) [1] using two additional
current mirrors.

The equivalent resistance of the entire structure can be defined as the ratio
between the $V_1 - V_2$ differential input voltage and the $I_1 - I_2$ differential current:

$$R_{ECH} = \frac{V_1 - V_2}{I_1 - I_2} = \frac{1}{G_m}, \qquad (7.1)$$

G_m being the equivalent transconductance of the linear differential amplifier.
The possibility of controlling the value of the equivalent resistance is given by the
dependence of the G_m equivalent transconductance on the I_O biasing current.

The replacing of the input–output connections from Fig. 7.1 by two input–output
cross-connections will change the sign of the equivalent resistance for the structure
presented in Fig. 7.2 [1]. The area of applications of controllable negative resistance

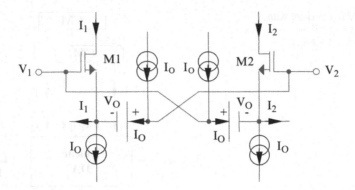

Fig. 7.3 Linear DA block – principle circuit

active resistors covers many domains, including the cancellation of amplifiers' gain load or the design of improved performances integrators.

$$R_{ECH} = \frac{V_1 - V_2}{I_2 - I_1} = -\frac{1}{G_m}. \tag{7.2}$$

The linearity of both positive and negative resistance active resistor circuits is, mainly, determined by the linearity of the constitutive differential amplifier. A possible method for linearizing the transfer characteristic of classical differential amplifier uses the principle based on the constant sum of gate-source voltages (Fig. 7.3) [2].

For a biasing in saturation of MOS transistors from Fig. 7.3, the $V_1 - V_2$ differential input voltage can be expressed as follows:

$$V_1 - V_2 = V_{GS1} - V_O \tag{7.3}$$

and:

$$V_1 - V_2 = V_O - V_{GS2}, \tag{7.4}$$

resulting the expressions of the sum and difference between gate-source voltages:

$$V_{GS1} + V_{GS2} = 2V_O \tag{7.5}$$

and:

$$V_{GS1} - V_{GS2} = 2(V_1 - V_2). \tag{7.6}$$

The $I_1 - I_2$ differential output current is:

$$I_1 - I_2 = \frac{K}{2}(V_{GS1} - V_T)^2 - \frac{K}{2}(V_{GS2} - V_T)^2, \tag{7.7}$$

equivalent with:

$$I_1 - I = \frac{K}{2}(V_{GS1} - V_{GS2})(V_{GS1} + V_{GS2} - 2V_T). \tag{7.8}$$

Replacing (7.5) and (7.6) in (7.8), it results a linear transfer characteristic of the differential amplifier presented in Fig. 7.3:

$$I_1 - I_2 = 2K(V_O - V_T)(V_1 - V_2). \tag{7.9}$$

Usually, the V_O voltage sources are implemented as current-controlled voltage sources. The simplest way to realize these sources, having the advantages of simplicity and of minimization the errors introduced by the bulk effect is to use the gate-source voltage of a MOS transistor in saturation, biased at a constant current, I_O:

$$V_O = V_{GSO} = V_T + \sqrt{\frac{2I_O}{K}}. \tag{7.10}$$

Replacing these particular expressions of V_O voltage sources in the general expression (7.9) of the output current of the differential amplifier, it results:

$$I_1 - I_2 = \sqrt{8KI_O}(V_1 - V_2), \tag{7.11}$$

so, an equivalent transconductance of the differential amplifier:

$$G_m = \sqrt{8KI_O}, \tag{7.12}$$

that can be controlled by the biasing current, I_O.

An active resistor circuit based on the first mathematical principle (PR 7.1), that uses the linearization technique of the composing differential amplifier, based on the constant sum of the gate-source voltages is presented in Fig. 7.4. The V_O voltage sources from Fig. 7.3 are practically implemented in Fig. 7.4, using the gate-source voltages of M3 and M4 transistors, biased at a constant current, I_O. In order to obtain a differential amplifier with two I_1 output currents and two I_2 output currents (as it is shown in Fig. 7.1), a parallel connection of two identical differential amplifiers is used in Fig. 7.4 (M1–M2 and M5–M6, respectively).

The expression of the V_O current-controlled voltage sources will be:

$$V_O = V_{GS3} = V_{GS4} = V_T + \sqrt{\frac{2I_O}{K}}. \tag{7.13}$$

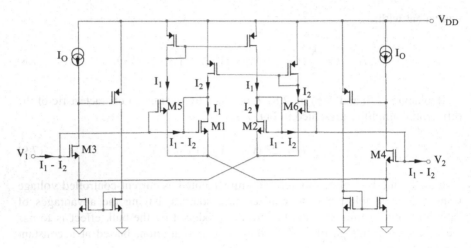

Fig. 7.4 Active resistor with positive equivalent resistance based on PR 7.1 – first implementation

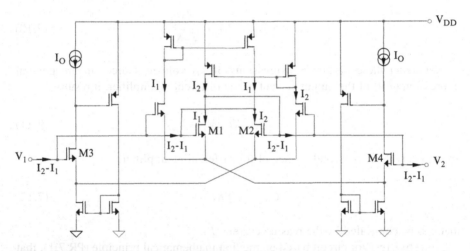

Fig. 7.5 Active resistor with negative equivalent resistance based on PR 7.1 – first implementation

The equivalent resistance of the entire structure can be set by choosing the value of the biasing current, I_O:

$$R_{ECH} = \frac{V_1 - V_2}{I_1 - I_2} = \frac{1}{G_m} = \frac{1}{\sqrt{8KI_O}}.$$ (7.14)

In order to obtain an active resistor circuit with negative equivalent resistance based on the previous active resistor structure, two input–output cross-connections can be used. The resulting active resistor circuit, using the block diagram presented in Fig. 7.2, is shown in Fig. 7.5.

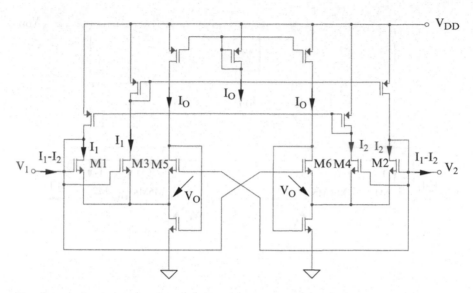

Fig. 7.6 Active resistor with positive equivalent resistance based on PR 7.1 – second implementation

The equivalent resistance of the active resistor structure is:

$$R_{ECH} = \frac{V_1 - V_2}{I_2 - I_1} = -\frac{1}{G_m} = -\frac{1}{\sqrt{8KI_O}}. \tag{7.15}$$

An alternative implementation of the active resistor circuit with positive equivalent resistance using the same principle is presented in Fig. 7.6. The V_O voltage sources from Fig. 7.3 are practically implemented in Fig. 7.6 using the gate-source voltages of M5 and M6 transistors, biased at a constant current, I_O. Similar with the active resistor presented in Fig. 7.4, two parallel-connected differential amplifiers (M1–M2 and M3–M4) have been used in order to generate two output currents I_1 and two output currents I_2. The expression of the equivalent resistance of the structure presented in Fig. 7.6 is also given by (7.14).

Similar input–output cross-connections allow to obtain a negative resistance active resistor circuit (Fig. 7.7).

The active resistor circuit presented in Fig. 7.8 [3, 4] is based on the same operation principle. The M3–M4 and M5–M6 pairs implement two current mirrors, necessary to extract I_1 and I_2 currents from the circuit, in order to compute $I_{12} = I_1 - I_2$ current. The V_C is a DC potential which fixes the value of the equivalent resistance between the input pins and sets I_{D7} and I_{D8} current to be equal with I_O current. The V_O generators are voltage sources, controlled by I_O current.

Because M1 and M3 transistors are identical and biased at the same drain current, it results $V_{GS1} = V_{GS3}$, so:

$$V_{GS1} = \frac{V_1 - V_X}{2}. \tag{7.16}$$

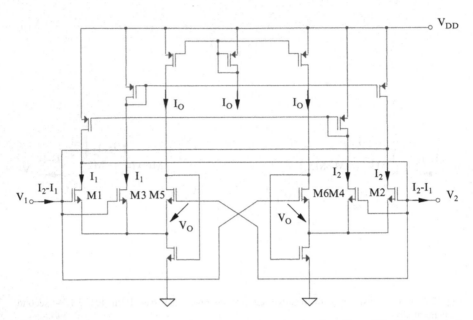

Fig. 7.7 Active resistor with negative equivalent resistance based on PR 7.1 – second implementation

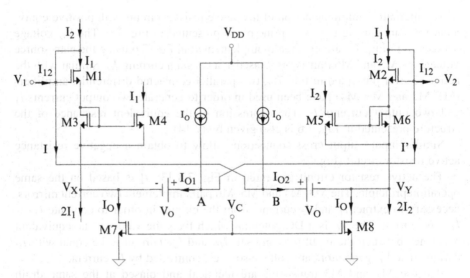

Fig. 7.8 Active resistor with positive equivalent resistance based on PR 7.1 – third principle implementation

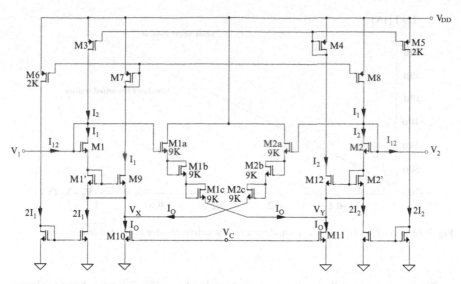

Fig. 7.9 Active resistor with positive equivalent resistance based on PR 7.1 – third complete implementation

where $V_X = V_2 - V_O$. It results:

$$V_{GS1} = \frac{V_1 - V_2 + V_O}{2} \tag{7.17}$$

and, similarly:

$$V_{GS2} = \frac{V_2 - V_1 + V_O}{2}. \tag{7.18}$$

So, the differential current, $I_{12} = I_1 - I_2$, can be expressed as follows:

$$I_{12} = \frac{K}{2}(V_{GS1} - V_T)^2 - \frac{K}{2}(V_{GS2} - V_T)^2 = \frac{K}{2}(V_1 - V_2)(V_O - 2V_T). \tag{7.19}$$

Thus, the active resistor presented in Fig. 7.8 has an equivalent resistance, expressed by:

$$R_{ECH} = \frac{V_1 - V_2}{I_{12}} = \frac{2}{K(V_O - 2V_T)}. \tag{7.20}$$

The complete circuit implementation is presented in Fig. 7.9 [3, 4]. The V_C is an externally applied potential, used to control the value of equivalent resistance (V_O depends on I_O current, which is fixed by V_C potential).

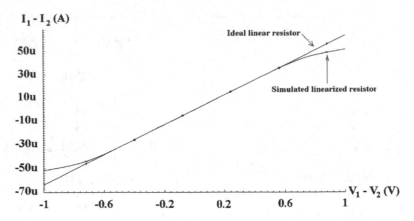

Fig. 7.10 The $(I_1 - I_2)(V_1 - V_2)$ simulation for the active resistor presented in Fig. 7.9

The current-controlled voltage generators, V_O from Fig. 7.8 have been replaced in Fig. 7.9 by two series connections of three MOS transistors (M1a, M1b, M1c and M2a, M2b, M2c, respectively). The V_O voltage can be expressed as follows:

$$V_O = 3V_{GS}(I_O) = 3\left(V_T + \sqrt{\frac{2I_O}{9K}}\right) = 3V_T + \sqrt{\frac{2}{K}}\sqrt{\frac{K}{2}}(V_C - V_T) = V_C + 2V_T.$$

$$(7.21)$$

From (7.20) and (7.21), it results an expression of the equivalent resistance of the circuit presented in Fig. 7.9, that does not depend on the threshold voltage:

$$R_{ECH} = \frac{1}{2KV_C}.$$

$$(7.22)$$

The $(I_1 - I_2)(V_1 - V_2)$ simulation for the active resistor presented in Fig. 7.9 is shown in Fig. 7.10.

In order to estimate the linearity of the circuit, the active resistor is compared with an ideal resistor, resulting the simulated linearity error presented in Fig. 7.11.

The maximum linearity error of the active resistor presented in Fig. 7.9 for a limited input voltage range ($|V_1 - V_2| \leq 500\,\text{mV}$) is smaller that 0.35%.

The circuit presented in Fig. 7.8 can be changed, in order to obtain an active resistor with negative equivalent resistance, resulting the circuit presented in Fig. 7.12 [3, 4].

It exists the possibility of implementing an active resistor circuit based on the same linearization principle, using FGMOS transistors (Fig. 7.13) [3–5]. The V_1 and V_2 are the output pins of the active resistor, I_{12} is the current passing between these terminals, V_C is a DC potential which controls the value of the equivalent

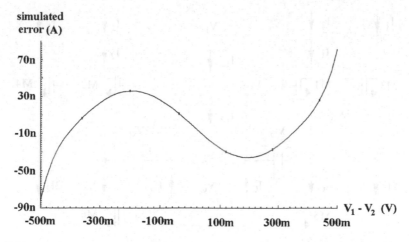

Fig. 7.11 The simulated linearity error for the active resistor circuit presented in Fig. 7.9

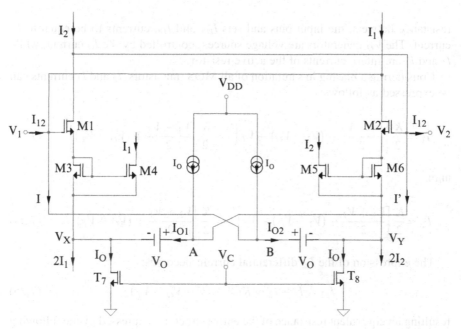

Fig. 7.12 Active resistor with negative equivalent resistance based on PR 7.1 – third principle implementation

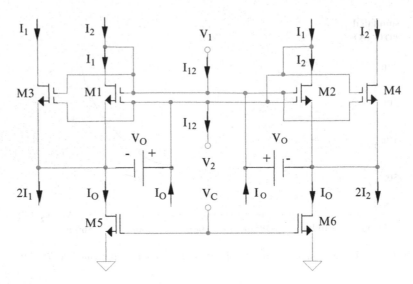

Fig. 7.13 Active resistor with positive equivalent resistance based on PR 7.1 – fourth principle implementation

resistance between the input pins and sets I_{D5} and I_{D6} currents to be equal to I_O current. The V_O generators are voltage sources, controlled by the I_O current, while I_1 and I_2 are intern currents of the active resistor.

Considering a biasing in saturation of all MOS transistors, I_1 and I_2 currents can be expressed as follows:

$$I_1 = \frac{K}{2}\left[\frac{V_1+V_2}{2} - (V_2-V_O) - V_T\right]^2 = \frac{K}{2}\left[\frac{V_1-V_2}{2} + (V_O-V_T)\right]^2 \quad (7.23)$$

and:

$$I_2 = \frac{K}{2}\left[\frac{V_1+V_2}{2} - (V_1-V_O) - V_T\right]^2 = \frac{K}{2}\left[\frac{V_2-V_1}{2} + (V_O-V_T)\right]^2. \quad (7.24)$$

The expression of the I_{12} differential current becomes:

$$I_{12} = I_1 - I_2 = K(V_1-V_2)(V_O-V_T), \quad (7.25)$$

resulting an equivalent resistance of the entire structure, expressed by the following relation:

$$R_{ECH} = \frac{V_1-V_2}{I_{12}} = \frac{1}{K(V_O-V_T)}. \quad (7.26)$$

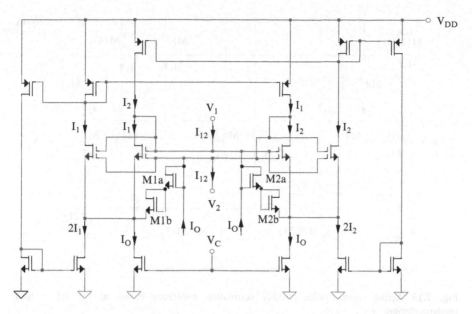

Fig. 7.14 Active resistor with positive equivalent resistance based on PR 7.1 – fourth complete implementation

In order to avoid the degradation of circuit linearity caused by the bulk effect, a proper implementation of the current-controlled voltage sources V_O from Fig. 7.13 has been realized in Fig. 7.14. For this particular choice, the V_O voltage can be expressed as follows:

$$V_O = V_{GS1a} + V_{GS1b} = V_{GS2a} + V_{GS2b} = 2\left(V_T + \sqrt{\frac{2I_O}{4K}}\right), \tag{7.27}$$

because M1a, M1b, M2a and M2b transistors (having the aspect ratio fourth time greater than the other transistors from the circuit) are biased at the same constant current, I_O. The precision of the entire structure presented in Fig. 7.14 [3–5] will be not affected by the bulk effect:

$$R_{ECH} = \frac{1}{K\left(V_T + 2\sqrt{\frac{2I_O}{4K}}\right)} = \frac{1}{KV_C}. \tag{7.28}$$

Another active resistor circuit having the possibility of applying between the input pins both positive and negative voltages, is presented in Fig. 7.15, this additional feature being possible using a double supply voltage.

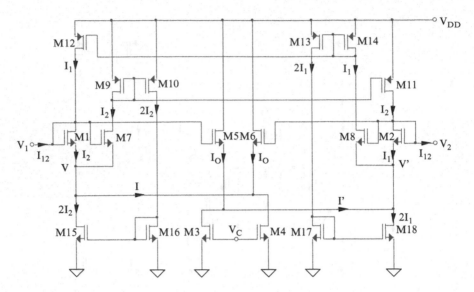

Fig. 7.15 Active resistor with positive equivalent resistance based on PR 7.1 – fifth implementation

Because of the multiple current mirrors, I and I' currents are zero and $I_{12} = I_2 - I_1$. The equivalent resistance of the circuit is:

$$R_{ECH} = \frac{V_1 - V_2}{I_{12}}. \tag{7.29}$$

The expression of I_1 current is:

$$I_1 = \frac{K}{2}(V_2 - V' - V_T)^2, \tag{7.30}$$

where $V' = V_1 - V_{GS5}$. As M3 and M5 transistors are identical and biased at the same drain current, their gate-source voltages will be also equal, so $V_{GS5} = V_{GS3} = V_C$. It results:

$$V' = V_1 - V_C. \tag{7.31}$$

The expression of I_1 current becomes:

$$I_1 = \frac{K}{2}[(V_2 - V_1) + (V_C - V_T)]^2 \tag{7.32}$$

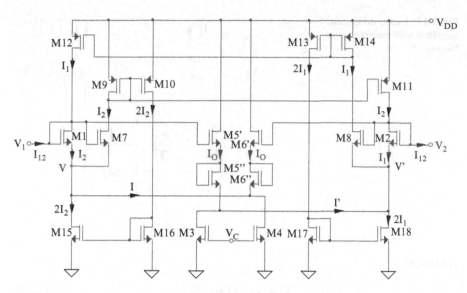

Fig. 7.16 Active resistor with positive equivalent resistance based on PR 7.1 – fifth improved implementation

and, similarly:

$$I_2 = \frac{K}{2}[(V_1 - V_2) + (V_C - V_T)]^2. \tag{7.33}$$

The I_{12} differential current can be expressed as follows:

$$I_{12} = I_2 - I_1 = 2K(V_1 - V_2)(V_C - V_T). \tag{7.34}$$

The equivalent resistance of the entire structure presented in Fig. 7.15 will have the following expression:

$$R_{ECH} = \frac{V_1 - V_2}{I_{12}} = \frac{1}{2K(V_C - V_T)}. \tag{7.35}$$

In order to remove the dependence of the equivalent resistance on the threshold voltage, it can be used an improved circuit, presented in Fig. 7.16.

The M5 and M6 transistors from Fig. 7.15 have been replaced by two series connections, M5'–M5'' and M6'–M6'', respectively, all these four transistors having aspect ratios fourth time greater than the other transistors from the circuit. Now, $V' = V_1 - 2V_{GS5'}$, where:

$$
\begin{aligned}
V_{GS5'} &= V_T + \sqrt{\frac{2I_{D5'}}{4K}} = V_T + \sqrt{\frac{2I_{D3}}{4K}} \\
&= V_T + \sqrt{\frac{1}{2K}\frac{K}{2}(V_C - V_T)^2} = \frac{V_C + V_T}{2},
\end{aligned} \tag{7.36}
$$

Fig. 7.17 Linear differential
amplifier with proper current
biasing

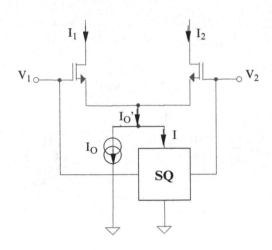

resulting:

$$V' = V_1 - V_T - V_C. \tag{7.37}$$

The expression of I_1 current becomes:

$$I_1 = \frac{K}{2}[(V_2 - V_1) + V_C]^2 \tag{7.38}$$

and, similarly:

$$I_2 = \frac{K}{2}[(V_1 - V_2) + V_C]^2. \tag{7.39}$$

The I_{12} differential current can be expressed as follows:

$$I_{12} = I_2 - I_1 = 2KV_C(V_1 - V_2). \tag{7.40}$$

The equivalent resistance of the entire structure presented in Fig. 7.16 will be independent on the threshold voltage:

$$R_{ECH} = \frac{V_1 - V_2}{I_{12}} = \frac{1}{2KV_C}. \tag{7.41}$$

Another possible implementation of a linear differential amplifier is based on a proper current biasing of a classical differential amplifier. The method (Fig. 7.17) [6] for obtaining a linear transfer characteristic of the differential amplifier is to obtain the I_O' biasing current of the entire differential structure as a sum of a main constant term, I_O and an additional term, proportional with the squaring of the differential input voltage, $I = K(V_1 - V_2)^2/4$:

$$I_O' = I_O + I = I_O + \frac{K}{4}(V_1 - V_2)^2, \tag{7.42}$$

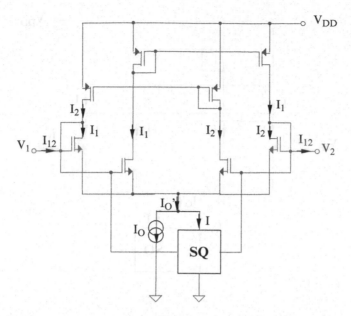

Fig. 7.18 Active resistor with positive equivalent resistance based on PR 7.1 – sixth implementation

resulting, in this case, a perfect linear behavior of the optimized differential amplifier:

$$
\begin{aligned}
I_1 - I_2 &= \frac{V_1 - V_2}{2} \sqrt{4KI_O' - K^2(V_1 - V_2)^2} \\
&= \sqrt{KI_O}(V_1 - V_2) = G_m(V_1 - V_2),
\end{aligned}
\tag{7.43}
$$

G_m being the equivalent transconductance of the proposed structure, that can be controlled by the biasing current, I_O.

Based on the block diagram presented in Fig. 7.1, the improved linearity differential cell presented in Fig. 7.17 can be re-used in order to obtain an original active resistor with excellent linearity, having the circuit presented in Fig. 7.18 [6].

For this circuit, the current passing through the input pins, I_{12}, can be expressed as follows:

$$
I_{12} = I_1 - I_2 = \frac{V_1 - V_2}{2} \sqrt{4KI_O' - K^2(V_1 - V_2)^2}.
\tag{7.44}
$$

Because the biasing current of the circuit core, I_O', was designed to be the sum of a main constant term I_O, and an additional term proportional with the squaring of the differential input voltage:

$$
I_O' = I_O + \frac{K}{4}(V_1 - V_2)^2,
\tag{7.45}
$$

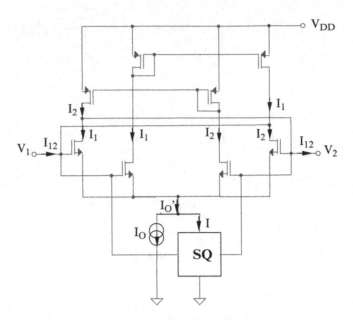

Fig. 7.19 Active resistor with negative equivalent resistance based on PR 7.1 – sixth implementation

the I_{12} current will have the following expression:

$$I_{12} = \sqrt{KI_O}(V_1 - V_2).\tag{7.46}$$

Defining the equivalent resistance of the circuit from Fig. 7.18 as the ratio between the $V_1 - V_2$ differential input voltage and the current passing through the input pins, I_{12}, it results:

$$R_{ECH} = \frac{V_1 - V_2}{I_{12}} = \frac{1}{\sqrt{KI_O}}.\tag{7.47}$$

The circuit presents the advantage of controllability (the active resistance R_{ECH} can be changed by modifying the I_O biasing current).

Using the active resistor with positive equivalent resistance presented in Fig. 7.18, in order to obtain a circuit with controllable negative resistance circuit, it is necessary to use two input–output cross-connections, resulting the circuit presented in Fig. 7.19 [6]. Because now $I_{12} = I_2 - I_1$, the equivalent resistance of the circuit from Fig. 7.19 will be:

$$R_{ECH}' = -R_{ECH} = -\frac{1}{\sqrt{KI_O}}.\tag{7.48}$$

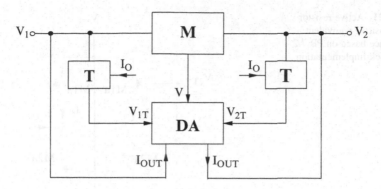

Fig. 7.20 Active resistor with positive equivalent resistance based on PR 7.2 – block diagram

7.2.2   Active Resistor Circuits Based on the Second Mathematical Principle (PR 7.2)

The linearization techniques based on the second mathematical principle uses a proper biasing of a classical differential amplifier, that modifies the transfer characteristic of this circuit, in order to obtain a linear behavior for the resulting active resistor structure.

The block diagram of active resistor circuit using the second mathematical principle is presented in Fig. 7.20. The output currents of the "DA" block from Fig. 7.20 [4, 7] are forces to pass through the input pins, implementing, in this way, a linear current–voltage characteristic, $I_{OUT}(V_1 - V_2)$, for the active resistor structure, so a constant equivalent resistance of the structure.

7.2.2.1   The "DA" (Differential Amplifier) Block

The "DA" block is implemented as a classical active-load differential amplifier, having the concrete realization presented in Fig. 7.21. Considering a biasing in saturation of MOS devices from Fig. 7.21 [4, 7, 8], the output current of the differential amplifier can be expressed as:

$$I_{OUT} = I_2 - I_1 = \frac{K}{2}(V_{1T} - V_{2T})(2V - V_{1T} - V_{2T} - 2V_T). \tag{7.49}$$

In order to obtain a linear transfer characteristic, $I_{OUT}(V_{1T} - V_{2T})$, it is necessary that the value from the second parenthesis from (7.49) to be constant with respect to the differential input voltage $V_{1T} - V_{2T}$:

$$2V - V_{1T} - V_{2T} - 2V_T = A = ct., \tag{7.50}$$

Fig. 7.21 Active resistor
with positive equivalent
resistance based on PR 7.2 –
DA block implementation

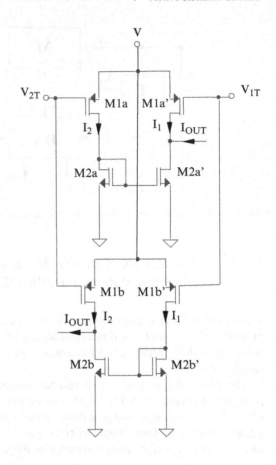

resulting the necessity of implementing a V voltage equal with:

$$V = \frac{V_{1T} + V_{2T}}{2} + V_T + \frac{A}{2}, \tag{7.51}$$

7.2.2.2 The "T" (Translation) Block

The translation of the V potential by $V_T + A/2$ can be obtained using "T" block,
having the implementation presented in Fig. 7.22 [4, 7].

Because the same I_O current is passing through both transistors from Fig. 7.22, it
can be obtained:

$$V_1 = V_{1T} + V_T + \sqrt{\frac{2I_O}{K}} \tag{7.52}$$

Fig. 7.22 Active resistor
with positive equivalent
resistance based on PR 7.2 –
T block implementation

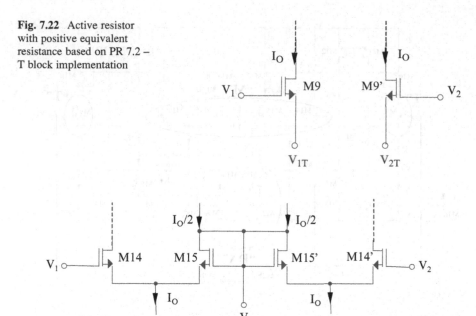

Fig. 7.23 Active resistor with positive equivalent resistance based on PR 7.2 – M block
implementation

and:

$$V_2 = V_{2T} + V_T + \sqrt{\frac{2I_O}{K}}. \tag{7.53}$$

So, both V_1 and V_2 input potentials are DC shifted with the same amount,
$V_T + \sqrt{2I_O/K}$.

7.2.2.3   The "M" (Arithmetic Mean) Block

In order to obtain the arithmetic mean of input potentials expressed by (7.51), an
arithmetical mean circuit must be used (Fig. 7.23) [4, 7], this particular implementation
having the advantage of containing only MOS transistors, biased in saturation region.

$$V = \frac{V_1 + V_2}{2}. \tag{7.54}$$

It can be obtained:

$$V = \frac{V_{1T} + V_{2T}}{2} + V_T + \sqrt{\frac{2I_O}{K}}. \tag{7.55}$$

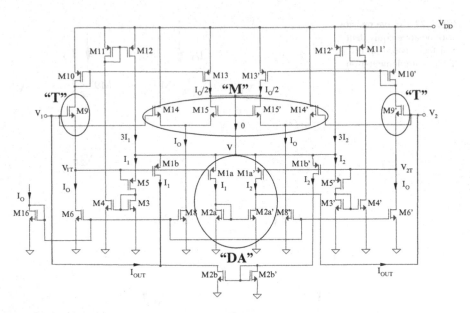

Fig. 7.24 Active resistor with positive equivalent resistance based on PR 7.2 – complete implementation

Comparing (7.51) and (7.55) relations, it results that $A = 2\sqrt{2I_O/K}$, so:

$$I_{OUT} = \frac{K}{2}(V_{1T} - V_{2T})2\sqrt{\frac{2I_O}{K}} = \sqrt{2KI_O}(V_{1T} - V_{2T}). \tag{7.56}$$

And, by using (7.52) and (7.53), equivalent with:

$$I_{OUT} = \sqrt{2KI_O}(V_1 - V_2) = G_m(V_1 - V_2), \tag{7.57}$$

$G_m = \sqrt{2KI_O}$ being the equivalent transconductance of the differential amplifier. As a result of using translation blocks, the G_m transconductance of the differential core will be not dependent on the threshold voltage, so the active resistor linearity will be not affected by the bulk effect.

Because the output currents of the "DA" block are forces to pass through the input pins (Fig. 7.20), a linear current–voltage characteristic, $I_{OUT}(V_1 - V_2)$, of the active resistor structure will be obtained. So, a constant equivalent resistance of the entire structure from Fig. 7.20 can be achieved:

$$R_{ECH} = \frac{V_1 - V_2}{I_{OUT}} = \frac{1}{G_m} = \frac{1}{\sqrt{2KI_O}}. \tag{7.58}$$

The full implementation of the active resistor circuit is presented in Fig. 7.24 [4, 7].

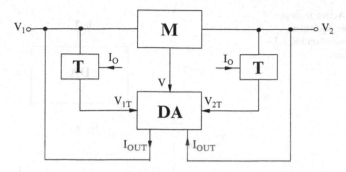

Fig. 7.25 Active resistor with negative equivalent resistance based on PR 7.2 – block diagram

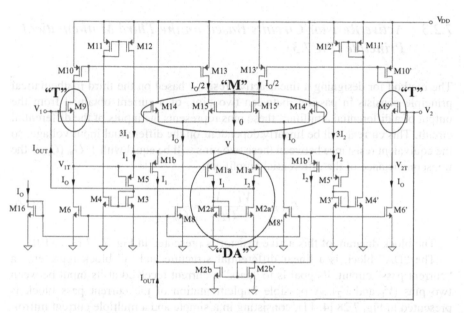

Fig. 7.26 Active resistor with negative equivalent resistance based on PR 7.2 – complete implementation

In order to obtain an active resistor circuit with negative equivalent resistance, the senses of the output currents of "DA" block must be inversed, resulting the block diagram presented in Fig. 7.25 [4].

The implementation of the active circuit with negative resistance is presented in Fig. 7.26 [4, 7], the equivalent resistance of this structure being:

$$R_{ECH}' = -\frac{1}{G_m} = -\frac{1}{\sqrt{2KI_O}}. \tag{7.59}$$

Fig. 7.27 Active resistor
with positive equivalent
resistance based on PR 7.3 –
block diagram

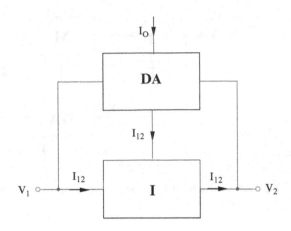

7.2.3 Active Resistor Circuits Based on the Third Mathematical Principle (PR 7.3)

The method for designing a linear active resistor based on the third mathematical principle consists in passing, between two pins, of a current obtained from the output of a differential amplifier; these pins represents the inputs of the differential circuit. This current will be linearly dependent on the differential input voltage, so the equivalent resistance between these two pins will be equal with $1/G_m$ (G_m is the transconductance of the differential amplifier).

$$R_{ECH.} = \frac{V_1 - V_2}{I_{12}} = \frac{1}{G_m}. \tag{7.60}$$

The block diagram of this active resistor is presented in Fig. 7.27 [4, 8–10].

The "DA" block is a linear differential structure and "I" block represents a "current pass" circuit. Its goal is to "pass" a current received at its input between two pins (V_1 and V_2). A possible implementation of the current pass block is presented in Fig. 7.28 [4, 11], consisting in a simple and a multiple current mirror.

As a result of the quadratic characteristic of a MOS transistor operating in saturation, the transfer characteristic of the classical CMOS differential amplifier will be strongly nonlinear, its linearity being in reasonable limits only for a very limited range of the differential input voltage amplitudes.

There are many possibilities of improving the linearity of the classical differential amplifier. The method used for increasing the linearity of the active resistor circuit having the block diagram presented in Fig. 7.27 is based on the compensation of quadratic characteristic of the MOS transistor working in saturation region by two identical current-mode square-root circuits. The result will be a more linear transfer characteristic of the circuit, quantitatively evaluated by an important reduction of the total harmonic distortions of the differential amplifier (Fig. 7.29) [4, 9, 10, 12].

Fig. 7.28 Active resistor
with positive equivalent
resistance based on PR 7.3 –
current-pass block
implementation

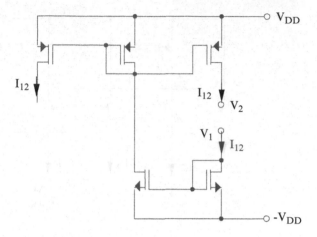

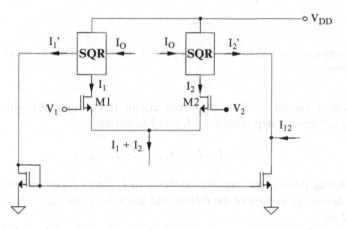

Fig. 7.29 Active resistor with positive equivalent resistance based on PR 7.3 – linear DA block
implementation

A possible implementation of the square-root circuits from Fig. 7.29 is shown in
Fig. 7.30 [4, 10, 12], representing a perfect symmetrical structure using MOS
transistors and a FGMOS device.

The expression of I current is:

$$I = \frac{4K}{2}\left[\frac{V_{GS}(I_O) + V_{GS}(I_{1,2})}{2} - V_T\right]^2$$

$$= \left(\sqrt{I_O} + \sqrt{I_{1,2}}\right)^2 = I_O + I_{1,2} + 2\sqrt{I_O I_{1,2}}.$$

(7.61)

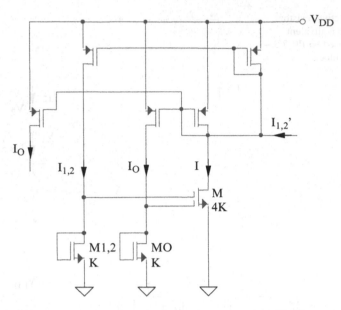

Fig. 7.30 Active resistor with positive equivalent resistance based on PR 7.3 – square-root block implementation

The output current of the square-root circuit from Fig. 7.30 will have the following square-root dependence on $I_{1,2}$ and I_O currents:

$$I_{1,2}' = I - I_O - I_{1,2} = 2\sqrt{I_{1,2}I_O}. \tag{7.62}$$

Considering this square-root dependencies of I_1' and I_2' currents on I_1 and I_2 currents, the output current of the differential amplifier from Fig. 7.29, I_{12}, can be expressed as:

$$I_{12} = I_1' - I_2' = 2\sqrt{I_O}\left(\sqrt{I_1} - \sqrt{I_2}\right), \tag{7.63}$$

resulting:

$$I_{12} = \sqrt{2KI_O}(V_{GS1} - V_{GS2}), \tag{7.64}$$

V_{GS1} and V_{GS2} being the gate-source voltages of M1 and M2 transistors from Fig. 7.29. It results a linear transfer characteristic of the differential amplifier:

$$I_{12} = G_m(V_1 - V_2), \tag{7.65}$$

where G_m is the circuit transconductance, $G_m = \sqrt{2KI_O}$. So, in a first-order analysis, the dependence of the differential circuit output current on its differential input voltage will be perfectly linear.

Fig. 7.31 Active resistor
with negative equivalent
resistance based on PR 7.3 –
current-pass block
implementation

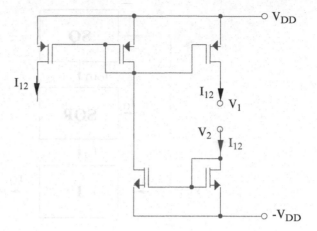

The equivalent resistance of the circuit presented in Fig. 7.27 will have the
following expression:

$$R_{ECH.} = \frac{1}{\sqrt{2KI_O}} \qquad (7.66)$$

and it can be controlled by the I_O biasing current.

An advantage of the active resistor presented in Fig. 7.27 is that a circuit with a
negative equivalent resistance can be obtained by a minor change in the current-
pass circuit presented in Fig. 7.28. The modified implementation of the current-pass
circuit that can be used for obtaining an active resistor with a negative equivalent
resistance is presented in Fig. 7.31 [13].

The resulting expression of the active resistance will be $R_{ECH}' = -R_{ECH} =$
$-1/2\sqrt{2KI_O}$.

7.2.4   Active Resistor Circuits Based on the Fourth
Mathematical Principle (PR 7.4)

The structure of the active resistor circuit based on the fourth mathematical
principle contains three important blocks (Fig. 7.32) [4, 14]: a voltage-current
squarer, "SQ", a current square-root circuit, "SQR", and a current-pass circuit, "I".

The I_{12} current is proportional with the square-root of I_{OUT} and I_O currents, while
I_{OUT} current is proportional with the square of the differential input voltage,
$V_1 - V_2$. So, the result will be a linear relation between the differential voltage
across the input pins of the active resistor and the current passing between them.

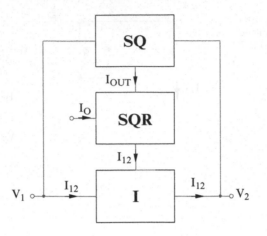

Fig. 7.32 Active resistor with positive equivalent resistance based on PR 7.4 – block diagram

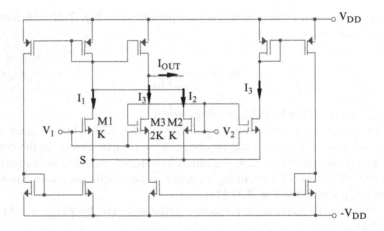

Fig. 7.33 Active resistor with positive equivalent resistance based on PR 7.4 – squaring circuit implementation

The current squaring circuit is based on the perfect symmetrical structure, presented in Fig. 7.33 [15]. The utilization of a FGMOS device decreases the silicon occupied area of this circuit.

The output current expression has a linear dependence on the drain currents of M1, M2 and M3 transistors:

$$I_{OUT} = I_1 + I_2 - I_3. \tag{7.67}$$

Considering a biasing in saturation of all MOS devices from Fig. 7.33, the previous currents will have the following expressions:

$$I_1 = \frac{K}{2}(V_1 - V_S - V_T)^2, \tag{7.68}$$

Fig. 7.34 Active resistor with positive equivalent resistance based on PR 7.4 – square-root circuit implementation

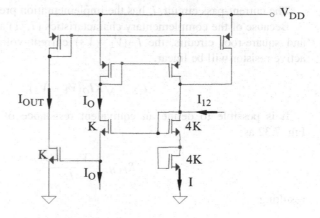

$$I_2 = \frac{K}{2}(V_2 - V_S - V_T)^2, \tag{7.69}$$

$$I_3 = K\left(\frac{V_1 + V_2}{2} - V_S - V_T\right)^2. \tag{7.70}$$

From the previous relations, it results a quadratic dependence of the output current, I_{OUT}, on the $V_1 - V_2$ differential input voltage:

$$I_{OUT} = \frac{K}{4}(V_1 - V_2)^2. \tag{7.71}$$

The square-root circuit is presented in Fig. 7.34, designed using exclusively MOS active devices biased in the saturation region.

The relation between the currents from the circuit is:

$$\left(V_T + \sqrt{\frac{2I_{OUT}}{K}}\right) + \left(V_T + \sqrt{\frac{2I_O}{K}}\right) = 2\left(V_T + \sqrt{\frac{2I}{4K}}\right), \tag{7.72}$$

equivalent with:

$$I = I_{OUT} + I_O + 2\sqrt{I_{OUT}I_O}. \tag{7.73}$$

Implementing the proper linear relation between the previous currents of the square-root circuit:

$$I_{12} = I - I_{OUT} - I_O, \tag{7.74}$$

the output current of the circuit from Fig. 7.34 will be proportional with the square-root of the input current:

$$I_{12} = 2\sqrt{I_{OUT}I_O}. \tag{7.75}$$

The current-pass circuit, I, has the implementation presented in Fig. 7.28.

Because of the complementary characteristics (7.71) and (7.75) of the squaring and square-root circuits, the $I_{12}(V_1 - V_2)$ current–voltage characteristic of the active resistor will be linear.

$$I_{12} = \sqrt{KI_O}(V_1 - V_2). \qquad (7.76)$$

It is possible to define an equivalent resistance of the circuit presented in Fig. 7.32 as:

$$R_{ECH.} = \frac{V_1 - V_2}{I_{12}}, \qquad (7.77)$$

resulting:

$$R_{ECH.} = \frac{1}{\sqrt{KI_O}}. \qquad (7.78)$$

The advantage of the active resistor circuit is that the value of the equivalent active resistance can be controlled by modifying the reference current I_O.

7.2.5   Active Resistor Circuits Based on the Fifth Mathematical Principle (PR 7.5)

The active resistor based on the fifth mathematical principle has the block diagram presented in Fig. 7.35, [4, 12] and it uses three important types of blocks:

- Two linearized differential amplifiers for converting the $V_1 - V_2$ input voltage and the V_O reference voltage in two currents that will be inserted into the multiplier circuit. The most important requirements for these differential amplifiers are referring to the linearity of the transfer characteristics, associated with the maximization of their input voltage ranges that allows a good linearity, to the common mode input range and to the independence of the circuit performances on the second-order effects. The currents generated by the differential amplifiers will be proportional with their input voltages, $I_1 = G_m V_O$ and $I_2 = G_m(V_1 - V_2)$;
- A current-mode multiplier circuit, "MULT", for "mirroring" the Ohm law, whose operation is described by the relation $I_{12} = I_O I_2 / I_1$;
- A current-pass circuit, which imposes the condition that the same current to pass between the V_1 and V_2 input pins

Consider that DA1 and DA2 differential amplifiers from Fig. 7.35 are implemented using the linearization technique proposed in Fig. 7.29. So:

$$I_2 = \sqrt{2KI}(V_1 - V_2), \qquad (7.79)$$

$$I_1 = \sqrt{2KI}\,V_O, \qquad (7.80)$$

Fig. 7.35 Active resistor (1) with positive equivalent resistance based on PR 7.5 – block diagram

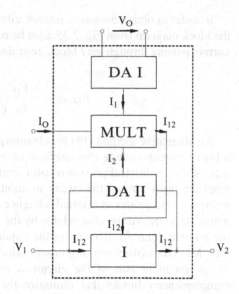

I biasing current replacing the I_O current from Fig. 7.29. Defining the equivalent resistance between V_1 and V_2 pins as the ratio between the $V_1 - V_2$ differential input voltage and the current passing through these pins, I_{12}, it results:

$$R_{ECH} = \frac{V_1 - V_2}{I_{12}} = \frac{V_O}{I_O}. \tag{7.81}$$

In conclusion, the active resistor presented in Fig. 7.35 will have an equivalent resistance that can be controlled by modifying the ratio of the V_O reference voltage and the I_O reference current.

A possible method for obtaining the multiplying function using the previous designed square-root circuit is to use two identical circuits presented in Fig. 7.29, implementing the following functions:

$$I_{OUT1} = 2\sqrt{I_O I_2} \tag{7.82}$$

and:

$$I_{OUT2} = 2\sqrt{I_{12} I_1}, \tag{7.83}$$

I_{OUT1} and I_{OUT2} being the output currents of these square-root circuits. Using a classical current mirror, it is possible to impose $I_{OUT1} = I_{OUT2}$, resulting the necessary multiplying function:

$$I_{12} = I_O \frac{I_2}{I_1}. \tag{7.84}$$

In order to obtain an active resistor circuit with negative equivalent resistance, the block diagram from Fig. 7.35 must be modified by inversing the sense of the I_{12} current passing through the I block, resulting an equivalent resistance expressed by:

$$R_{ECH}' = -\frac{V_O}{I_O}\sqrt{\frac{(W/L)_1}{(W/L)_2}\frac{I_{O1}}{I_{O2}}}. \tag{7.85}$$

An alternative method [15] for obtaining, using the fifth mathematical principle, a linear current–voltage characteristic of the active resistor, similar with the characteristic of a classical passive resistor, consists in the "mirroring" of the Ohm law from the input pins of the circuit to another pins, used for applying an external reference voltage and an external reference current. The equivalent resistance of the active structure will be controllable by the ratio between the reference voltage and the reference current. Because of the requirements for a good frequency response, only MOS transistors working in saturation are used and a current-mode operation of an important part of the circuit is implemented. A possible choice of the complementary blocks that minimize the complexity of the entire structure is referring to the squaring and square-rooting functions. In order to further reduce the silicon occupied area, classical MOS devices have been replaced by FGMOS transistors.

The structure of the active resistor is based on four important blocks: two voltage-current squarers, a current square-root circuit, a current divider circuit and a current-pass circuit, named SQ, SQR, DIV and I, respectively on the block diagram from Fig. 7.36 [15].

The I_{12} current is proportional with the square-root of the product between I and I_O currents, while I is proportional with the ratio of I_2 and I_1 currents. Each of these two last currents is proportional with the squaring of $V_1 - V_2$ and V_O voltages, respectively. The result of this implementation of the circuit will be a linear relation between the $V_1 - V_2$ differential voltage across the input pins of the active resistor and the current passing between these pins, I_{12}.

The current squaring circuits can be implemented using the circuit presented in Fig. 7.33, resulting a quadratic dependence of the I_2 output current on the $V_1 - V_2$ differential input voltage:

$$I_2 = \frac{K}{4}(V_1 - V_2)^2. \tag{7.86}$$

In a similar way, the other squarer from the block diagram will compute the following expression of I_1 current:

$$I_1 = \frac{K}{4}V_O^2. \tag{7.87}$$

Fig. 7.36 Active resistor (2) with positive equivalent resistance based on PR 7.5 – block diagram

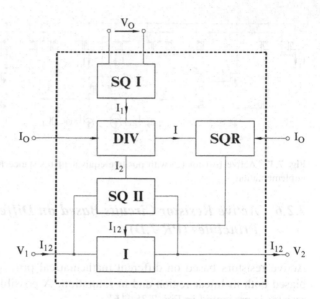

The current square-root block from Fig. 7.36 can be realized using the circuit presented in Fig. 7.34, the output current being proportional with the square-root of the input current:

$$I_{12} = 2\sqrt{I_O I}. \tag{7.88}$$

The divider circuit can be obtained using two square-root circuits from Fig. 7.34, connected as it is shown in Fig. 7.37. The computed functions are $I_{O1} = 2\sqrt{I_1 I}$ and $I_{O2} = 2\sqrt{I_2 I_O}$. Because $I_{O1} = I_{O2}$, the function implemented by the circuit from Fig. 7.37 [15] will be:

$$I = I_O \frac{I_2}{I_1}. \tag{7.89}$$

The current-pass circuit is similar with the structure presented in Fig. 7.28. Using previous relations, it results that the equivalent resistance of the active resistor having the block diagram presented in Fig. 7.36 is:

$$R_{ECH.} = \frac{V_1 - V_2}{I_{12}} = \frac{V_O}{2I_O}. \tag{7.90}$$

An important advantage of the previous presented circuit is that the value of the equivalent active resistance can be controlled by modifying the ratio of a reference voltage, V_O and a reference current, I_O.

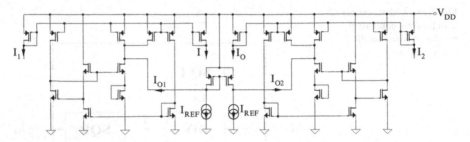

Fig. 7.37 Active resistor (2) with positive equivalent resistance based on PR 7.5 – DIV block implementation

7.2.6   Active Resistor Circuits Based on Different Mathematical Principles (PR 7.D)

Active resistors based on different mathematical principles use MOS transistors biased both in linear region and in saturation. A possible realization of an active resistor is presented in Fig. 7.38 [16].

The M1 and M2 transistors are biased in linear region, while all other MOS transistors are operated in saturation region. The I_1 and I_2 currents can be expressed as follows:

$$I_1 = \frac{K}{2}\left[V_{DS1}(V_{GS1} - V_T) - \frac{V_{DS1}^2}{2}\right] \tag{7.91}$$

and:

$$I_2 = \frac{K}{2}\left[V_{DS2}(V_{GS2} - V_T) - \frac{V_{DS2}^2}{2}\right]. \tag{7.92}$$

The M5–M6 current mirror imposes the identity between I_1 and I_2 currents, so $V_{GS3} = V_{GS4}$, resulting $V_{DS1} = V_{DS2}$. Using (7.91) and (7.92), it results:

$$I_{IN} = I_2 - I_1 = \frac{K}{2}V_{DS2}(V_{GS2} - V_{GS1})$$
$$= \frac{K}{2}V_{DS2}[(V_{G2} + V_{DD}) - (0 + V_{DD})] = \frac{K}{2}V_{DS2}V_{G2}. \tag{7.93}$$

The equivalent input impedance can be defined as follows:

$$R_{ECH} = \frac{V_{DS2}}{I_{IN}} = \frac{2}{KV_{G2}}. \tag{7.94}$$

Fig. 7.38 Active resistor (1) with positive equivalent resistance based on PR 7.D

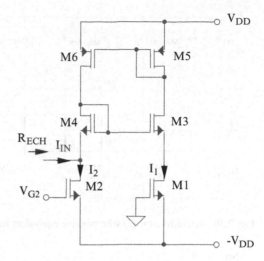

The circuit simulates in the input pin an equivalent resistance, R_{ECH}, having a value that can be controlled by a biasing potential V_{G2}.

Another possible realization of an active resistor circuit is presented in Fig. 7.39 [17]. The expressions of $I_1 - I_4$ currents are:

$$I_1 = \frac{K}{2}(V_{C1} - V_1 - V_T)^2, \qquad (7.95)$$

$$I_2 = \frac{K}{2}(V_{C1} - V_2 - V_T)^2, \qquad (7.96)$$

$$I_3 = \frac{K}{2}(V_{C2} - V_1 - V_T)^2, \qquad (7.97)$$

$$I_4 = \frac{K}{2}(V_{C2} - V_2 - V_T)^2. \qquad (7.98)$$

Because of M16–M17 and M20–M21 current mirrors, $I_7 = 2I_1$ and $I_{10} = 2I_2$. The I and I' currents can be expressed using a linear relation containing the previous currents:

$$I = I' = I_1 + I_4 - I_2 - I_3 \qquad (7.99)$$

resulting:

$$I = I' = \frac{K}{2}(V_2 - V_1)(2V_{C1} - V_1 - V_2 - 2V_T)$$
$$+ \frac{K}{2}(V_1 - V_2)(2V_{C2} - V_1 - V_2 - 2V_T). \qquad (7.100)$$

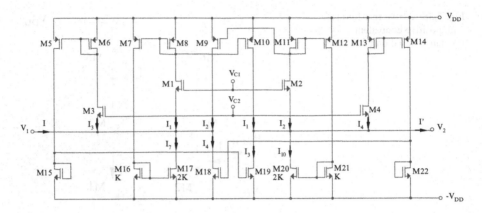

Fig. 7.39 Active resistor (2) with positive equivalent resistance based on PR 7.D

So:

$$I = I' = K(V_{C2} - V_{C1})(V_1 - V_2). \qquad (7.101)$$

The equivalent resistance of the active structure presented in Fig. 7.39 can be expressed as follows:

$$R_{ECH} = \frac{V_1 - V_2}{I} = \frac{V_1 - V_2}{I'} = \frac{1}{K(V_{C2} - V_{C1})} \qquad (7.102)$$

and it can be controlled using $V_{C2} - V_{C1}$ differential voltage.

Another implementation of an active resistor circuit using exclusively MOS transistors biased in saturation is presented in Fig. 7.40 .

The I current can be expressed as a linear function on the currents from the circuit:

$$I = I_2 - I_1 - I_5 + I_6. \qquad (7.103)$$

Because $I_6 = 2I_1$ and $I_5 = 2I_2$, it results:

$$I = I_1 - I_2. \qquad (7.104)$$

Similarly, the I' current will have the following expression:

$$I' = I_2 - I_1 + I_7 - I_8. \qquad (7.105)$$

Using $I_7 = 2I_1$ and $I_8 = 2I_2$ relations (implemented using different aspect ratios MOS transistors), it can be obtained:

$$I' = I_1 - I_2. \qquad (7.106)$$

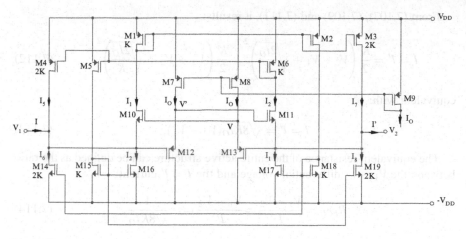

Fig. 7.40 Active resistor (3) with positive equivalent resistance based on PR 7.D

So, the same current $I = I'$ will pass between the inputs, being expressed by the following relation:

$$I = I' = \frac{K}{2}(V - V_1 - V_T)^2 - \frac{K}{2}(V' - V_2 - V_T)^2. \qquad (7.107)$$

As the drain current of M12 transistor is imposed to be I_O, their gate-source voltage will be equal with:

$$V' - V_1 = V_T + \sqrt{\frac{2I_O}{K}}. \qquad (7.108)$$

Thus:

$$V' = V_1 + V_T + \sqrt{\frac{2I_O}{K}}. \qquad (7.109)$$

Similarly:

$$V - V_2 = V_T + \sqrt{\frac{2I_O}{K}}. \qquad (7.110)$$

So:

$$V = V_2 + V_T + \sqrt{\frac{2I_O}{K}}. \qquad (7.111)$$

From (7.107), (7.109) and (7.111), it results:

$$I = I' = \frac{K}{2}\left(V_2 - V_1 + \sqrt{\frac{2I_O}{K}}\right)^2 - \frac{K}{2}\left(V_1 - V_2 + \sqrt{\frac{2I_O}{K}}\right)^2, \qquad (7.112)$$

equivalent with:

$$I = I' = \sqrt{8KI_O}(V_2 - V_1), \qquad (7.113)$$

The equivalent resistance of the entire active structure can be defined as the ratio between the $V_1 - V_2$ differential voltage and the $I = I'$ current:

$$R_{ECH} = \frac{V_1 - V_2}{I} = \frac{V_1 - V_2}{I'} = -\frac{1}{\sqrt{8KI_O}}. \qquad (7.114)$$

The R_{ECH} equivalent resistance can be controlled by the I_O reference current.

7.3   Conclusions

Chapter presents a multitude of active resistor structures implemented in CMOS technology. Functionally equivalent with a classical resistor, active resistor circuits have the most important advantage of reducing the silicon area, especially for large values of the simulated resistances. Additionally, both positive and negative equivalent resistances are available by small changing of the design. The possibility of controlling the value of the equivalent resistance using a control current or a control voltage extends the area of utilization of this class of VLSI circuits.

References

1. Popa C (2010) Improved linearity CMOS differential amplifiers with applications in VLSI designs. In: International symposium on electronics and telecommunications, Athens, pp 29–32
2. Popa C (2006) Multifunctional linear structure with applications in VLSI designs. In: International semiconductor conference, Romania, pp 433–436
3. Popa C (2007) Improved linearity active resistors using MOS and floating-gate MOS transistors. In: The international conference on "computer as a tool", Warsaw, pp 224–230
4. Manolescu AM, Popa C (2009) Low-voltage low-power improved linearity CMOS active resistor circuits. Springer J Analog Integr Circuits Signal Process 62:373–387
5. Popa C (2004) A new FGMOS active resistor with improved linearity and frequency response. In: International semiconductor conference, pp 295–298, Sinaia, Romania
6. Popa C (2008) Programmable CMOS active resistor using computational circuits. In: International semiconductor conference, pp 389–392, Sinaia, Romania

7. Popa C (2010) Tunable CMOS resistor circuit with improved linearity based on the arithmetical mean computation. In: IEEE Mediterranean electrotechnical conference, pp 1379–1382, Valletta, Malta

8. Popa C (2006) Improved linearity differential structure with applications in VLSI designs. In: International conference on optimization of electric and electronic equipment, pp 24–27, Brasov, Romania

9. Popa C (2005) Linear active resistor based on CMOS square-root circuits for VLSI applications1. In: International conference "computer as a tool", pp 894–897, Belgrade, Serbia and Montenegro

10. Popa C (2009) Negative resistance active resistor with improved linearity and frequency response. J Circuits Syst Comput 18:1–10

11. Popa C (2006) Improved linearity active resistor using equivalent FGMOS devices. In: International conference on microelectronics, pp 396–399, Nis, Serbia

12. Popa C, Manolescu AM, Manolescu A (2006) Improved linearity CMOS active resistor with increased frequency response and controllable equivalent resistance. In: International semiconductor conference, Sinaia, pp 355–358

13. Popa C (2006) Improved linearity active resistor with controllable negative resistance. In: IEEE international conference on integrated circuit design and technology, Padova, pp 1–4

14. Popa C (2005) A new improved linearity active resistor using complementary functions. In: International semiconductor conference, Sinaia, pp 391–394

15. Popa C (2010) Improved linearity CMOS active resistor based on complementary computational circuits. In: IEEE international conference on electronics, circuits, and systems, pp 455–458, Athens, Greece

16. Weihsing L, Shen-Iuan L, Shui-Ken W (2005) CMOS current-mode divider and its applications. IEEE Trans Circuits Syst II, Exp Briefs 52:145–148

17. Oura T, Yoneyama T, Tantry S, Asai H (2002) A threshold voltage independent floating resistor circuit exhibiting both positive and negative resistance values. In: IEEE international symposium on circuits and systems, vol III, pp 739–742, Arizona, USA

18. Tantry S, Yoneyama T, Asai H (2001) Two floating resistor circuits and their applications to synaptic weights in analog neural networks. In: IEEE international symposium on circuits and systems, pp 564–567, Sydney, Australia

19. Sakurai S, Ismail M (1992) A CMOS square-law programmable floating resistor independent of the threshold voltage. IEEE Trans Circuits Systems II, Analog Digit Signal Process 39:565–574

20. Popa C, Mitrea O, Manolescu AM, Glesner M (2002) Linearization technique for a CMOS active resistor. In: International conference on optimization of electric and electronic equipment, pp 613–616, Brasov, Romania

21. Popa C (2007) Low-voltage low-power curvature-corrected voltage reference circuit using DTMOSTs. Lecture notes in computer science, Springer, pp 117–124

22. De La Cruz-Blas CA, Lopez-Martin A, Carlosena A (2003) 1.5-V MOS translinear loops with improved dynamic range and their applications to current-mode signal processing. IEEE Trans Circuits Syst II, Analog Digit Signal Process 50:918–927

23. Desheng M, Wilamowski BM, Dai FF (2009) A tunable CMOS resistor with wide tuning range for low pass filter application. In: IEEE topical meeting on silicon monolithic integrated circuits in RF systems, pp 1–4, San Diego, USA

24. Torralba A et al (2009) Tunable linear MOS resistors using quasi-floating-gate techniques. IEEE Trans Circuits Syst II, Exp Briefs 56:41–45

25. Tadić N, Zogović M (2010) A low-voltage CMOS voltage-controlled resistor with wide resistance dynamic range. In: International conference on microelectronics proceedings, pp 341–344, Nis, Serbia

Chapter 8
Multifunctional Structures

8.1 Mathematical Analysis for Synthesis of Multifunctional Structures

An important goal in VLSI designs is represented by the possibility of a multiple utilization of the same cell, the increased modularity that can be achieved being reflected in an important reduction of power consumption and of design costs per circuit function. Many fundamental linear or nonlinear analog blocks can be realized starting from the same core, the optimization technique implemented for the nucleus being efficient for all the derived circuits. The presented multifunctional structures are based on four different elementary mathematical principles, each of them being illustrated by concrete implementations in CMOS technology of their functional relations.

The multifunctional structures [2, 3, 5, 9, 35, 46] that can be realized starting from an improved performances multifunctional core are: differential amplifiers [1, 11, 14, 16–18, 20–31, 34, 59, 64], multiplier circuits [32, 33, 36–45, 47, 48, 50], active resistors [10, 12, 13, 19, 60–63] (with both positive and negative controllable equivalent resistance), squaring [4, 38, 49–54], square-rooting [8, 15, 55–58] or exponential [6, 7] circuits.

8.1.1 First Mathematical Principle (PR 8.1)

The output currents of the multifunctional circuit core (Fig. 8.1) have the following general expressions:

$$I_{OUT1} = I_O + a\sqrt{KI_O}(V_1 - V_2) + bK(V_1 - V_2)^2 \qquad (8.1)$$

and:

$$I_{OUT2} = I_O - a\sqrt{KI_O}(V_1 - V_2) + bK(V_1 - V_2)^2, \qquad (8.2)$$

C.R. Popa, *Synthesis of Computational Structures for Analog Signal Processing*,
DOI 10.1007/978-1-4614-0403-3_8, © Springer Science+Business Media, LLC 2011

Fig. 8.1 MFC core based on
PR 8.1 – symbolical
representation

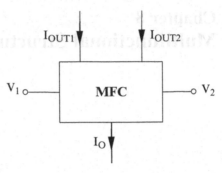

Fig. 8.2 Differential
amplifier based on PR 8.1 –
block diagram

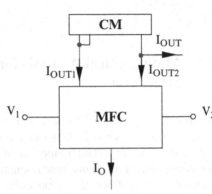

a and b being constants, depending on the particular implementation of the multifunctional circuit core.

8.1.1.1   Principle of Operation of a Linear Differential Amplifier

The block diagram of the differential amplifier using the multifunctional core from Fig. 8.1 is shown in Fig. 8.2.

The output current of the differential amplifier circuit is obtained as the difference between the individual output currents, I_{OUT1} and I_{OUT2}:

$$I_{OUT} = I_{OUT1} - I_{OUT2} = 2a\sqrt{KI_O}(V_1 - V_2). \tag{8.3}$$

The equivalent transconductance of the structure can be defined as follows:

$$G_m = \frac{I_{OUT}}{V_1 - V_2} = 2a\sqrt{KI_O} \tag{8.4}$$

and it can be controlled using the I_O biasing current.

Fig. 8.3 Active resistor with positive equivalent resistance based on PR 8.1 – block diagram

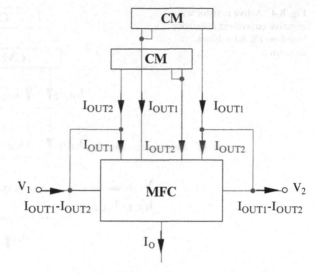

8.1.1.2 Principle of Operation of an Active Resistor with Positive Equivalent Resistance

In order to obtain an active resistor with positive equivalent resistance (Fig. 8.3) [1], the multifunctional core must be modified for generating two I_{OUT1} output currents and two I_{OUT2} output currents.

Additionally, two input–output connections have been added, forcing between the input pins the same current, $I_{OUT1} - I_{OUT2}$. The equivalent resistance of the entire structure can be defined as the ratio between the differential input voltage, $V_1 - V_2$ and the differential current, $I_{OUT1} - I_{OUT2}$:

$$R_{ECH} = \frac{V_1 - V_2}{I_{OUT1} - I_{OUT2}} = \frac{1}{G_m} = \frac{1}{2a\sqrt{KI_O}}. \tag{8.5}$$

The possibility of controlling the value of the equivalent resistance is fulfilled using the dependence of the equivalent transconductance, G_m on the biasing current, I_O.

8.1.1.3 Principle of Operation of an Active Resistor with Negative Equivalent Resistance

The replacing of the input–output connections from Fig. 8.3 with two input–output cross-connections will change the sign of the equivalent resistance for the structure presented in Fig. 8.4 [1].

$$R_{ECH} = \frac{V_1 - V_2}{I_{OUT2} - I_{OUT1}} = -\frac{1}{G_m} = -\frac{1}{2a\sqrt{KI_O}}. \tag{8.6}$$

Fig. 8.4 Active resistor with
negative equivalent resistance
based on PR 8.1 – block
diagram

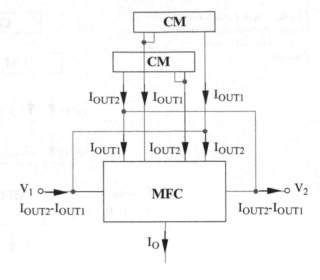

Fig. 8.5 Squaring circuit (1)
based on PR 8.1 – block
diagram

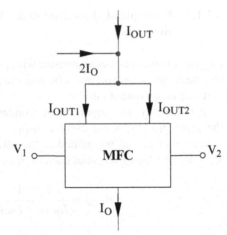

8.1.1.4 Principle of Operation of a Voltage Squaring Circuit

The output current of the squaring circuit presented in Fig. 8.5 [2] can be obtained as a
linear relation, containing the sum of the individual output currents, I_{OUT1} and I_{OUT2}:

$$I_{OUT} = I_{OUT1} + I_{OUT2} - 2I_O = 2bK(V_1 - V_2)^2. \tag{8.7}$$

8.1.1.5 Principle of Operation of a Multiplier Circuit (First Method)

For implementing a voltage multiplier circuit, using the multifunctional core
presented in Fig. 8.1, a possible method consists in the replacing of the I_O constant

Fig. 8.6 Multiplier circuit
(1) based on PR 8.1 – block
diagram

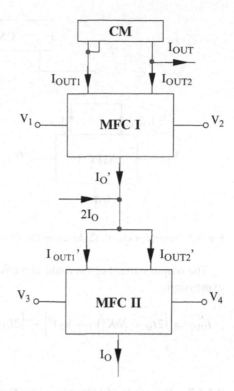

biasing current with a current, $I_O{}'$, representing the output current of a squaring circuit from Fig. 8.5, having as input the differential voltage $V_3 - V_4$ (Fig. 8.6):

$$I_O{}' = I_{OUT1}{}' + I_{OUT2}{}' - 2I_O = 2bK(V_3 - V_4)^2. \qquad (8.8)$$

It results:

$$\begin{aligned} I_{OUT} = I_{OUT1} - I_{OUT2} &= 2a\sqrt{KI_O{}'}\,(V_1 - V_2) \\ &= 2aK\sqrt{2b}\,(V_1 - V_2)(V_3 - V_4). \end{aligned} \qquad (8.9)$$

8.1.1.6 Principle of Operation of a Multiplier Circuit (Second Method)

In order to obtain the multiplication function using two squaring circuits from Fig. 8.5, a possible method is presented in Fig. 8.7 [2] (the consideration of the difference between the output currents of two squaring circuits, the first circuit having as input potentials, V_1 and $-V_2$, while the second-one, V_1 and V_2).

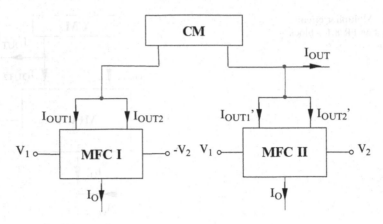

Fig. 8.7 Multiplier circuit (2) based on PR 8.1 – block diagram

The output current of the multiplier circuit from Fig. 8.7 will have the following expression:

$$I_{OUT} = \left[2I_O + 2bK(V_1 + V_2)^2\right] - \left[2I_O + 2bK(V_1 - V_2)^2\right] = 8bKV_1V_2. \quad (8.10)$$

8.1.1.7 Principle of Operation of a Square-Root Circuit

A possible method for obtaining the square-root function using the squaring circuit from Fig. 8.5 is presented in Fig. 8.8 [2]. The input potentials, V_1 and V_2, are obtained using four gate-drain connected MOS transistors, biased at I_1 and I_2 input currents.

The sum of the output currents of the multifunctional core is:

$$I_{OUT1} + I_{OUT2} = 2I_O + 2bK(V_1 - V_2)^2. \quad (8.11)$$

The $V_1 - V_2$ differential input voltage can be expressed as a function of the gate-source voltages as follows:

$$V_1 - V_2 = 2V_{GS}(I_1) - 2V_{GS}(I_2). \quad (8.12)$$

Replacing (8.12) in (8.11) and using the square-root dependence of the drain current on its gate-source voltage for a MOS transistor biased in saturation, it results:

$$I_{OUT1} + I_{OUT2} = 2I_O + 2bK\left(2\sqrt{\frac{2I_1}{K}} - 2\sqrt{\frac{2I_2}{K}}\right)^2, \quad (8.13)$$

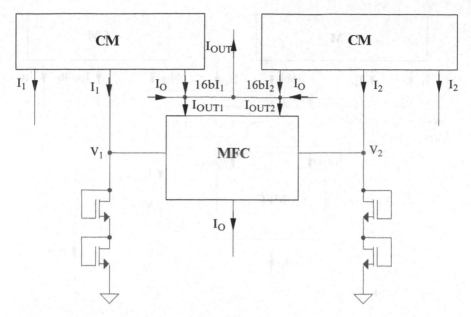

Fig. 8.8 Square-root circuit based on PR 8.1 – block diagram

equivalent with:

$$I_{OUT1} + I_{OUT2} = 2I_O + 16bI_1 + 16bI_2 - 32b\sqrt{I_1 I_2}. \tag{8.14}$$

The output current can be expressed using a linear relation between the currents from the circuit:

$$I_{OUT} = 16bI_1 + 16bI_2 + 2I_O - (I_{OUT1} + I_{OUT2}), \tag{8.15}$$

resulting a square-root dependence of the output current, I_{OUT}, on the input currents, I_1 and I_2:

$$I_{OUT} = 32b\sqrt{I_1 I_2}. \tag{8.16}$$

8.1.1.8   Principle of Operation of a Current Squaring Circuit

The method for obtaining the current squaring function is based on the modifying of the previous square-root circuit by changing the positions of current mirrors from Fig. 8.8 (Fig. 8.9) [2].

Similar with the previous circuit, the sum of the output currents, I_{OUT1} and I_{OUT2}, is:

$$I_{OUT1} + I_{OUT2} = 2I_O + 16bI_1 + 16bI_2 - 32b\sqrt{I_1 I_2}. \tag{8.17}$$

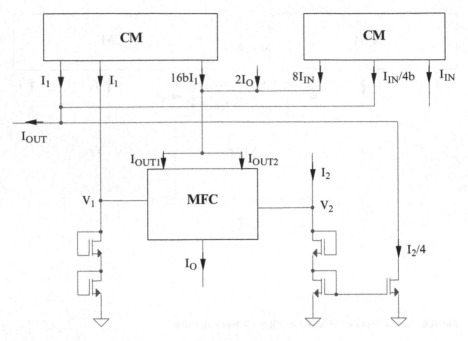

Fig. 8.9 Squaring circuit (2) based on PR 8.1 – block diagram

The current mirrors and the connections from Fig. 8.9 implement the following relation between the currents from the circuit:

$$I_{OUT1} + I_{OUT2} = 16bI_1 + 2I_O + 8I_{IN}. \tag{8.18}$$

From the previous relations, it results:

$$I_1 = \frac{(2bI_2 - I_{IN})^2}{16b^2 I_2} = \frac{I_2}{4} - \frac{I_{IN}}{4b} + \frac{I_{IN}^2}{16b^2 I_2}. \tag{8.19}$$

The expression of the output current of the current squaring circuit will be:

$$I_{OUT} = I_1 + \frac{I_{IN}}{4b} - \frac{I_2}{4} = \frac{I_{IN}^2}{16b^2 I_2}, \tag{8.20}$$

where I_{IN} is considered to be the input current and I_2 represents the reference current. For simplicity, the I_2 current can be considered to be equal with the reference current, I_O, that biases the differential amplifier. Thus:

$$I_{OUT} = \frac{I_{IN}^2}{16b^2 I_O}. \tag{8.21}$$

Fig. 8.10 MFC core based on PR 8.2

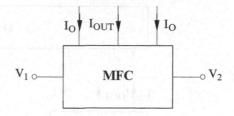

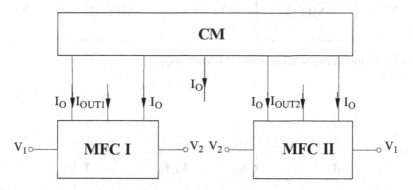

Fig. 8.11 Differential amplifier based on PR 8.2 – block diagram

8.1.2 Second Mathematical Principle (PR 8.2)

The output current of the multifunctional circuit core (Fig. 8.10) has the following expression:

$$I_{OUT} = -\sqrt{2KI_O}(V_1 - V_2) + \frac{K}{2}(V_1 - V_2)^2. \tag{8.22}$$

8.1.2.1 Principle of Operation of a Linear Differential Amplifier

The block diagram of the differential amplifier using the multifunctional core from Fig. 8.10 is shown in Fig. 8.11 [3].

Using the previous relation for the multifunctional cores from Fig. 8.11, it results:

$$I_{OUT1} = -\sqrt{2KI_O}(V_1 - V_2) + \frac{K}{2}(V_1 - V_2)^2, \tag{8.23}$$

$$I_{OUT2} = \sqrt{2KI_O}(V_1 - V_2) + \frac{K}{2}(V_1 - V_2)^2, \tag{8.24}$$

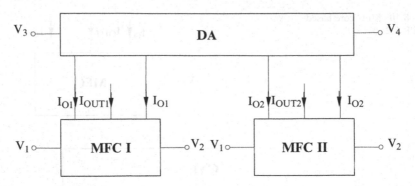

Fig. 8.12 Multiplier circuit based on PR 8.2 – block diagram

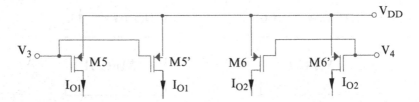

Fig. 8.13 Implementation of DA block

I_O being the reference current. For obtaining the amplifying function with theoretical null distortions, it is necessary to consider the difference of the previous output currents:

$$I_{OUT2} - I_{OUT1} = 2\sqrt{2KI_O}(V_1 - V_2). \tag{8.25}$$

8.1.2.2 Principle of Operation of a Multiplier Circuit

In order to implement the multiplying function, the first linear dependent on the differential input voltage term from relation (8.22) is used. The block diagram of the multiplier circuit, based on the second mathematical principle is presented in Fig. 8.12 [3].

The method for removing the last quadratic term from (8.22) consists in the utilization of two identical multifunctional cores from Fig. 8.1, having identical differential input voltage, but different biasing currents, I_{O1} and I_{O2} (MFC 1 and MFC 2 in Fig. 8.12). As the quadratic term does not depend on I_{O1} and I_{O2} currents, the consideration of the difference between the output currents, I_{OUT1} and I_{OUT2}, will cancel out this undesired term.

The I_{O1} and I_{O2} currents are generated by a classical differential amplifier, DA (implemented in Fig. 8.13 [3]), having $V_3 - V_4$ as differential input voltage. In order to obtain a double current of the output of this differential amplifier, its practical implementation is realized using a parallel connection of two identical classical differential amplifiers.

The expressions of the output currents are:

$$I_{OUT1} = -\sqrt{2KI_{O1}}(V_1 - V_2) + \frac{K}{2}(V_1 - V_2)^2, \tag{8.26}$$

$$I_{OUT2} = -\sqrt{2KI_{O2}}(V_1 - V_2) + \frac{K}{2}(V_1 - V_2)^2. \tag{8.27}$$

The difference between the output currents, I_{OUT1} and I_{OUT2}, will be:

$$I_{OUT} = I_{OUT1} - I_{OUT2} = \sqrt{2K}\left(\sqrt{I_{O2}} - \sqrt{I_{O1}}\right)(V_1 - V_2). \tag{8.28}$$

For the circuit presented in Fig. 8.13, considering that its composing transistors are biased in saturation, it can write that:

$$\sqrt{I_{O2}} - \sqrt{I_{O1}} = \sqrt{\frac{K}{2}}(V_3 - V_4). \tag{8.29}$$

Replacing (8.29) in (8.28), it results that the circuit proposed in Fig. 8.12 implements the multiplying function:

$$I_{OUT} = K(V_1 - V_2)(V_3 - V_4). \tag{8.30}$$

8.1.2.3   Principle of Operation of a Squaring Circuit

The proposed method for implementing the squaring function is derived from the realization of the differential amplifier circuit with increased linearity, presented in Fig. 8.11. In order to obtain an output current proportional with the square of the differential input voltage, the second term from (8.22) must be used. Practically, the sum of I_{OUT1} and I_{OUT2} output currents from Fig. 8.11 will contain only the quadratic term:

$$I_{OUT1} + I_{OUT2} = K(V_1 - V_2)^2. \tag{8.31}$$

8.1.3 Third Mathematical Principle (PR 8.3)

The third mathematical principle can be written as follows:

$$I_{OUT} = \frac{V_1 - V_2}{2} \sqrt{4K \left[I_O + \frac{K}{4}(V_1 - V_2)^2 \right] - K^2(V_1 - V_2)^2} = \sqrt{KI_O}(V_1 - V_2).$$

(8.32)

8.1.4 Fourth Mathematical Principle (PR 8.4)

8.1.4.1 Superior-Order Approximation of a Continuous Mathematical Function

A possible method for obtaining any continuous function using current squaring circuits consists in the consideration of the superior-order approximation of this function using the limited Taylor series expansion. The input variable is represented by the ratio between the input current and the reference current, $x = I_{IN}/I_O$.

A $f(x)$ continuous function can be expand in Taylor series as follows:

$$f(x) = f(x)|_{x=0} + \sum_{k=1}^{\infty} \frac{1}{k!} f^{(k)}(x) \bigg|_{x=0} x^k,$$

(8.33)

$f^{(k)}(x)$ being the kth order derivate of $f(x)$ function. The previous relation is equivalent with a polynomial expression of $f(x)$ function with constant coefficients a_k:

$$f(x) = a_0 + \sum_{k=1}^{\infty} a_k x^k,$$

(8.34)

where:

$$a_O = f(x)|_{x=0}$$

(8.35)

and:

$$a_k = \frac{f^{(k)}(x)}{k!} \bigg|_{x=0}.$$

(8.36)

Any continuous function can be approximated using a nth order limited Taylor expansion, the approximation error being proportional with the number of the neglected terms:

$$f(x) \cong a_O + \sum_{k=1}^{n} a_k x^k.$$ (8.37)

Because x variable is non-dimensional, the current-mode evaluation of $f(x)$ function can be made by generating a current $I_{OUT}(x) = I_O f(x)$:

$$I_{OUT}(x) \cong a_O I_O + I_O \sum_{k=1}^{n} a_k \left(\frac{I_{IN}}{I_O}\right)^k.$$ (8.38)

The block diagram of a function generator circuit is presented in Fig. 8.14 [4–7], consisting in $n - 1$ identical current squaring circuits for a n - th order polynomial series expansion of $f(x)$ function and in a block that computes a_k coefficients for $k = 0, 1, ..., n$. The advantage of this implementation is that a very good precision of the circuit can be achieved by increasing the value of n.

The implemented currents using the previous circuits are:

$$I_{OUT(0)} = I_{IN},$$ (8.39)

$$I_{OUT(1)} = \frac{I_{IN}^2}{I_O},$$ (8.40)

$$I_{OUT(2)} = \frac{I_{OUT(1)}^2}{I_{IN}} = \frac{I_{IN}^3}{I_O^2},$$ (8.41)

$$I_{OUT(n)} = \frac{I_{OUT(n-1)}^2}{I_{OUT(n-2)}} = \frac{I_{IN}^{n+1}}{I_O^n}.$$ (8.42)

The "a_k" block is implemented using simple and multiple current mirrors and must be able to compute the a_k coefficients from (8.35) and (8.36). The output current of this block, $I_{OUT}(x)$, will be proportional with the superior-order approximated function, $f(x)$. In order to increase the circuit accuracy, the number of squaring blocks from Fig. 8.14 can be increased. So, a compromise between the circuit complexity and its precision must be made. The circuit accuracy is also increased because of the independence of the output current on technological parameters.

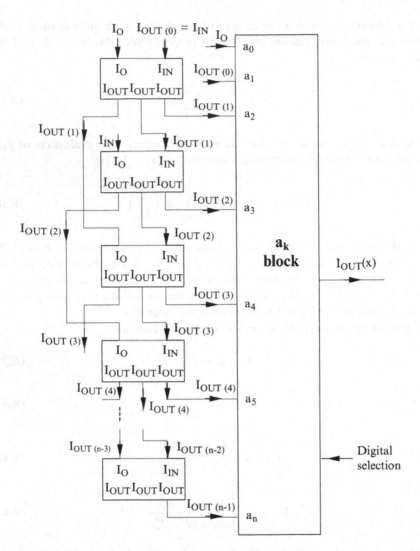

Fig. 8.14 Function generator circuit based on PR 8.2 – block diagram

8.1.4.2 Second-Order Approximation of a Continuous Mathematical Function

Another possible implementation of the same mathematical principle is also based on the approximation of any continuous mathematical function by its superior-order limited Taylor series expansion:

$$f(x) = a_O + a_1 x + a_2 x^2 + a_3 x^3 + \cdots, \tag{8.43}$$

Table 8.1 Coefficients of usual functions for second-order Taylor approximation

Function	a_0	a_1	a_2
$\exp(x)$	1	1	1/2
$\cos(x)$	1	0	$-1/2$
$\cosh(x)$	1	0	1/2
$(1-x)^{-1}$	1	1	1
$(1+x)^{1/2}$	1	1/2	$-1/8$
$(1-x)^{1/2}$	1	$-1/2$	$-1/8$
$(1+x)^{1/3}$	1	1/3	$-1/9$
$(1-x)^{1/3}$	1	$-1/3$	$-1/9$
$(1+x)^{1/4}$	1	1/4	$-3/32$
$(1-x)^{1/4}$	1	$-1/4$	$-3/32$
$(1+x)^{-2}$	1	-2	3
$(1-x)^{-2}$	1	-2	3
$\ln(1-x)$	0	-1	$-1/2$
$\ln(1+x)$	0	1	$-1/2$
$(1+x)^2$	1	2	1
$(1-x)^2$	1	-2	1

where $x = I_{IN}/I_O$ (the ratio of the input current and the reference current).

For some usual continuous mathematical functions, the values of a_0, a_1 and a_2 coefficients used for a second-order approximation are centralized in Table 8.1.

8.1.4.3 Third-Order Approximation of a Continuous Mathematical Function

A $g(x)$ function that can approximate many $f(x)$ continuous mathematical functions could have the following expression:

$$g(x) = \frac{a_1 x}{1 + a_2 x} + a_3 x + a_4,\qquad(8.44)$$

the a_k constants having the following expressions: $a_1 = p^2/q$, $a_2 = -q/p$, $a_3 = n - p^2/q$, $a_4 = m$. In order to evaluate the capability of $g(x)$ function to approximate a $f(x)$ continuous function, the Taylor series expansion for $g(x)$ must be determined:

$$g'(x) = \frac{a_1}{(1 + a_2 x)^2} + a_3,\qquad(8.45)$$

$$g''(x) = \frac{-2a_1 a_2}{(1 + a_2 x)^3},\qquad(8.46)$$

$$g'''(x) = \frac{6a_1 a_2^2}{(1 + a_2 x)^4},\qquad(8.47)$$

$$g''''(x) = \frac{-24a_1 a_2^3}{(1 + a_2 x)^5},$$ (8.48)

resulting:

$$g(x) = m + nx + px^2 + qx^3 + \frac{q^2}{p}x^4 + \cdots.$$ (8.49)

So, the $g(x)$ function third-order approximates a continuous $f(x)$ function, having the Taylor series expansion expressed by:

$$f(x) = m + nx + px^2 + qx^3 + rx^4 + \cdots.$$ (8.50)

The approximation error is mainly caused by the fourth-order terms from the previous expansions:

$$\varepsilon_{f(x)}^{g(x)}(x) \cong \frac{(r + a_1 a_2^3)x^4}{f(x)}.$$ (8.51)

The expressions of $g(x)$ functions and of the approximation errors for 13 usual mathematical functions are presented in Appendix 2.

8.1.4.4 Third-Order Approximation of a Continuous Mathematical Function Using Two Primitive Functions

A method for obtaining a multitude of continuous mathematical functions using their third-order Taylor series expansion is to use two primitive functions and to express the approximation of each required mathematical function as a linear combination of these primitives. The continuous function that will be implemented is noted with $f(x)$, its third-order approximation – with $g(x)$, while the primitive functions are noted with $f_1(x)$ and $f_2(x)$. The approximation function $g(x)$ can be expressed as follows:

$$g(x) = af_1(x) + bf_2(x) + cx + d,$$ (8.52)

a, b, c and d being constant coefficients associated with each implemented function $f(x)$. The following analysis will be made for a particular choosing of these primitive functions that generate relatively simple mathematical relations and reasonable values of the coefficients:

$$f_1(x) = \frac{1}{1 - x}$$ (8.53)

and:

$$f_2(x) = \frac{1}{2-x}. \tag{8.54}$$

The Taylor series expansions of $f_1(x)$ and $f_2(x)$ are:

$$f_1(x) = 1 + x + x^2 + x^3 + x^4 + \cdots \tag{8.55}$$

and:

$$f_2(x) = \frac{1}{2} + \frac{x}{2^2} + \frac{x^2}{2^3} + \frac{x^3}{2^4} + \frac{x^4}{2^5} + \cdots. \tag{8.56}$$

So, the function $g(x)$ can be expressed as follows:

$$g(x) = \left(a + \frac{b}{2} + d\right) + \left(a + \frac{b}{4} + c\right)x + \left(a + \frac{b}{8}\right)x^2$$
$$+ \left(a + \frac{b}{16}\right)x^3 + \left(a + \frac{b}{32}\right)x^4 + \cdots. \tag{8.57}$$

Considering that the Taylor expansion of $f(x)$ function is:

$$f(x) = m + nx + px^2 + qx^3 + rx^4 + \cdots, \tag{8.58}$$

m, n, p, q and r being constant coefficients of the expansion (known, because $f(x)$ function is also known), the condition that $g(x)$ function must represent the third-order approximation of $f(x)$ function can be written as follows:

$$m = a + \frac{b}{2} + d, \tag{8.59}$$

$$n = a + \frac{b}{4} + c, \tag{8.60}$$

$$p = a + \frac{b}{8} \tag{8.61}$$

and:

$$q = a + \frac{b}{16}, \tag{8.62}$$

resulting:

$$a = 2q - p, \tag{8.63}$$

$$b = 16(p - q), \tag{8.64}$$

$$c = n + 2q - 3p \tag{8.65}$$

and:

$$d = m + 6q - 7p. \tag{8.66}$$

So, $g(x)$ function that third-order approximates $f(x)$ will have the following expression:

$$g(x) = (2q - p)\frac{1}{1 - x} + 16(p - q)\frac{1}{2 - x}$$
$$+ (n + 2q - 3p)x + (m + 6q - 7p). \tag{8.67}$$

The approximation error is mainly caused by the fourth-order terms from the previous expansions, (8.57) and (8.58):

$$\varepsilon_{f(x)}^{g(x)}(x) \cong \frac{\left(a + \frac{b}{32}\right) - r}{f(x)} x^4 = \frac{3q - p - r}{2f(x)} x^4. \tag{8.68}$$

Table 8.2 centralizes the values of constants m, n, p, q and r and also the expressions of the approximation errors (8.68) and of the approximation function $g(x)$ for the previous 11 usual continuous mathematical functions:

8.1.4.5 Fifth-Order Approximation of a Continuous Mathematical Function Using Four Primitive Functions

A similar method for obtaining a multitude of continuous mathematical functions using their fifth-order Taylor series expansion is to use four primitive functions and to express the approximation of each required mathematical function as a linear combination of these primitives. The continuous function that will be implemented is noted with $f(x)$, its fifth-order approximation – with $g(x)$, while the primitive functions are noted with $f_1(x), f_2(x), f_3(x)$ and $f_4(x)$. The approximation function $g(x)$ can be expressed as follows:

$$g(x) = a_1 f_1(x) + a_2 f_2(x) + a_3 f_3(x) + a_4 f_4(x) + a_5 x + a_6, \tag{8.69}$$

a_1, a_2, a_3, a_4, a_5 and a_6 being constant coefficients associated with each implemented function $f(x)$. The following analysis will be made for a particular choosing of these primitive functions that generate relatively simple mathematical relations and reasonable values of the coefficients:

$$f_1(x) = \frac{1}{1 - x}, \tag{8.70}$$

Table 8.2 Coefficients, $\varepsilon(x)$ and $g(x)$ of eleven usual functions for third-order Taylor approximation using two primitive functions

$f(x)$	m	n	p	q	r	ε	$g(x)$
$\exp(x)$	1	1	$\frac{1}{2}$	$\frac{1}{6}$	$\frac{1}{24}$	$\dfrac{x^4}{48\exp(x)}$	$-\dfrac{1}{6}\dfrac{1}{1-x}+\dfrac{16}{3}\dfrac{1}{2-x}-\dfrac{x}{6}-\dfrac{3}{2}$
$\sqrt{1+x}$	1	$\frac{1}{2}$	$-\frac{1}{8}$	$\frac{1}{16}$	$-\frac{5}{128}$	$\dfrac{45x^4}{256\sqrt{1+x}}$	$\dfrac{1}{4}\dfrac{1}{1-x}-\dfrac{1}{2}\dfrac{1}{2-x}+x+\dfrac{5}{2}$
$\sqrt{1-x}$	1	$-\frac{1}{2}$	$-\frac{1}{8}$	$-\frac{1}{16}$	$-\frac{5}{128}$	$\dfrac{3x^4}{256\sqrt{1-x}}$	$\dfrac{1}{2}\dfrac{1}{2-x}-\dfrac{x}{4}+\dfrac{3}{2}$
$\sqrt[3]{1+x}$	1	$\frac{1}{3}$	$-\frac{1}{9}$	$\frac{5}{81}$	$-\frac{10}{243}$	$\dfrac{41x^4}{243\sqrt[3]{1+x}}$	$\dfrac{19}{81}\dfrac{1}{1-x}-\dfrac{224}{81}\dfrac{1}{2-x}+\dfrac{64x}{81}+\dfrac{58}{27}$
$\sqrt[3]{1-x}$	1	$-\frac{1}{3}$	$-\frac{1}{9}$	$-\frac{5}{81}$	$\frac{10}{243}$	$\dfrac{4x^4}{243\sqrt[3]{1-x}}$	$\dfrac{1}{81}\dfrac{1}{1-x}-\dfrac{64}{81}\dfrac{1}{2-x}-\dfrac{10x}{81}+\dfrac{38}{27}$
$\sqrt[4]{1+x}$	1	$\frac{1}{4}$	$-\frac{3}{32}$	$\frac{7}{128}$	$-\frac{77}{2048}$	$\dfrac{605x^4}{4096\sqrt[4]{1+x}}$	$\dfrac{13}{64}\dfrac{1}{1-x}-\dfrac{19}{64}\dfrac{1}{2-x}+\dfrac{41x}{64}+\dfrac{127}{64}$
$\sqrt[4]{1-x}$	1	$-\frac{1}{4}$	$-\frac{3}{32}$	$-\frac{7}{128}$	$-\frac{77}{2048}$	$\dfrac{67x^4}{4096\sqrt[4]{1-x}}$	$\dfrac{1}{64}\dfrac{1}{1-x}-\dfrac{5}{64}\dfrac{1}{2-x}-\dfrac{5x}{64}+\dfrac{85}{64}$
$\dfrac{1}{(1+x)^2}$	1	-2	3	-4	5	$10x^4(1+x)^2$	$-\dfrac{11}{112}\dfrac{1}{1-x}+\dfrac{1}{112}\dfrac{1}{2-x}-19x-44$

(continued)

Table 8.2 (continued)

$f(x)$	m	n	p	q	r	ε	$g(x)$
$\dfrac{1}{(1-x)^2}$	1	2	3	4	5	$2x^4(1-x)^2$	$\dfrac{5}{1-x}-\dfrac{16}{2-x}+x+4$
$\ln(1+x)$	0	1	$-\dfrac{1}{2}$	$\dfrac{1}{3}$	$-\dfrac{1}{4}$	$\dfrac{7x^4}{8\ln(1+x)}$	$\dfrac{7}{6}\dfrac{1}{1-x}-\dfrac{40}{3}\dfrac{1}{2-x}\\+\dfrac{19x}{6}+\dfrac{11}{2}$
$\ln(1-x)$	0	-1	$-\dfrac{1}{2}$	$-\dfrac{1}{3}$	$-\dfrac{1}{4}$	$\dfrac{x^4}{8\ln(1-x)}$	$-\dfrac{1}{6}\dfrac{1}{1-x}-\dfrac{8}{3}\dfrac{1}{2-x}\\-\dfrac{x}{6}+\dfrac{3}{2}$

$$f_2(x) = \frac{1}{1+x}, \tag{8.71}$$

$$f_3(x) = \frac{1}{2-x}, \tag{8.72}$$

$$f_4(x) = \frac{1}{2+x}, \tag{8.73}$$

The Taylor series expansions of $f_1(x), f_2(x), f_3(x)$ and $f_4(x)$ are:

$$f_1(x) = 1 + x + x^2 + x^3 + x^4 + x^5 + x^6 + \cdots, \tag{8.74}$$

$$f_2(x) = 1 - x + x^2 - x^3 + x^4 - x^5 + x^6 - \cdots, \tag{8.75}$$

$$f_3(x) = \frac{1}{2} + \frac{x}{2^2} + \frac{x^2}{2^3} + \frac{x^3}{2^4} + \frac{x^4}{2^5} + \frac{x^5}{2^6} + \frac{x^6}{2^7} + \cdots, \tag{8.76}$$

$$f_4(x) = \frac{1}{2} - \frac{x}{2^2} + \frac{x^2}{2^3} - \frac{x^3}{2^4} + \frac{x^4}{2^5} - \frac{x^5}{2^6} + \frac{x^6}{2^7} + \cdots, \tag{8.77}$$

So, $g(x)$ function can be expressed as follows:

$$g(x) = \left(a_1 + a_2 + \frac{a_3 + a_4}{2} + a_6\right) + \left(a_1 - a_2 + \frac{a_3 - a_4}{4} + a_5\right)x$$
$$+ \left(a_1 + a_2 + \frac{a_3 + a_4}{8}\right)x^2 + \left(a_1 - a_2 + \frac{a_3 - a_4}{16}\right)x^3$$
$$+ \left(a_1 + a_2 + \frac{a_3 + a_4}{32}\right)x^4 + \left(a_1 - a_2 + \frac{a_3 - a_4}{64}\right)x^5$$
$$+ \left(a_1 + a_2 + \frac{a_3 + a_4}{128}\right)x^6 + \cdots. \tag{8.78}$$

Considering that the Taylor expansion of $f(x)$ function is:

$$f(x) = b_0 + b_1 x + b_2 x^2 + b_3 x^3 + b_4 x^4 + b_5 x^5 + b_6 x^6 + \cdots, \tag{8.79}$$

$b_0 - b_6$ being constant coefficients of the expansion (known, because $f(x)$ function is also known), the condition that $g(x)$ function must represent the fifth-order approximation of $f(x)$ function can be written as follows:

$$b_0 = a_1 + a_2 + \frac{a_3 + a_4}{2} + a_6, \tag{8.80}$$

$$b_1 = a_1 - a_2 + \frac{a_3 - a_4}{4} + a_5, \tag{8.81}$$

$$b_2 = a_1 + a_2 + \frac{a_3 + a_4}{8}, \tag{8.82}$$

$$b_3 = a_1 - a_2 + \frac{a_3 - a_4}{16}, \tag{8.83}$$

$$b_4 = a_1 + a_2 + \frac{a_3 + a_4}{32} \tag{8.84}$$

and:

$$b_5 = a_1 - a_2 + \frac{a_3 - a_4}{64}, \tag{8.85}$$

resulting:

$$a_1 = \frac{2}{3}(b_4 + b_5) - \frac{1}{6}(b_2 + b_3), \tag{8.86}$$

$$a_2 = \frac{1}{6}(b_3 - b_2) - \frac{2}{3}(b_5 - b_4), \tag{8.87}$$

$$a_3 = \frac{16}{3}(b_2 - b_4) + \frac{32}{3}(b_3 - b_5), \tag{8.88}$$

$$a_4 = \frac{16}{3}(b_2 - b_4) - \frac{32}{3}(b_3 - b_5), \tag{8.89}$$

$$a_5 = b_1 + 4b_5 - 5b_3 \tag{8.90}$$

and:

$$a_6 = b_0 + 4b_4 - 5b_2. \tag{8.91}$$

So, $g(x)$ function that fifth-order approximates $f(x)$ will have the following expression:

$$g(x) = \frac{a_1}{1 - x} + \frac{a_2}{1 + x} + \frac{a_3}{2 - x} + \frac{a_4}{2 + x} + a_5 x + a_6. \tag{8.92}$$

8.2 Analysis and Design of Multifunctional Structures

8.2.1 Multifunctional Structures Based on the First Mathematical Principle

8.2.1.1 First Implementation of PR 8.1

The first implementation of PR 8.1 is based on the multifunctional core presented in Fig. 8.15 [2]. This circuit uses the principle of the constant sum of gate-source voltages in order to obtain the required function. It presents the advantage of a symmetrical structure that strongly reduce the intrinsic nonlinearity of the entire circuit.

The output currents of the multifunctional core presented in Fig. 8.15 have the following expressions:

$$I_{OUT1} = \frac{K}{2}(V_O - V_T)^2 + K(V_O - V_T)(V_1 - V_2) + \frac{K}{2}(V_1 - V_2)^2 \qquad (8.93)$$

and:

$$I_{OUT2} = \frac{K}{2}(V_O - V_T)^2 - K(V_O - V_T)(V_1 - V_2) + \frac{K}{2}(V_1 - V_2)^2. \qquad (8.94)$$

The V_O sources are usually implemented as current-controlled voltage sources.

An example of a possible realization of V_O uses the gate-source voltages of MOS transistors biased in saturation region (V_{GS3} and V_{GS5} from Fig. 8.16) [8, 9]. Comparing with other circuits, the structure presented in Fig. 8.16 has an extremely high input impedance.

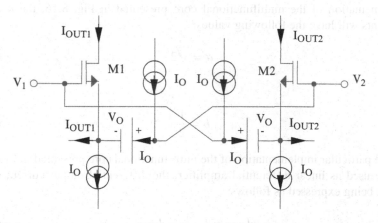

Fig. 8.15 First implementation of the MFC core based on PR 8.1 – principle circuit

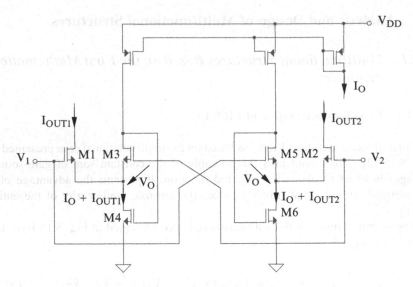

Fig. 8.16 MFC core based on PR 8.1 – complete implementation (1)

For this particular implementation of V_O sources, the expressions of I_{OUT1} and I_{OUT2} currents become:

$$I_{OUT1} = I_O + \sqrt{2KI_O}(V_1 - V_2) + \frac{K}{2}(V_1 - V_2)^2 \qquad (8.95)$$

and:

$$I_{OUT2} = I_O - \sqrt{2KI_O}(V_1 - V_2) + \frac{K}{2}(V_1 - V_2)^2. \qquad (8.96)$$

Comparing these relations with the general relations (8.1) and (8.2), for the implementation of the multifunctional core presented in Fig. 8.16, the a and b constants will have the following values:

$$a = \sqrt{2} \qquad (8.97)$$

and:

$$b = \frac{1}{2}. \qquad (8.98)$$

The particular implementation of the multifunctional core presented in Fig. 8.16 can be used as linear differential amplifier, the differential output current of the circuit being expressed as follows:

$$I_{OUT} = I_{OUT1} - I_{OUT2}. \qquad (8.99)$$

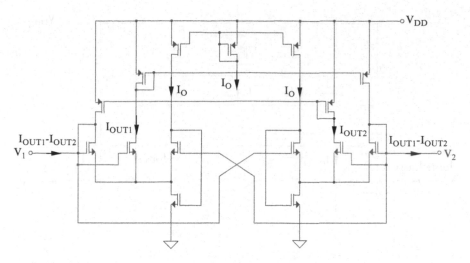

Fig. 8.17 Active resistor with positive equivalent resistance based on PR 8.1 – complete implementation (1)

So:

$$I_{OUT} = \sqrt{8KI_O}\,(V_1 - V_2). \tag{8.100}$$

resulting an equivalent transconductance of the circuit having the following expression:

$$G_m = \frac{I_{OUT}}{V_1 - V_2} = \sqrt{8KI_O}. \tag{8.101}$$

In order to obtain two active resistors having positive and negative equivalent resistances, the concrete implementation of the multifunctional core shown in Fig. 8.16 must be used in the blocks diagrams presented in Figs. 8.3 and 8.4, the complete realizations of the active resistor circuits being shown in Figs. 8.17 and 8.18. As a result on their excellent linearity and of their relative small complexity, the following structures find a multitude of applications in analog signal processing.

The equivalent resistances of the circuits presented in Figs. 8.17 and 8.18 are expressed by the following relations, respectively:

$$R_{ECH} = \frac{V_1 - V_2}{I_{OUT1} - I_{OUT2}} = \frac{1}{\sqrt{8KI_O}} \tag{8.102}$$

and:

$$R_{ECH} = \frac{V_1 - V_2}{I_{OUT2} - I_{OUT1}} = -\frac{1}{\sqrt{8KI_O}}. \tag{8.103}$$

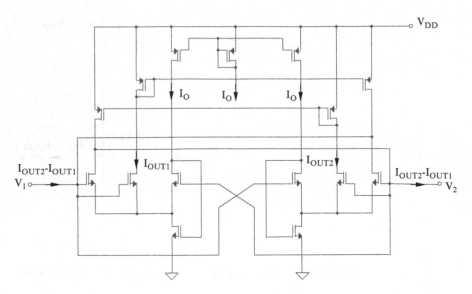

Fig. 8.18 Active resistor with negative equivalent resistance based on PR 8.1 – complete implementation (1)

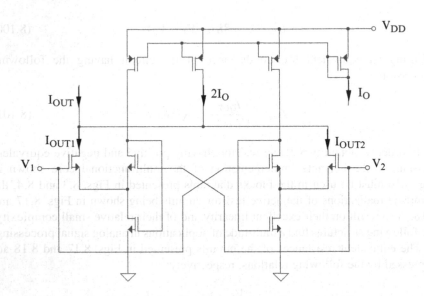

Fig. 8.19 Squaring circuit (1) based on PR 8.1 – complete implementation (1)

A squaring circuit can be obtained replacing in the block diagram presented in Fig. 8.5 the particular realization of the multifunctional core shown in Fig. 8.16, the implementation of the squaring circuit being presented in Fig. 8.19 [8, 9]. The design effort for this circuit is relatively small, as the only changing with respect to the MFC core from Fig. 8.16 is the consideration of the sum of its output currents.

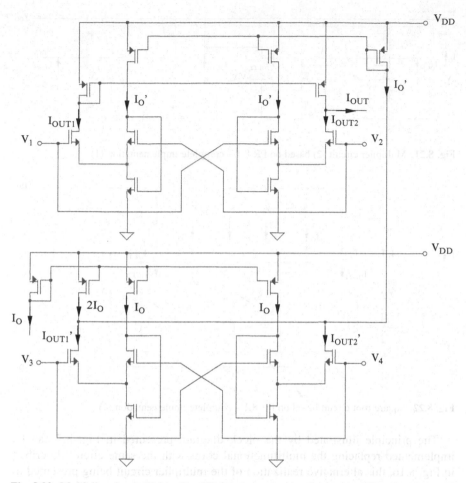

Fig. 8.20 Multiplier circuit (1) based on PR 8.1 – complete implementation (1)

The output current of the circuit from Fig. 8.19 is proportional with the square of the differential input voltage:

$$I_{OUT} = I_{OUT1} + I_{OUT2} - 2I_O = K(V_1 - V_2)^2. \qquad (8.104)$$

A multiplier circuit based on the block diagram presented in Fig. 8.6 can be obtained using the particular implementation of the multifunctional core shown in Fig. 8.16, the complete multiplier circuit being presented in Fig. 8.20. The structure can be used for applications that require differential input voltages.

The output current of the circuit from Fig. 8.20 is proportional with the product between the differential input voltages:

$$I_{OUT} = 2\sqrt{2}K(V_1 - V_2)(V_3 - V_4). \qquad (8.105)$$

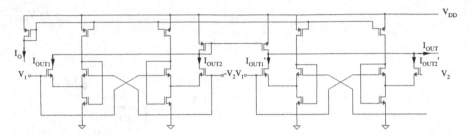

Fig. 8.21 Multiplier circuit (2) based on PR 8.1 – complete implementation (1)

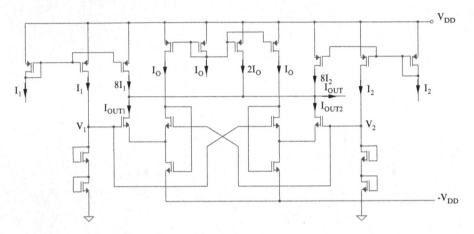

Fig. 8.22 Square-root circuit based on PR 8.1 – complete implementation (1)

The principle illustrated by the block diagram presented in Fig. 8.7 can be implemented replacing the multifunctional cores with the same circuit described in Fig. 8.16, this alternative realization of the multiplier circuit being presented in Fig. 8.21 [2].

The output current of the alternative realization of the multiplier circuit presented in Fig. 8.21 is proportional with the product between the input voltages, $I_{OUT} = 4KV_1V_2$.

A current-mode square-root circuit, having many applications in analog signal processing, can be obtained combining the block diagram from Fig. 8.8 with the multifunctional core from Fig. 8.16 (Fig. 8.22) [2].

The expression of the output current is:

$$I_{OUT} = 16\sqrt{I_1 I_2}. \tag{8.106}$$

The current squaring circuit based on the block diagram shown in Fig. 8.9 and on the multifunctional core from Fig. 8.16 is shown in Fig. 8.23 [2]. This circuit is useful for a current-mode signal processing, presenting relatively small errors as a result of the independence of the output variable on technological parameters.

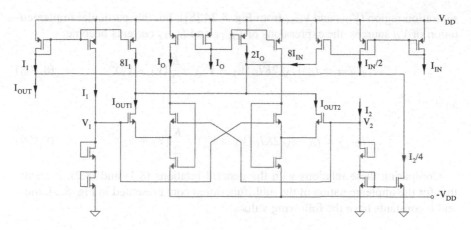

Fig. 8.23 Squaring circuit (2) based on PR 8.1 – complete implementation (1)

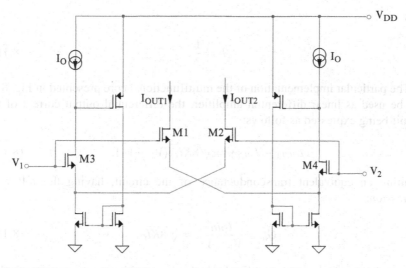

Fig. 8.24 MFC core based on PR 8.1 – complete implementation (2)

The expression of the output current is:

$$I_{OUT} = \frac{I_{IN}^2}{4I_O}.$$
(8.107)

An alternative realization of V_O voltage sources from Fig. 8.15, presenting the advantage of simplicity, uses gate-source voltages of MOS transistors biased in saturation region (V_{GS3} and V_{GS4} from Fig. 8.24 [8]). For this particular implementation of V_O sources, the expressions of I_{OUT1} and I_{OUT2} currents become:

$$I_{OUT1} = I_O + \sqrt{2KI_O}(V_1 - V_2) + \frac{K}{2}(V_1 - V_2)^2 \qquad (8.108)$$

and:

$$I_{OUT2} = I_O - \sqrt{2KI_O}(V_1 - V_2) + \frac{K}{2}(V_1 - V_2)^2. \qquad (8.109)$$

Comparing these relations with the general relations (8.1) and (8.2), it results that for the implementation of the multifunctional core presented in Fig. 8.24, the a and b constants have the following values:

$$a = \sqrt{2} \qquad (8.110)$$

and:

$$b = \frac{1}{2}. \qquad (8.111)$$

The particular implementation of the multifunctional core presented in Fig. 8.24 can be used as linear differential amplifier, the differential output current of the circuit being expressed as follows:

$$I_{OUT1} - I_{OUT2} = \sqrt{8KI_O}\,(V_1 - V_2), \qquad (8.112)$$

resulting an equivalent transconductance of the circuit, having the following expression:

$$G_m = \frac{I_{OUT}}{V_1 - V_2} = \sqrt{8KI_O}. \qquad (8.113)$$

In order to obtain two active resistors having positive or negative equivalent resistances, the concrete implementation of the multifunctional core shown in Fig. 8.24 must be used in the blocks diagrams presented in Figs. 8.3 and 8.4, the complete realizations of the active resistor circuits being shown in Figs. 8.25 and 8.26. The replacing of classical input–output connections (Fig. 8.25) with two input–output cross-connections (Fig. 8.26) offers the possibility of obtaining a negative resistance active resistor circuit, finding a multitude of applications in VLSI designs.

The equivalent resistances of the circuits presented in Figs. 8.25 and 8.26 are expressed by the following relations, respectively:

$$R_{ECH} = \frac{V_1 - V_2}{I_{OUT1} - I_{OUT2}} = \frac{1}{\sqrt{8KI_O}} \qquad (8.114)$$

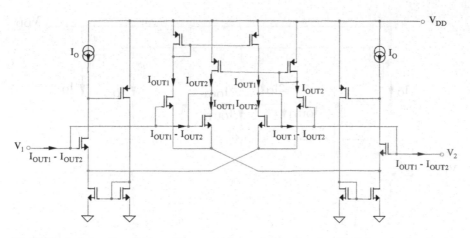

Fig. 8.25 Active resistor with positive equivalent resistance based on PR 8.1 – complete implementation (2)

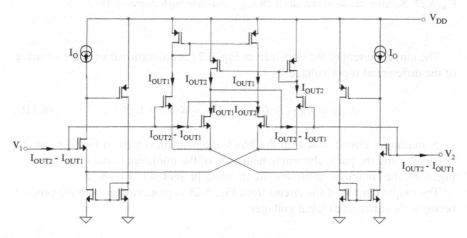

Fig. 8.26 Active resistor with negative equivalent resistance based on PR 8.1 – complete implementation (2)

and:

$$R_{ECH} = \frac{V_1 - V_2}{I_{OUT2} - I_{OUT1}} = -\frac{1}{\sqrt{8KI_O}}. \qquad (8.115)$$

A squaring circuit can be obtained replacing in the block diagram presented in Fig. 8.5 the particular realization of the multifunctional core shown in Fig. 8.24, the realization of the squaring circuit being presented in Fig. 8.27 [8, 9].

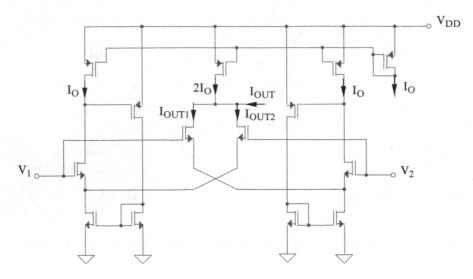

Fig. 8.27   Squaring circuit (1) based on PR 8.1 – complete implementation (2)

The output current of the circuit from Fig. 8.27 is proportional with the squaring of the differential input voltage:

$$I_{OUT} = I_{OUT1} + I_{OUT2} - 2I_O = K(V_1 - V_2)^2. \qquad (8.116)$$

A multiplier circuit based on the block diagram presented in Fig. 8.6 can be obtained using the particular implementation of the multifunctional core shown in Fig. 8.24, the complete multiplier circuit being presented in Fig. 8.28.

The output current of the circuit from Fig. 8.28 is proportional with the product between the differential input voltages:

$$I_{OUT} = 2\sqrt{2}K(V_1 - V_2)(V_3 - V_4). \qquad (8.117)$$

The principle illustrated by the block diagram presented in Fig. 8.7 can be implemented replacing the multifunctional cores with the same circuit described in Fig. 8.24, this alternative realization of the multiplier circuit being presented in Fig. 8.29 [2].

The output current of the alternative realization of the multiplier circuit presented in Fig. 8.29 is proportional with the product between the input voltages:

$$I_{OUT} = 4KV_1V_2. \qquad (8.118)$$

The square-root circuit obtained combining the block diagram from Fig. 8.8 and the multifunctional core from Fig. 8.24 is shown in Fig. 8.30 [2].

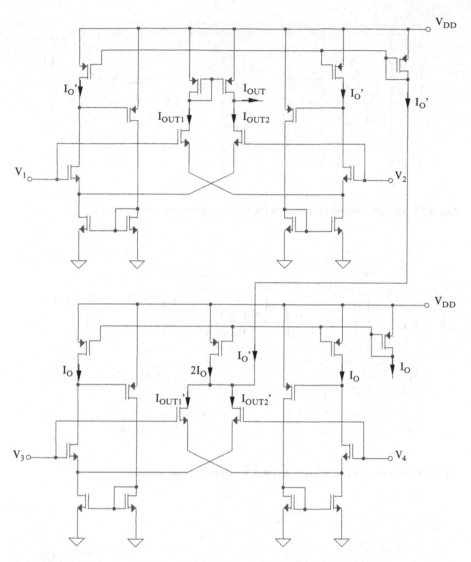

Fig. 8.28 Multiplier circuit (1) based on PR 8.1 – complete implementation (2)

The expression of the output current is:

$$I_{OUT} = 16\sqrt{I_1 I_2}. \tag{8.119}$$

The current squaring circuit based on the block diagram shown in Fig. 8.9 and on the multifunctional core from Fig. 8.24 is shown in Fig. 8.31 [2].

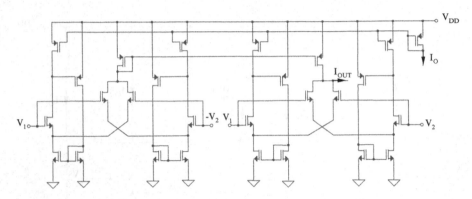

Fig. 8.29 Multiplier circuit (2) based on PR 8.1 – complete implementation (2)

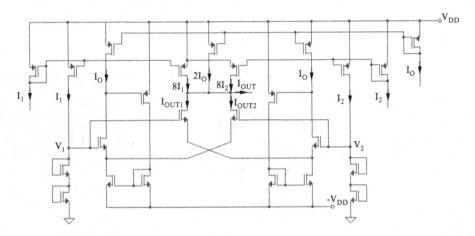

Fig. 8.30 Square-root circuit based on PR 8.1 – complete implementation (2)

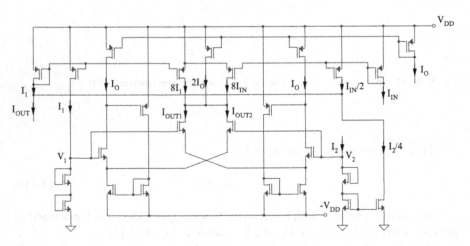

Fig. 8.31 Squaring circuit (2) based on PR 8.1 – complete implementation (2)

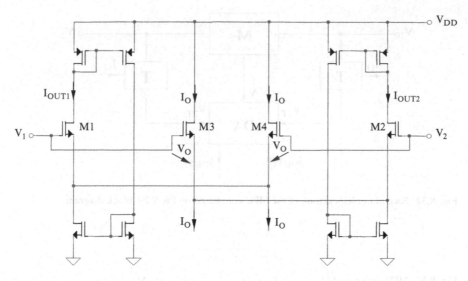

Fig. 8.32 MFC core based on PR 8.1 – complete implementation (3)

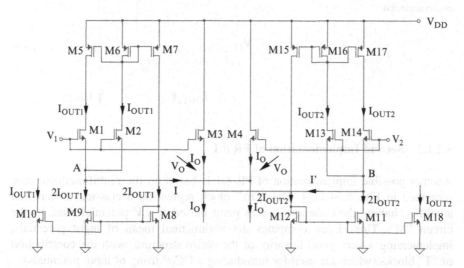

Fig. 8.33 MFC core based on PR 8.1 – complete implementation (4)

The expression of the output current is:

$$I_{OUT} = \frac{I_{IN}^2}{4I_O}. \tag{8.120}$$

Two other alternative implementations of the multifunctional core based on the same principle are presented in Fig. 8.32 [2] and Fig. 8.33 [10].

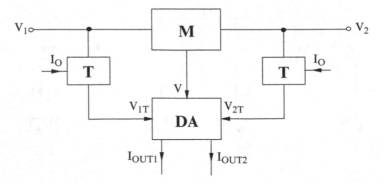

Fig. 8.34 Second implementation of the MFC core based on PR 8.1 – block diagram

Fig. 8.35 MFC core based
on PR 8.1 – DA block
implementation

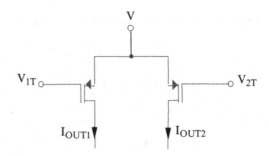

8.2.1.2 Second Implementation of PR 8.1

Another possible implementation of PR 8.1 is based on the multifunctional core
presented in Fig. 8.34 [11]. The "DA" block represents a classical differential
amplifier, having the common-sources point biased at a V potential fixed by the
circuit "M". This circuit computes the arithmetical mean of input potentials,
implementing a very good linearity of the entire structure, with the contribution
of "T" blocks (which are used for introducing a DC shifting of input potentials).

The "DA" block has the concrete realization presented in Fig. 8.35.

The DC shifting of the V potential could be obtained using "T" blocks, with an
implementation proposed in Fig. 8.36 [12, 13].

Because the same current I_O is passing through all transistors from Fig. 8.36, it
can write that:

$$V_1 = V_{1T} + V_T + \sqrt{\frac{2I_O}{K}} \tag{8.121}$$

Fig. 8.36 MFC core based
on PR 8.1 – T block
implementation

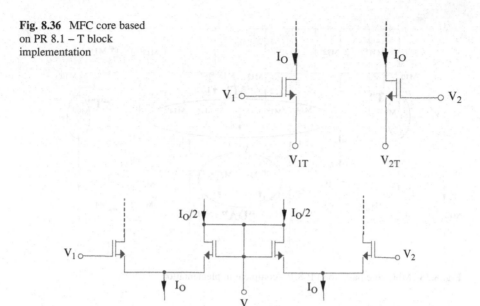

Fig. 8.37 MFC core based on PR 8.1 – M block implementation

and:

$$V_2 = V_{2T} + V_T + \sqrt{\frac{2I_O}{K}}. \tag{8.122}$$

So, both V_1 and V_2 input potentials are shifted with the same amount, $V_T + \sqrt{2I_O/K}$.

In order to obtain the arithmetic mean of input potentials, the circuit from Fig. 8.37 [12, 13] can be used.

$$V = \frac{V_1 + V_2}{2}. \tag{8.123}$$

The complete implementation of the multifunctional circuit is presented in Fig. 8.38 [11].

The expressions of I_{OUT1} and I_{OUT2} currents are:

$$I_{OUT1} = \frac{K}{2}(V - V_{1T} - V_T)^2 = \frac{K}{2}\left[V - \left(V_1 - V_T - \sqrt{\frac{2I_O}{K}}\right) - V_T\right]^2$$

$$= \frac{K}{2}\left(\frac{V_1 + V_2}{2} - V_1 + \sqrt{\frac{2I_O}{K}}\right)^2 = \frac{K}{2}\left(-\frac{V_1 - V_2}{2} + \sqrt{\frac{2I_O}{K}}\right)^2$$

$$= I_O - \sqrt{\frac{KI_O}{2}}(V_1 - V_2) + \frac{K}{8}(V_1 - V_2)^2 \tag{8.124}$$

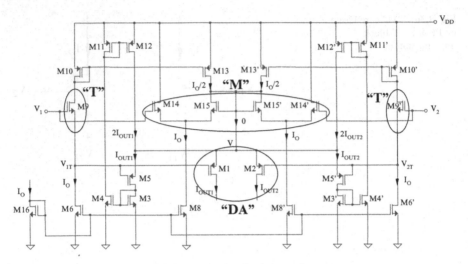

Fig. 8.38 MFC core based on PR 8.1 – complete implementation

and:

$$I_{OUT2} = I_O + \sqrt{\frac{KI_O}{2}}(V_1 - V_2) + \frac{K}{8}(V_1 - V_2)^2. \qquad (8.125)$$

Comparing the previous relations with (8.1) and (8.2), it results:

$$a = -\frac{1}{\sqrt{2}} \qquad (8.126)$$

and:

$$b = \frac{1}{8}. \qquad (8.127)$$

The implementation of a linear differential amplifier using the multifunctional core shown in Fig. 8.34 is identical with the structure presented in Fig. 8.38, the equivalent transconductance of the differential structure being defined as follows:

$$G_m = \frac{I_{OUT1} - I_{OUT2}}{V_1 - V_2} = -\sqrt{2KI_O}. \qquad (8.128)$$

In order to obtain two active resistors having positive and negative equivalent resistances, the concrete implementation of the multifunctional core shown in Fig. 8.38 must be used in the blocks diagrams presented in Figs. 8.3 and 8.4, the complete realizations of the active resistor circuits being shown in Fig. 8.39 [12, 13] and Fig. 8.40 [12, 13].

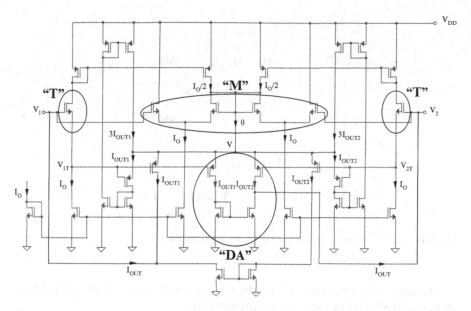

Fig. 8.39 Active resistor with positive equivalent resistance based on PR 8.1 – complete implementation

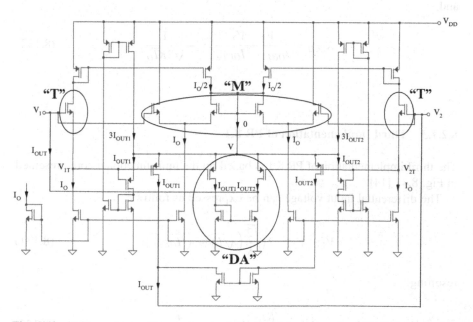

Fig. 8.40 Active resistor with negative equivalent resistance based on PR 8.1 – complete implementation

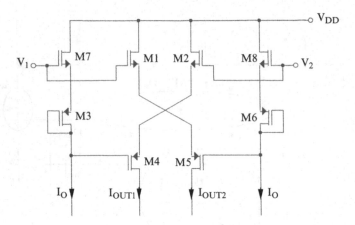

Fig. 8.41 Third implementation of the MFC core based on PR 8.1

The equivalent resistances of the circuits presented in Figs. 8.39 and 8.40 are expressed by the following relations, respectively:

$$R_{ECH} = \frac{V_1 - V_2}{I_{OUT2} - I_{OUT1}} = \frac{1}{\sqrt{2KI_O}} \qquad (8.129)$$

and:

$$R_{ECH} = \frac{V_1 - V_2}{I_{OUT1} - I_{OUT2}} = -\frac{1}{\sqrt{2KI_O}}. \qquad (8.130)$$

8.2.1.3 Third Implementation of PR 8.1

The third implementation of PR 8.1 is based on the multifunctional core presented in Fig. 8.41 [14].

The differential input voltage can be expressed as follows:

$$V_1 - V_2 = 2\sqrt{\frac{2}{K}}\left(\sqrt{I_O} - \sqrt{I_{OUT1}}\right), \qquad (8.131)$$

resulting:

$$I_{OUT1} = I_O - \sqrt{\frac{KI_O}{2}}(V_1 - V_2) + \frac{K}{8}(V_1 - V_2)^2 \qquad (8.132)$$

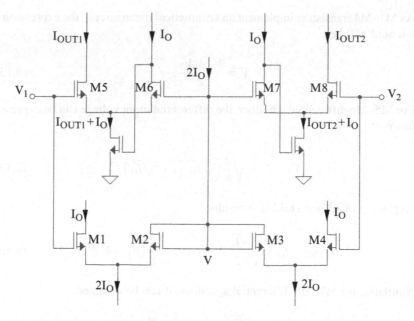

Fig. 8.42 Fourth implementation of the MFC core based on PR 8.1

and:

$$I_{OUT2} = I_O + \sqrt{\frac{KI_O}{2}}(V_1 - V_2) + \frac{K}{8}(V_1 - V_2)^2. \qquad (8.133)$$

Comparing the previous relations with (8.1) and (8.2), it results:

$$a = -\frac{1}{\sqrt{2}} \qquad (8.134)$$

and:

$$b = \frac{1}{8}. \qquad (8.135)$$

8.2.1.4 Fourth Implementation of PR 8.1

The fourth implementation of PR 8.1 is based on the multifunctional core presented in Fig. 8.42.

As M1–M4 transistors implement an arithmetical mean circuit, the expression of V potential will be:

$$V = \frac{V_1 + V_2}{2}.$$

(8.136)

For M5–M6 differential amplifier, the differential input voltage can be expressed as follows:

$$V_1 - V = \sqrt{\frac{2}{K}}\left(\sqrt{I_{OUT1}} - \sqrt{I_O}\right).$$

(8.137)

Replacing (8.136) in (8.137), it results:

$$I_{OUT1} = I_O + \sqrt{\frac{KI_O}{2}}(V_1 - V_2) + \frac{K}{8}(V_1 - V_2)^2.$$

(8.138)

Similarly, for M7–M8 differential amplifier, it can be obtained:

$$I_{OUT2} = I_O - \sqrt{\frac{KI_O}{2}}(V_1 - V_2) + \frac{K}{8}(V_1 - V_2)^2.$$

(8.139)

Comparing the previous relations with (8.1) and (8.2), it results:

$$a = \frac{1}{\sqrt{2}}$$

(8.140)

and:

$$b = \frac{1}{8}.$$

(8.141)

8.2.1.5 Fifth Implementation of PR 8.1

The fifth implementation of PR 8.1 is based on the multifunctional core presented in Fig. 8.43 [15].

For M1–M2 differential amplifier, the differential input voltage can be expressed as follows:

$$V_1 - V_2 = \sqrt{\frac{2}{K}}\left(\sqrt{I_{OUT1}} - \sqrt{I_O}\right).$$

(8.142)

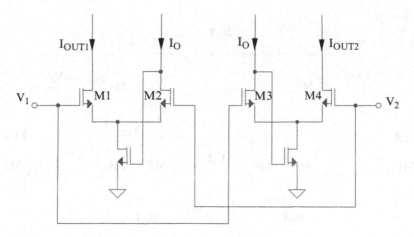

Fig. 8.43 Fifth implementation of the MFC core based on PR 8.1

So:

$$I_{OUT1} = I_O + \sqrt{2KI_O}(V_1 - V_2) + \frac{K}{2}(V_1 - V_2)^2. \qquad (8.143)$$

Similarly, for M3–M4 differential amplifier, the expression of I_{OUT2} current will be:

$$I_{OUT2} = I_O - \sqrt{2KI_O}(V_1 - V_2) + \frac{K}{2}(V_1 - V_2)^2. \qquad (8.144)$$

Comparing the previous relations with (8.1) and (8.2), it results:

$$a = \sqrt{2} \qquad (8.145)$$

and:

$$b = \frac{1}{2}. \qquad (8.146)$$

8.2.1.6 Sixth Implementation of PR 8.1

The sixth implementation of PR 8.1 is based on the multifunctional core presented in Fig. 8.44 [16].

It was demonstrated in Chap. 3 ("Squaring Circuits") – Fig. 3.17 that:

$$I_{OUT1} = I_O + \sqrt{2KI_O}(V_1 - V_2) + \frac{K}{2}(V_1 - V_2)^2 \qquad (8.147)$$

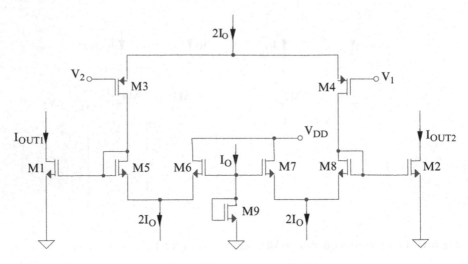

Fig. 8.44 Sixth implementation of the MFC core based on PR 8.1

and:

$$I_{OUT2} = I_O - \sqrt{2KI_O}(V_1 - V_2) + \frac{K}{2}(V_1 - V_2)^2. \qquad (8.148)$$

Comparing the previous relations with (8.1) and (8.2), it results:

$$a = \sqrt{2} \qquad (8.149)$$

and:

$$b = \frac{1}{2}. \qquad (8.150)$$

8.2.1.7   Seventh Implementation of PR 8.1

The seventh implementation of PR 8.1 is based on the multifunctional core presented in Fig. 8.45 [17].

The differential input voltage can be expressed as follows:

$$V_1 - V_2 = 2\sqrt{\frac{2}{K}}\left(\sqrt{I_O} - \sqrt{I_{OUT1}}\right), \qquad (8.151)$$

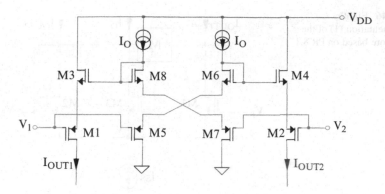

Fig. 8.45 Seventh implementation of the MFC core based on PR 8.1

resulting:

$$I_{OUT1} = I_O - \sqrt{\frac{KI_O}{2}}(V_1 - V_2) + \frac{K}{8}(V_1 - V_2)^2 \qquad (8.152)$$

and, similarly:

$$I_{OUT2} = I_O + \sqrt{\frac{KI_O}{2}}(V_1 - V_2) + \frac{K}{8}(V_1 - V_2)^2. \qquad (8.153)$$

Comparing the previous relations with (8.1) and (8.2), it results:

$$a = -\frac{1}{\sqrt{2}} \qquad (8.154)$$

and:

$$b = \frac{1}{8}. \qquad (8.155)$$

8.2.1.8   Eighth Implementation of PR 8.1

The eighth implementation of PR 8.1 uses the multifunctional core presented in Fig. 8.46 [18].

An alternative realization of the same circuit is presented in Fig. 8.47 [18] and it uses FGMOS transistors for implementing the arithmetical mean of the input potentials.

Fig. 8.46 Eighth
implementation (1) of the
MFC core based on PR 8.1

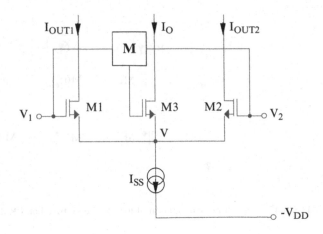

Fig. 8.47 Eighth (2)
implementation of the MFC
core based on PR 8.1

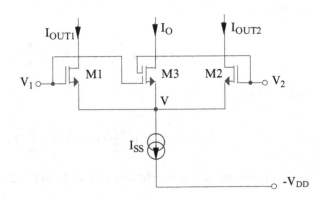

The expressions of I_{OUT1} and I_{OUT2} currents are:

$$I_{OUT1} = \frac{K}{2}(V_1 - V - V_T)^2 \tag{8.156}$$

and:

$$I_{OUT2} = \frac{K}{2}(V_2 - V - V_T)^2. \tag{8.157}$$

The drain current of M3 transistor can be expressed as follows:

$$I_O = \frac{K}{2}\left(\frac{V_1 + V_2}{2} - V - V_T\right)^2. \tag{8.158}$$

Thus:

$$V = \frac{V_1 + V_2}{2} - V_T - \sqrt{\frac{2I_O}{K}}.$$ (8.159)

It results:

$$I_{OUT1} = \frac{K}{2}\left(\frac{V_1 - V_2}{2} + \sqrt{\frac{2I_O}{K}}\right)^2$$ (8.160)

and:

$$I_{OUT1} = \frac{K}{2}\left(\frac{V_1 - V_2}{2} + \sqrt{\frac{2I_O}{K}}\right)^2,$$ (8.161)

or:

$$I_{OUT1} = I_O + \sqrt{\frac{KI_O}{2}}(V_1 - V_2) + \frac{K}{8}(V_1 - V_2)^2$$ (8.162)

and:

$$I_{OUT2} = I_O - \sqrt{\frac{KI_O}{2}}(V_1 - V_2) + \frac{K}{8}(V_1 - V_2)^2.$$ (8.163)

Comparing the previous relations with (8.1) and (8.2), it can be obtained:

$$a = \frac{1}{\sqrt{2}}$$ (8.164)

and:

$$b = \frac{1}{8}.$$ (8.165)

8.2.2 Multifunctional Structures Based on the Second Mathematical Principle

The implementation of the multifunctional core using the second mathematical principle is presented in Fig. 8.48 [3].

Fig. 8.48 MFC core based
on PR 8.2

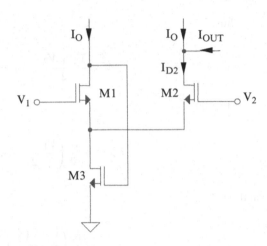

Replacing the square-root dependence of the gate-source voltage on the drain
current for a MOS transistor biased in saturation and considering identical
transistors, it results:

$$V_1 - V_2 = \sqrt{\frac{2I_O}{K}} - \sqrt{\frac{2I_{D2}}{K}}, \qquad (8.166)$$

equivalent with:

$$I_{D2} = I_O - \sqrt{2KI_O}(V_1 - V_2) + \frac{K}{2}(V_1 - V_2)^2. \qquad (8.167)$$

Thus, the output current of the multifunctional core presented in Fig. 8.48, I_{OUT},
will have the following expression:

$$I_{OUT} = I_{D2} - I_O = -\sqrt{2KI_O}(V_1 - V_2) + \frac{K}{2}(V_1 - V_2)^2. \qquad (8.168)$$

The complete implementation of a linear differential amplifier circuit based on
the second mathematical principle is presented in Fig. 8.11 [3]. MFC 1 from
Fig. 8.49 is realized using M3 and M4 transistors, while MFC 2 from the same
figure is composed from M1 and M2 transistors.

The complete circuit of the multiplier circuit based on the second mathematical
principle is presented in Fig. 8.50 [3].

The implementation in CMOS technology of the squaring circuit based on the
second mathematical principle is also presented in Fig. 8.49, the difference between
the differential amplifier and the squaring circuit being the consideration of the
difference, respectively the sum of the output currents.

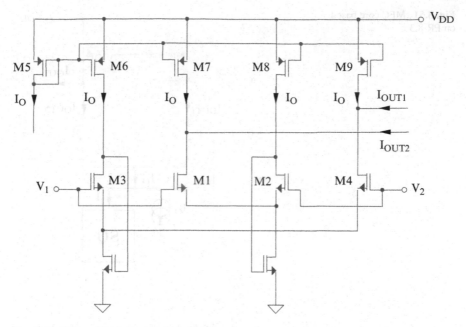

Fig. 8.49 Differential amplifier based on PR 8.2 – complete implementation

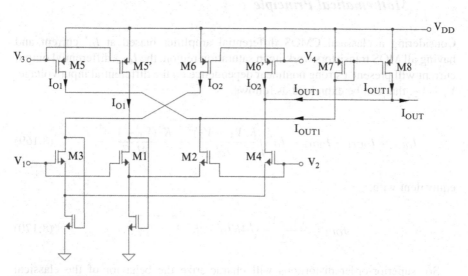

Fig. 8.50 Multiplier circuit based on PR 8.2 – complete implementation

Fig. 8.51 MFC core based
on PR 8.3

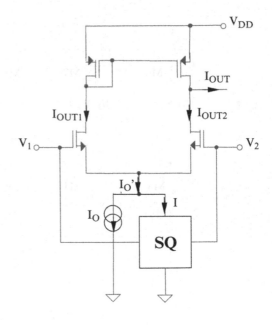

8.2.3 Multifunctional Structures Based on the Third Mathematical Principle

Considering a classical CMOS differential amplifier biased at I_O' current and having all MOS transistors working in saturation region, the I_{OUT} differential output current will present a strong nonlinear dependence on the differential input voltage, $V_1 - V_2$, that can be expressed as follows:

$$I_{OUT} = I_{OUT1} - I_{OUT2} = I_O'\sqrt{\frac{K(V_1 - V_2)^2}{I_O'} - \frac{K^2(V_1 - V_2)^4}{4I_O'^2}}, \qquad (8.169)$$

equivalent with:

$$I_{OUT} = \frac{V_1 - V_2}{2}\sqrt{4KI_O' - K^2(V_1 - V_2)^2}. \qquad (8.170)$$

So, superior-order distortions will characterize the behavior of the classical differential structure, imposing the design of a linearization technique for removing the superior-order terms from the transfer characteristic.

The method illustrated in Fig. 8.51 [19] (based on the third mathematical principle) for obtaining a linear transfer characteristic of the differential amplifier is to obtain the bias current, I_O', of the entire differential structure as a sum of a

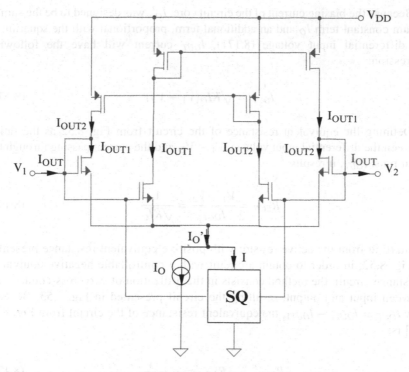

Fig. 8.52 Active resistor with positive equivalent resistance based on PR 8.3

main constant term, I_O and an additional term proportional with the square of the differential input voltage, $I = K(V_1 - V_2)^2/4$:

$$I_O' = I_O + I = I_O + \frac{K}{4}(V_1 - V_2)^2, \qquad (8.171)$$

resulting, in this case, a perfect linear behavior of the optimized differential amplifier:

$$I_{OUT} = I_{OUT1} - I_{OUT2} = \sqrt{KI_O}(V_1 - V_2) = G_m(V_1 - V_2), \qquad (8.172)$$

G_m being the equivalent transconductance of the proposed structure, that can be controlled by the biasing current, I_O.

The improved linearity differential amplifier presented in Fig. 8.51 can be reused in order to obtain a linear active resistor, having the circuit presented in Fig. 8.52 [19].

For this circuit, the current passing through the input pins, I_{OUT}, can be expressed as:

$$I_{OUT} = I_{OUT1} - I_{OUT2} = \frac{V_1 - V_2}{2}\sqrt{4KI_O' - K^2(V_1 - V_2)^2}. \qquad (8.173)$$

Because the biasing current of the circuit core, I_O', was designed to be the sum of a main constant term I_O and an additional term, proportional with the squaring of the differential input voltage (8.171), I_{OUT} current will have the following expression:

$$I_{OUT} = \sqrt{KI_O}(V_1 - V_2).\tag{8.174}$$

Defining the equivalent resistance of the circuit from Fig. 8.52 as the ration between the differential input voltage, $V_1 - V_2$, and the current passing through the input pins, I_{OUT}, it results:

$$R_{ECH} = \frac{V_1 - V_2}{I_{OUT}} = \frac{1}{\sqrt{KI_O}}.\tag{8.175}$$

Starting from the active resistor with positive equivalent resistance presented in Fig. 8.52, in order to obtain a circuit with a controllable negative equivalent resistance circuit, the method consists in the utilization of two cross-connections between input and output, resulting the circuit presented in Fig. 8.53. Because now $I_{OUT} = I_{OUT2} - I_{OUT1}$, the equivalent resistance of the circuit from Fig. 8.53 [19] is:

$$R_{ECH}' = -R_{ECH} = -\frac{1}{\sqrt{KI_O}}.\tag{8.176}$$

In order to obtain a voltage multiplier starting from the differential linearized structure presented in Fig. 8.51, a similar squaring circuit will be re-used for generating a current proportional with the squaring of another differential voltage, $V_3 - V_4$, resulting the circuit presented in Fig. 8.54 [19].

This current, named I_{OX}, will be further used for replacing the I_O constant current, which was biasing the differential amplifier presented in Fig. 8.51. Thus:

$$I_{OX} = \frac{K}{4}(V_3 - V_4)^2.\tag{8.177}$$

Replacing I_O from (8.174) with I_{OX} given by (8.177), it results that the circuit presented in Fig. 8.54 implements the multiplying function:

$$I_{OUT} = \frac{K}{2}(V_1 - V_2)(V_3 - V_4).\tag{8.178}$$

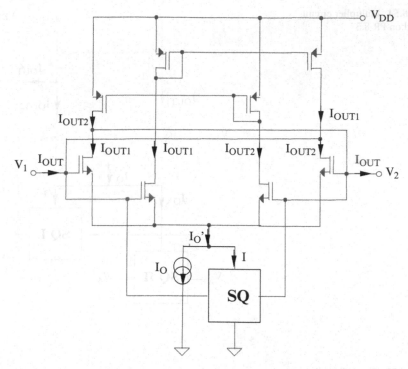

Fig. 8.53 Active resistor with negative equivalent resistance based on PR 8.3

8.2.4 Multifunctional Structures Based on the Fourth Mathematical Principle

8.2.4.1 Implementation of the Multifunctional Circuit Based on the Second-Order Approximation of a Continuous Mathematical Function

In order to obtain the second-order approximation of a function, a "C" coefficient block (Fig. 8.55) can be used, a_0, a_1 and a_2 coefficients corresponding to the Taylor series expansion (8.38).

The output current of the circuit presented in Fig. 8.55 will be:

$$I_{OUT} = a_0 I_O + a_1 I_{IN} + a_2 \frac{I_{IN}^2}{I_O}$$

$$= I_O \left[a_0 + a_1 \left(\frac{I_{IN}}{I_O} \right) + a_2 \left(\frac{I_{IN}}{I_O} \right)^2 \right] \cong I_O f \left(\frac{I_{IN}}{I_O} \right) = I_O f(x). \tag{8.179}$$

Fig. 8.54 Multiplier circuit
based on PR 8.3

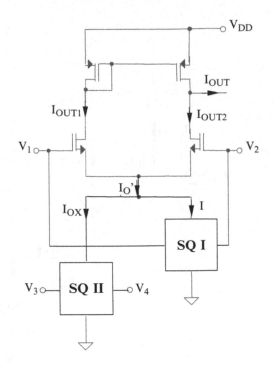

Fig. 8.55 The "C" block for
PR 8.4

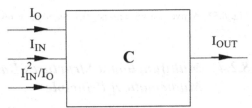

8.2.4.2   Implementation of the Multifunctional Circuit Based on the Third-Order Approximation of a Continuous Mathematical Function

The block diagram of the multifunctional circuit is presented in Fig. 8.56. The
MULT/DIV circuit has the implementation presented in Fig. 8.59.

The expression of I_{OUT}' current of MULT/DIV circuit is:

$$I_{OUT}' = I_0 \frac{I_1}{I_2}. \tag{8.180}$$

So:

$$I_{OUT}' = I_0 \frac{(p^2/q)I_{IN}}{I_0 - qI_{IN}/p} = I_0 \frac{\frac{p^2}{q}\left(\frac{I_{IN}}{I_0}\right)}{1 - \frac{q}{p}\left(\frac{I_{IN}}{I_0}\right)}. \tag{8.181}$$

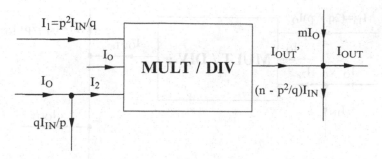

Fig. 8.56 MFC core for second-order approximation based on PR 8.4 – block diagram

The output current of the circuit having the block diagram presented in Fig. 8.56 will have the following expression:

$$
\begin{aligned}
I_{OUT} &= I_{OUT}' - \left(n - \frac{p^2}{q}\right)I_{IN} + mI_O \\
&= I_O\left[\frac{\frac{p^2}{q}\left(\frac{I_{IN}}{I_O}\right)}{1 - \frac{q}{p}\left(\frac{I_{IN}}{I_O}\right)} - \left(n - \frac{p^2}{q}\right)\left(\frac{I_{IN}}{I_O}\right) + m\right] \\
&= I_O\left[\frac{\frac{p^2}{q}x}{1 - \frac{q}{p}x} - \left(n - \frac{p^2}{q}\right)x + m\right].
\end{aligned}
\tag{8.182}
$$

Using the notation $x = I_{IN}/I_O$ and (8.44) relation, it results that I_{OUT} current represents the third-order approximation of $f(x)$ function:

$$
I_{OUT} = I_O g(x) = I_O g\left(\frac{I_{IN}}{I_O}\right) \cong I_O f(x).
\tag{8.183}
$$

8.2.4.3 Implementation of the Multifunctional Circuit Based on the Third-Order Approximation of a Continuous Mathematical Function Using Two Primitive Functions

The block diagram of the multifunctional circuit is presented in Fig. 8.57. The MULT/DIV circuit has the implementation presented in Fig. 8.59. The expressions of I_{OUTa} and I_{OUTb} currents are:

$$
I_{OUTa} = I_O\frac{I_{1a}}{I_{2a}}
\tag{8.184}
$$

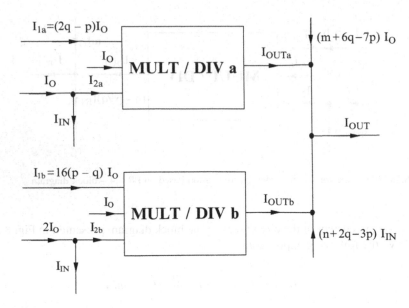

Fig. 8.57 MFC core for third-order approximation based on PR 8.4 – block diagram

and:

$$I_{OUTb} = I_O \frac{I_{1b}}{I_{2b}}. \tag{8.185}$$

So:

$$I_{OUTa} = I_O \frac{(2q - p)I_O}{I_O - I_{IN}} = I_O \frac{2q - p}{1 - \left(\frac{I_{IN}}{I_O}\right)} \tag{8.186}$$

and:

$$I_{OUTb} = I_O \frac{16(p - q)I_O}{2I_O - I_{IN}} = I_O \frac{16(p - q)}{2 - \left(\frac{I_{IN}}{I_O}\right)}. \tag{8.187}$$

The output current of the circuit having the block diagram presented in Fig. 8.57 will have the following expression:

$$I_{OUT} = I_{OUTa} + I_{OUTb} + (n + 2q - 3p)I_{IN} + (m + 6q - 7p)I_O. \tag{8.188}$$

Thus:

$$I_{OUT} = I_O \left[\frac{2q - p}{1 - \left(\frac{I_{IN}}{I_O}\right)} + \frac{16(p - q)}{2 - \left(\frac{I_{IN}}{I_O}\right)} + (n + 2q - 3p)\left(\frac{I_{IN}}{I_O}\right) + (m + 6q - 7p) \right].$$

(8.189)

Using the notation $x = I_{IN}/I_O$ and (8.44) relation, it results that I_{OUT} current represents the third-order approximation of $f(x)$ function:

$$I_{OUT} = I_O \left[\frac{2q - p}{1 - x} + \frac{16(p - q)}{2 - x} + (n + 2q - 3p)x + (m + 6q - 7p) \right]$$

$$= I_O g(x) \cong I_O f(x).$$

(8.190)

8.2.4.4 Implementation of the Multifunctional Circuit Based on the Fifth-Order Approximation of a Continuous Mathematical Function Using Four Primitive Functions

The block diagram of the multifunctional circuit is presented in Fig. 8.58. The MULT/DIV circuit has the implementation presented in Fig. 8.59.

The expressions of I_{OUTa}, I_{OUTb}, I_{OUTc} and I_{OUTd} currents are:

$$I_{OUTa} = I_O \frac{I_{1a}}{I_{2a}},$$

(8.191)

$$I_{OUTb} = I_O \frac{I_{1b}}{I_{2b}},$$

(8.192)

$$I_{OUTc} = I_O \frac{I_{1c}}{I_{2c}}$$

(8.193)

and:

$$I_{OUTd} = I_O \frac{I_{1d}}{I_{2d}}$$

(8.194)

So:

$$I_{OUTa} = I_O \frac{a_1 I_O}{I_O - I_{IN}} = I_O \frac{a_1}{1 - \left(\frac{I_{IN}}{I_O}\right)},$$

(8.195)

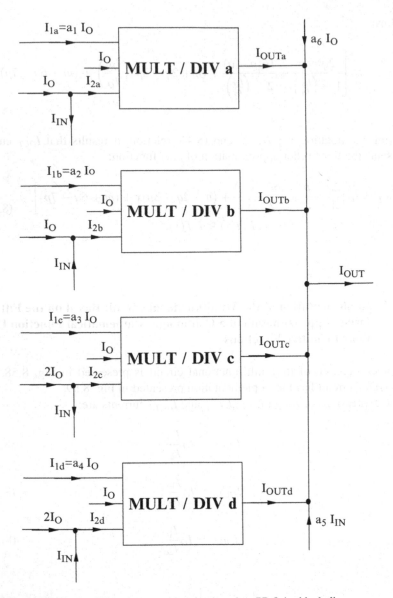

Fig. 8.58 MFC core for fifth-order approximation based on PR 8.4 – block diagram

$$I_{OUTb} = I_O \frac{a_2 I_O}{I_O + I_{IN}} = I_O \frac{a_2}{1 + \left(\frac{I_{IN}}{I_O}\right)}, \qquad (8.196)$$

$$I_{OUTc} = I_O \frac{a_3 I_O}{2 I_O - I_{IN}} = I_O \frac{a_3}{2 - \left(\frac{I_{IN}}{I_O}\right)} \qquad (8.197)$$

and:

$$I_{OUTd} = I_O \frac{a_4 I_O}{2I_O + I_{IN}} = I_O \frac{a_4}{2 + \left(\frac{I_{IN}}{I_O}\right)}. \tag{8.198}$$

The output current of the circuit having the block diagram presented in Fig. 8.58 will have the following expression:

$$I_{OUT} = I_{OUTa} + I_{OUTb} + I_{OUTc} + I_{OUTd} + a_5 I_{IN} + a_6 I_O. \tag{8.199}$$

Thus:

$$I_{OUT} = I_O \left[\frac{a_1}{1 - \left(\frac{I_{IN}}{I_O}\right)} + \frac{a_2}{1 + \left(\frac{I_{IN}}{I_O}\right)} + \frac{a_3}{2 - \left(\frac{I_{IN}}{I_O}\right)} + \frac{a_4}{2 + \left(\frac{I_{IN}}{I_O}\right)} + a_5 \left(\frac{I_{IN}}{I_O}\right) + a_6 \right]. \tag{8.200}$$

Using the notation $x = I_{IN}/I_O$ and (8.44) relation, it results that I_{OUT} current represents the fifth-order approximation of $f(x)$ function:

$$I_{OUT} = I_O \left(\frac{a_1}{1 - x} + \frac{a_2}{1 + x} + \frac{a_3}{2 - x} + \frac{a_4}{2 + x} + a_5 x + a_6 \right) = I_O g(x) \cong I_O f(x). \tag{8.201}$$

A possible implementation of the MULT/DIV circuit from the previous block diagrams uses as circuit cores two current squaring circuits, having the realization shown in Fig. 8.59.

For the MOS transistors from Fig. 8.59a, the equation of the translinear loop can be expressed as follows:

$$2V_{GS}(I_O) = V_{GS}(I_{D1}) + V_{GS}(I_{D1} + I_{IN}), \tag{8.202}$$

resulting:

$$2\sqrt{I_O} = \sqrt{I_{D1}} + \sqrt{I_{D1} + I_{IN}}. \tag{8.203}$$

So:

$$I_{D1} = I_O - \frac{I_{IN}}{2} + \frac{I_{IN}^2}{16 I_O}. \tag{8.204}$$

The expression of the output current will be:

$$I_{OUT} = I_{D1} + \frac{I_{IN}}{2} - I_O = \frac{I_{IN}^2}{16 I_O}. \tag{8.205}$$

a

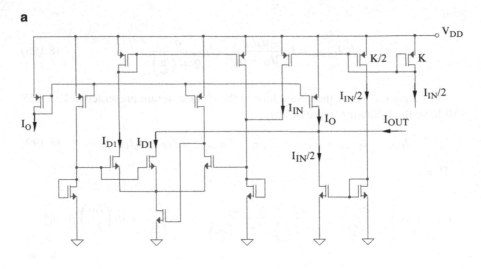

b

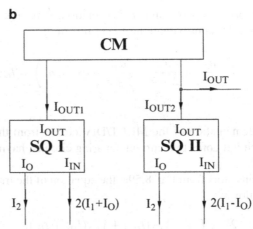

Fig. 8.59 (a) The functional core of the MULT/DIV circuit and (b) the block diagram of the MULT/DIV circuit

The output current of the MULT/DIV circuit from Fig. 8.59b has the following expression:

$$I_{OUT} = I_{OUT1} - I_{OUT2}, \qquad (8.206)$$

resulting:

$$I_{OUT} = \frac{(I_1 + I_O)^2}{4I_2} - \frac{(I_1 - I_O)^2}{4I_2} = I_O \frac{I_1}{I_2}. \qquad (8.207)$$

I_{OUT}[uA]

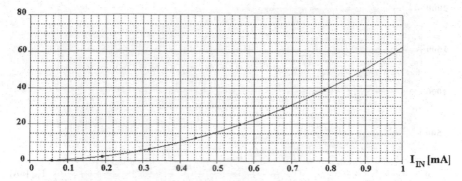

Fig. 8.60 The $I_{OUT}(I_{IN})$ simulation for the squaring circuit

	I_{IN} (μA)	$I_{OUT\ th.}$ (μA)	$I_{OUT\ sim}$ (μA)
Table 8.3 Comparison between the simulated and the theoretical estimated results for the current squarer presented in Fig. 8.59a	50	0.156	0.110
	100	0.625	0.660
	150	1.406	1.461
	200	2.500	2.574
	250	3.906	3.991
	300	5.625	5.711
	350	7.656	7.719
	400	10.000	10.089
	450	12.656	12.713
	500	15.625	15.714
	550	18.906	18.956
	600	22.500	22.587
	650	26.406	26.446
	700	30.625	30.707
	750	35.156	35.184
	800	40.000	40.077
	850	45.156	45.171
	900	50.625	50.596
	950	56.406	56.407
	1,000	62.500	62.564

The $I_{OUT}(I_{IN})$ simulation of the squaring circuit presented in Fig. 8.59a is shown in Fig. 8.60. The I_O current is equal with 1 mA, while the range of I_{IN} current was chosen to be between 0 and 1 mA.

A comparison between the simulated and the theoretical estimated results for the current squarer presented in Fig. 8.59a is shown in Table 8.3.

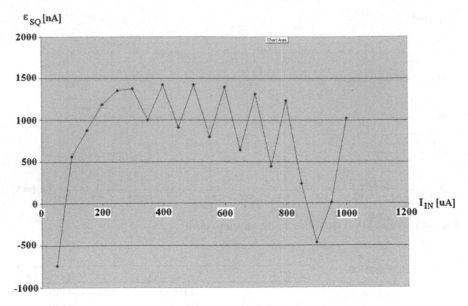

Fig. 8.61 The simulated approximation error $\varepsilon_{SQ}(I_{IN})$ for the squaring circuit

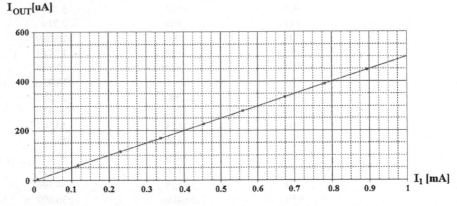

Fig. 8.62 The $I_{OUT}(I_1)$ simulation for the MULT/DIV circuit from Fig. 8.59b

The simulated approximation error, $\varepsilon_{SQ}(I_{IN})$, for the squaring circuit from Fig. 8.59a is shown in Fig. 8.61. The error is smaller than 0.0049% for an extended range of the input current.

The $I_{OUT}(I_1)$ simulation for the MULT/DIV circuit presented in Fig. 8.59b is shown in Fig. 8.62. The I_O and I_2 currents have the following values: $I_O = 500$ μA and $I_2 = 1$ mA, while the range of I_1 current was chosen to be between 0 and 1 mA.

ε_{MULT} [nA]

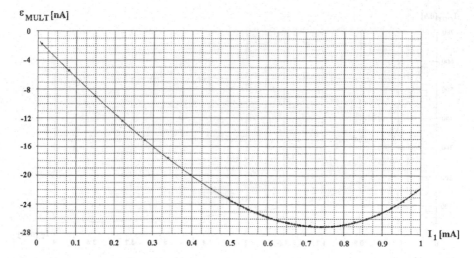

Fig. 8.63 The simulated linearity error of $I_{OUT}(I_1)$ characteristic for the MULT/DIV circuit from Fig. 8.59b

I_{OUT}[uA]

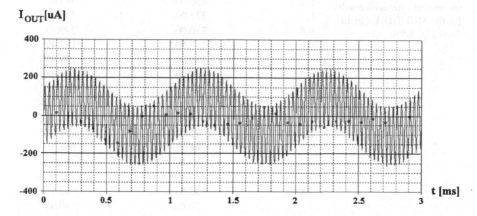

Fig. 8.64 The $I_{OUT}(t)$ simulation for the MULT/DIV circuit from Fig. 8.59b

The simulated linearity error of $I_{OUT}(I_1)$ characteristic for the MULT/DIV circuit from Fig. 8.59b is shown in Fig. 8.63. The linearity error is smaller than 0.006% for an extended range of the input currents.

For the same MULT/DIV circuit presented in Fig. 8.59b, a transient analysis was performed. The I_O current is a sinusoidal current with an amplitude of 50 μA and a frequency equal with 1 kHz, while I_1 current is a sinusoidal current having an amplitude of 0.3 mA and a frequency of 30 kHz. The I_2 current is a continuous current, equal with 1 mA. The simulation of the output current is presented in Fig. 8.64.

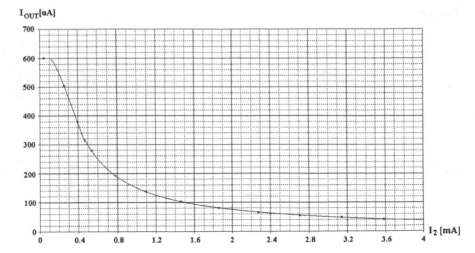

Fig. 8.65 The $I_{OUT}(I_2)$ simulation for the MULT/DIV circuit presented in Fig. 8.59b

	I_2 (mA)	$I_{OUT\ th.}$ (μA)	I_{OUTsim} (μA)
Table 8.4 Comparison between the simulated and the theoretical estimated results for the MULT/DIV circuit from Fig. 8.59b	0.2	750.00	545.45
	0.4	375.00	375.01
	0.6	250.00	249.99
	0.8	187.50	187.49
	1	150.00	149.983
	1.2	125.00	124.984
	1.4	107.14	107.932
	1.6	93.75	93.734
	1.8	83.33	83.318
	2	75.00	74.985
	2.2	68.18	68.167
	2.4	62.50	62.451
	2.6	57.69	57.678
	2.8	53.57	53.557
	3	50.00	49.987
	3.2	46.87	46.862
	3.4	44.12	44.105
	3.6	41.67	41.654
	3.8	39.47	39.460
	4	37.50	37.487

The $I_{OUT}(I_2)$ simulation for the MULT/DIV circuit presented in Fig. 8.59b is presented in Fig. 8.65. The I_O and I_1 currents have the following values: $I_O = 0.5$ mA and $I_1 = 0.3$ mA, while the range of I_2 current was chosen to be between 0 and 4 mA.

A comparison between the simulated and the theoretical estimated results for the previously presented MULT/DIV circuit is shown in Table 8.4.

8.3 Conclusions

Chapter introduces the original concept of multifunctional structures – active circuits that are able to implement, starting from the same circuit core, both linear and nonlinear mathematical functions. The approach of analog signal processing from the perspective of using multifunctional cores presents the very important advantages of reducing the power consumption and silicon area per implemented function. As the design effort is mainly concentrated for improving the performances of the functional core, the design costs for a circuit function can be strongly reduced using this design method.

References

1. Popa C (2010) Improved linearity CMOS differential amplifiers with applications in VLSI designs. International symposium on electronics and telecommunications, pp 29–32, Timisoara, Romania
2. Popa C (2009) High accuracy CMOS multifunctional structure for analog signal processing. International semiconductor conference, pp 427–430, Sinaia, Romania
3. Popa C (2010) CMOS multifunctional computational structure with improved performances. International semiconductors conference, pp 471–474, Sinaia, Romania
4. Popa C (2006) CMOS quadratic circuits with applications in VLSI designs. International conference on signals and electronic systems, pp 117–120, Lodz, Poland
5. Popa C (2004) A digital-selected current-mode function generator for analog signal processing applications. International semiconductor conference, pp 495–498, Sinaia, Romania
6. Popa C (2005) Improved accuracy pseudo-exponential function generator with applications in analog signal processing. International conference on computer as a tool, 1594–1597, Belgrade, Serbia and Montenegro
7. Popa C (2008) Improved accuracy pseudo-exponential function generator with applications in analog signal processing. IEEE Trans Very Large Scale Integr Syst 16:318–321
8. De La Cruz Blas CA, Feely O (2008) Limit cycle behavior in a class-AB second-order square root domain filter. IEEE international conference on electronics, circuits and systems, pp 117–120, St. Julians, Malta
9. Zarabadi SR, Ismail M, Chung-Chih H (1998) High performance analog VLSI computational circuits. IEEE J Solid-State Circuits 33:644–649
10. Sakurai S, Ismail M (1992) A CMOS square-law programmable floating resistor independent of the threshold voltage. IEEE Trans Circuits Syst II: Analog Digit Signal Process 39:565–574
11. Popa C, Manolescu AM (2007) CMOS differential structure with improved linearity and increased frequency response. International semiconductor conference, pp 517–520, Sinaia, Romania
12. Popa C (2010) Tunable CMOS resistor circuit with improved linearity based on the arithmetical mean computation. IEEE Mediterranean electrotechnical conference, pp 1379–1382, Valletta, Malta

13. Manolescu AM, Popa C (2009) Low-voltage low-power improved linearity CMOS active resistor circuits. Springer J Analog Integr Circuits Signal Process 62:373–387
14. Lee BW, Sheu BJ (1990) A high slew-rate CMOS amplifier for analog signal processing. IEEE J Solid-State Circuits 25:885–889
15. Kumar JV, Rao KR (2002) A low-voltage low power square-root domain filter. Asia-Pacific conference on circuits and systems, pp 375–378, Bali, Indonesia
16. Klumperink E, van der Zwan E, Seevinck E (1989) CMOS variable transconductance circuit with constant bandwidth. Electron Lett 25:675–676
17. Zele RH, Allstot DJ, Fiez TS (1991) Fully-differential CMOS current-mode circuits and applications. IEEE international symposium on circuits and systems, pp 1817–1820, Raffles City, Singapore
18. El Mourabit A, Sbaa MH, Alaoui-Ismaili Z, Lahjomri F (2007) A CMOS transconductor with high linear range. IEEE international conference on electronics, circuits and systems, pp 1131–1134, Marrakech, Morocco
19. Popa C (2008) Programmable CMOS active resistor using computational circuits. International semiconductor conference, pp 389–392, Sinaia, Romania
20. Farshidi E (2009) A low-voltage class-AB linear transconductance based on floating-gate MOS technology. European conference on circuit theory and design, pp 437–440, Antalya, Turkey
21. Abbasi M, Kjellberg T et al (2010) A broadband differential cascode power amplifier in 45 nm CMOS for high-speed 60 GHz system-on-chip. IEEE radio frequency integrated circuits symposium, pp 533–536, Anaheim, USA
22. Yonghui J, Ming L et al (2010) A low power single ended input differential output low noise amplifier for L1/L2 band. IEEE international symposium on circuits and systems, pp 213–216, Paris, France
23. Ong GT, Chan PK (2010) A micropower gate-bulk driven differential difference amplifier with folded telescopic cascode topology for sensor applications. IEEE international midwest symposium on circuits and systems, pp 193–196, Seattle, USA
24. Vaithianathan V, Raja J, Kavya R, Anuradha N (2010) A 3.1 to 4.85 GHz differential CMOS low noise amplifier for lower band of UWB applications. International conference on wireless communication and sensor computing, pp 1–4, Chennai, India
25. Figueiredo M, Santin E, Goes J, Santos-Tavares R, Evans G (2010) Two-stage fully-differential inverter-based self-biased CMOS amplifier with high efficiency. IEEE international symposium on circuits and systems, pp 2828–2831, Paris, France
26. Enche Ab, Rahim SAE, Ismail MA et al (2010) A wide gain-bandwidth CMOS fully-differential folded cascode amplifier. International conference on electronic devices, systems and applications, pp 165–168, Kuala Lumpur, Malaysia
27. Chanapromma C, Daoden K (2010) A CMOS fully differential operational transconductance amplifier operating in sub-threshold region and its application. International conference on signal processing systems, pp V2-73–V2-7728, Yantai, China
28. Rajput KK, Saini AK, Bose SC (2010) DC offset modeling and noise minimization for differential amplifier in subthreshold operation. IEEE computer society annual symposium on VLSI, pp 247–252, Greece
29. Bajaj N, Vermeire B, Bakkaloglu B (2010) A 10 MHz to 100 MHz bandwidth scalable, fully differential current feedback amplifier. IEEE international symposium on circuits and systems, pp 217–220, Paris, France
30. Harb A (2010) A rail-to-rail full clock fully differential rectifier and sample-and-hold amplifier. IEEE international symposium on circuits and systems, pp 1571–1574, Paris, France
31. Lili C, Zhiqun L et al (2010) A 10-Gb/s CMOS differential transimpedance amplifier for parallel optical receiver. International symposium on signals systems and electronics, pp 1–4, Nanjing, China
32. Popa C (2009) Computational circuits using bulk-driven MOS devices. IEEE international conference on computer as a tool, pp 246–251, St. Petersburg, Russia

33. Popa C (2009) Multiplier circuit with improved linearity using FGMOS transistors. International symposium ELMAR, pp 159–162, Zadar, Croatia
34. Popa C (2001) Low-power rail-to-rail CMOS linear transconductor. International semiconductor conference, pp 557–560, Sinaia, Romania
35. Wallinga H, Bult K (1989) Design and analysis of CMOS analog signal processing circuits by means of a graphical MOST model. IEEE J Solid-State Circuits 24:672–680
36. Sawigun C, Serdijn WA (2009) Ultra-low-power, class-AB, CMOS four-quadrant current multiplier. Electron Lett 45:483–484
37. Akshatha BC, Akshintala VK (2009) Low voltage, low power, high linearity, high speed CMOS voltage mode analog multiplier. International conference on emerging trends in engineering and technology, pp 149–154, Nagpur, India
38. Hidayat R, Dejhan K, Moungnoul P, Miyanaga Y (2008) OTA-based high frequency CMOS multiplier and squaring circuit. International symposium on intelligent signal processing and communications systems, pp 1–4, Bangkok, Thailand
39. Naderi A et al (2009) Four-quadrant CMOS analog multiplier based on new current squarer circuit with high-speed. IEEE international conference on computer as a tool, pp 282–287, St. Petersburg, USA
40. Khateb F, Biolek D, Khatib N, Vavra J (2010) Utilizing the bulk-driven technique in analog circuit design. IEEE international symposium on design and diagnostics of electronic circuits and systems, pp 16–19, Vienna, Austria
41. Machowski W, Kuta S, Jasielski J, Kolodziejski W (2010) Quarter-square analog four-quadrant multiplier based on CMOS invertes and using low voltage high speed control circuits. International conference on mixed design of integrated circuits and systems, pp 333–336, Wroklaw, Poland
42. Ehsanpour M, Moallem P, Vafaei A (2010) Design of a novel reversible multiplier circuit using modified full adder. International conference on computer design and applications, pp V3-230–V3-234, Hebei, China
43. Parveen T, Ahmed MT (2009) OFC based versatile circuit for realization of impedance converter, grounded inductance, FDNR and component multipliers. International multimedia, signal processing and communication technologies, pp 81–84, Aligarh, India
44. Feldengut T, Kokozinski R, Kolnsberg S (2009) A UHF voltage multiplier circuit using a threshold-voltage cancellation technique. Research in microelectronics and electronics, pp 288–291, Cork, Ireland
45. Popa C (2009) Logarithmic compensated voltage reference. Spanish conference on electron devices, pp 215–218, Santiago de Compostela, Spain
46. Popa C (2007) Improved accuracy function generator circuit for analog signal processing. International conference on computer as a tool, pp 231–236, Warsaw, Poland
47. Cheng-Chieh C, Shen-Iuan L (2000) Current-mode full-wave rectifier and vector summation circuit. Electron Lett 36:1599–1600
48. Hidayat R, Dejhan K, Moungnoul P, Miyanaga Y (2008) OTA-based high frequency CMOS multiplier and squaring circuit. International symposium on intelligent signal processing and communications systems, pp 1–4, Bangkok, Thailand
49. Kumbun J, Lawanwisut S, Siripruchyanun M (2009) A temperature-insensitive simple current-mode squarer employing only multiple-output CCTAs. IEEE region 10 conference TENCON, pp 1–4, Singapore
50. Naderi A, Mojarrad H, Ghasemzadeh H, Khoei A, Hadidi K (2009) Four-quadrant CMOS analog multiplier based on new current squarer circuit with high-speed. IEEE international conference on computer as a tool, pp 282–287, St Petersburg, Russia
51. Machowski W, Kuta S, Jasielski J, Kolodziejski W (2010) Quarter-square analog four-quadrant multiplier based on CMOS invertes and using low voltage high speed control circuits. International conference on mixed design of integrated circuits and systems, pp 333–336, Wroclaw, Poland

52. Raikos G, Vlassis S (2009) Low-voltage CMOS voltage squarer. IEEE international on electronics, circuits, and systems, pp 159–162, Medina, Tunisia
53. Muralidharan R, Chip-Hong C (2009) Fixed and variable multi-modulus squarer architectures for triple moduli base of RNS. IEEE international conference on circuits and systems, pp 441–444, Taipei, Taiwan
54. Garofalo V et al (2010) A novel truncated squarer with linear compensation function. IEEE international symposium on circuits and systems, pp 4157–4160, Paris, France
55. Kircay A, Keserlioglu MS (2009) Novel current-mode second-order square-root-domain highpass and allpass filter. International conference on electrical and electronics engineering, pp II-242–II-246, Bursa, Turkey
56. Kircay A, Keserlioglu MS, Cam U (2009) A new current-mode square-root-domain notch filter. European conference on circuit theory and design, pp 229–232, Antalya, Turkey
57. Popa C (2007) Improved linearity active resistors using MOS and floating-gate MOS transistors. The international conference on computer as a tool, pp 224–230, Warsaw, Poland
58. Popa C (2007) Low-voltage low-power curvature-corrected voltage reference circuit using DTMOSTs. Lecture notes in computer science, Springer, pp 117–124
59. Dermentzoglou LE, Arapoyanni A, Tsiatouhas Y (2010) A built-in-test circuit for RF differential low noise amplifiers. IEEE Trans Circuits Syst I: Regul Pap 57:1549–1558
60. De La Cruz-Blas CA, Lopez-Martin A, Carlosena A (2003) 1.5-V MOS translinear loops with improved dynamic range and their applications to current-mode signal processing. IEEE Trans Circuits Syst II: Analog Digit Signal Process 50:918–927
61. Desheng M, Wilamowski BM, Dai FF (2009) A tunable CMOS resistor with wide tuning range for low pass filter application. IEEE topical meeting on silicon monolithic integrated circuits in RF systems, pp 1–4, San Diego, USA
62. Torralba A et al (2009) Tunable linear MOS resistors using quasi-floating-gate techniques. IEEE Trans Circuits Syst II: Exp Briefs 56:41–45
63. Tadić N, Zogović M (2010) A low-voltage CMOS voltage-controlled resistor with wide resistance dynamic range. International conference on microelectronics proceedings, pp 341–344, Nis, Serbia
64. Mandai S, Nakura T, Ikeda M, Asada K (2010) Cascaded time difference amplifier using differential logic delay cell. Asia and South Pacific design automation conference, pp 355–356, Taipei, Taiwan

Appendix 1

Error Mechanisms in Computational Circuits

The Deviation of the MOS Transistor Characteristic from the Square-Law, Bulk Effect, Leakage

The first-order squaring characteristic of the MOS transistor biased in saturation is affected by the second-order effects: mobility degradation (A1.1), channel-length modulation (A1.2) and bulk effect (A1.3):

$$K = \frac{K_O}{[1 + \theta_G(V_{GS} - V_T)](1 + \theta_D V_{DS})}, \tag{A1.1}$$

$$I_D = \frac{K}{2}(V_{GS} - V_T)^2(1 + \lambda V_{DS}), \tag{A1.2}$$

$$V_T = V_{TO} + \gamma\left(\sqrt{\Phi - V_{BS}} - \sqrt{\Phi}\right). \tag{A1.3}$$

The errors introduced by the bulk effect can be minimized by proper designs that avoid the dependence of the circuit overall parameters (voltage gain, transconductance, equivalent resistance) on the threshold voltage. Alternatively, designing circuits in such a way that the bulk-source voltage to be equal with zero also cancels out the errors introduced by the bulk effect. Important errors produced by the second-order effects are given by the dependence of the K transconductance parameter on V_{GS} voltage. Considering only the $K(V_{GS})$ dependence, it results a small changing of V_{GS} voltage comparing with the analysis based on the first-order model of MOS transistors:

$$V_{GS} = V_T + \sqrt{\frac{2I_D}{K}} + \frac{\theta_G}{K}I_D. \tag{A1.4}$$

C.R. Popa, *Synthesis of Computational Structures for Analog Signal Processing*, DOI 10.1007/978-1-4614-0403-3, © Springer Science+Business Media, LLC 2011

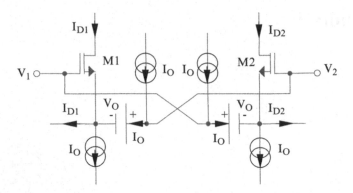

Fig. A1.1 Multifunctional circuit

In order to evaluate the total error introduced by this second-order effect in the ideal operation of a CMOS circuit, the multifunctional circuit presented in Fig. A1.1 will be considered.

The differential input voltage can be expressed as follows:

$$V_1 - V_2 = V_{GS1} - V_O = V_O - V_{GS2}. \tag{A1.5}$$

Using the first-order squaring characteristic of the MOS transistor biased in saturation, the circuit can act as a differential amplifier or as a squaring circuit, by taking the difference, respectively the sum of the I_{D1} and I_{D2} output currents:

$$I_{D1} - I_{D2} = \sqrt{8KI_O}(V_1 - V_2) \tag{A1.6}$$

and:

$$I_{D1} + I_{D2} = 2I_O + K(V_1 - V_2)^2. \tag{A1.7}$$

For a second-order analysis, considering only the $K(V_{GS})$ dependence illustrated by (A1.1), the expression of I_{D1} current will be:

$$I_{D1} = \frac{K}{2} \frac{(V_{GS1} - V_T)^2}{1 + \theta_G(V_{GS1} - V_T)} = \frac{K}{2} \frac{[(V_O - V_T) + (V_1 - V_2)]^2}{1 + \theta_G[(V_O - V_T) + (V_1 - V_2)]}. \tag{A1.8}$$

For an usual implementation of V_O voltage sources using gate-source voltages of MOS transistors biased in saturation region, it results:

$$I_{D1} = I_O + \left(\sqrt{2KI_O} - 2\theta_G I_O\right)(V_1 - V_2)$$

$$+ \frac{K}{2}(V_1 - V_2)^2 \left(1 - 3\theta_G\sqrt{\frac{2I_O}{K}}\right) - \frac{K\theta_G}{2}(V_1 - V_2)^3 \tag{A1.9}$$

and, similarly:

$$I_{D2} = I_O - \left(\sqrt{2KI_O} - 2\theta_G I_O\right)(V_1 - V_2)$$
$$+ \frac{K}{2}(V_1 - V_2)^2 \left(1 - 3\theta_G \sqrt{\frac{2I_O}{K}}\right) + \frac{K\theta_G}{2}(V_1 - V_2)^3. \tag{A1.10}$$

The amplifying function of the circuit can be expressed considering the difference between the previous currents:

$$I_{D1} - I_{D2} = 2\left(\sqrt{2KI_O} - 2\theta_G I_O\right)(V_1 - V_2) - K\theta_G(V_1 - V_2)^3. \tag{A1.11}$$

The main linear term is slightly affected by the second-order effects, but the most important problem is represented by the third-order term from (A1.11) that will introduce third-order distortions in the transfer characteristic of the differential amplifier. It is possible to minimize the impact of this undesired term by using a proper connection of two quasi-identical and different-biased opposite-excited differential structures.

The squaring function can be achieved using the previous circuit considering the sum of I_{D1} and I_{D2} currents:

$$I_{D1} + I_{D2} = 2I_O + K(V_1 - V_2)^2 \left(1 - 3\theta_G \sqrt{\frac{2I_O}{K}}\right). \tag{A1.12}$$

The overall performance degradation for the squaring function is relatively small, as only the amplitude of the output signal is slightly affected as a result of the second-order effects.

Especially for ultimate CMOS VLSI nanometer designs, the leakage becomes an important problem, the leakage current depending on the properties of the layout and also on the device structure (dimensions of channel, gate oxide thickness, doping). Unfortunately, the computation of the total leakage cannot be made summing the leakage currents considered individually, existing a correlation between the leakage currents.

The subthreshold leakage is represented by the current produced by minority electrons flowing through p substrate from source to drain, the mathematical law that models this process being an exponential function:

$$I_{SL} = I_{DO} \frac{W}{L} \exp\left(\frac{V_{GS} - V_T}{nV_{TH}}\right)\left[1 - \exp\left(-\frac{V_{DS}}{V_{TH}}\right)\right]. \tag{A1.13}$$

W/L represents the aspect ratio of the MOS transistor, V_{TH} is the thermal voltage, n is a parameter, while I_{DO} current is determined from the condition of the continuity of the MOS transistor model between weak inversion and strong inversion regions.

The expression of the subthreshold-off current (obtained for $V_{GS} = 0$ and $V_{DS} >> V_T$) will be:

$$I_{SLoff} = I_{DO} \frac{W}{L} \exp\left(\frac{-V_T}{nV_{TH}}\right). \tag{A1.14}$$

So, it exponentially increases when the threshold voltage is decreasing and also increases with temperature and with the decreasing of the channel length (because of the direct mathematical dependence, of the short-channel-effect and of the drain induced barrier lowering effect).

The gate leakage current can be carried by tunneling electrons or holes, the carriers leaking to source, drain and channel. The electrons can pass potential barriers, higher than their energy, the tunneling current exponentially depending on barrier height and width, as well as on electron's mass.

For junction leakage, low currents are carried by minority carriers drifting across the junction, by the electron–hole generation in junction or by the impact ionization at high reverse bias, the electrons being able to pass the barrier by tunneling through it – BTBT (Band To Band Tunneling) current.

Mismatches in Current Mirrors

As the I_O biasing currents from the previous circuit are generated by a current mirror, considering the inherent mismatches of this structure, the left-side I_O current becomes $I_O + \Delta I_O$, while the right-side one will be $I_O - \Delta I_O$ (ΔI_O represents the variation of the current source from the ideal value, I_O). Because $\Delta I_O/I_O << 1$, the difference and the sum of I_{D1} and I_{D2} currents will have the following expressions:

$$I_{D1} - I_{D2} = 2\Delta I_O + \sqrt{8KI_O}(V_1 - V_2) \tag{A1.15}$$

and:

$$I_{D1} + I_{D2} = 2I_O + K(V_1 - V_2)^2 + \sqrt{\frac{2K}{I_O}} \Delta I_O(V_1 - V_2). \tag{A1.16}$$

The operation of the differential amplifier is not affected by distortions as a result of the mismatches in current mirrors, only a DC offset $2\Delta I_O$ will appear in the expression of the differential output current.

The error introduced in the squaring circuit operation is represented by the appearance of a linear additional term, that can be compensated by design techniques.

Different Threshold Voltages for NMOS and PMOS Active Devices

Considering that the V_O voltage sources from the previous circuit are implemented using two gate-source voltages of PMOS transistors, having different threshold voltages comparing with M1 and M2 NMOS transistors, the expressions of I_{D1} and I_{D2} output currents become:

$$I_{D1} = I_O + \frac{K}{2}\left(V_{Tp} - V_{Tn}\right)^2 + \sqrt{2KI_O}\left(V_{Tp} - V_{Tn}\right)$$
$$+ \frac{K}{2}(V_1 - V_2)^2 + (V_1 - V_2)\left[\sqrt{2KI_O} + K\left(V_{Tp} - V_{Tn}\right)\right] \qquad (A1.17)$$

and:

$$I_{D2} = I_O + \frac{K}{2}\left(V_{Tp} - V_{Tn}\right)^2 + \sqrt{2KI_O}\left(V_{Tp} - V_{Tn}\right)+$$
$$+ \frac{K}{2}(V_1 - V_2)^2 - (V_1 - V_2)\left[\sqrt{2KI_O} + K\left(V_{Tp} - V_{Tn}\right)\right]. \qquad (A1.18)$$

The differential output current will have the following expression:

$$I_{D1} - I_{D2} = 2(V_1 - V_2)\left[\sqrt{2KI_O} + K\left(V_{Tp} - V_{Tn}\right)\right], \qquad (A1.19)$$

while the sum of these currents will be:

$$I_{D1} + I_{D2} = 2I_O + K(V_1 - V_2)^2 + K\left(V_{Tp} - V_{Tn}\right)^2 + \sqrt{8KI_O}\left(V_{Tp} - V_{Tn}\right). \quad (A1.20)$$

The equivalent transconductance of the differential amplifier is affected by the difference between the threshold voltages of PMOS and NMOS transistors, the linearity of the structure remaining the same comparing with the circuit based on the first-order model of MOS transistors.

The modification of the squaring circuit operation as a result of different threshold voltages is represented by the changing of the constant term $2I_O$. This error can be easily compensated by inserting an additional current generator in the output terminal of the circuit.

Variations over the Process, Temperature and Supply Voltage

Expressions (A1.6) and (A1.7) characterize the fundamental operation of the circuit as differential amplifier and, respectively, as squaring circuit. As the K transconductance parameter is strongly temperature dependent, both circuit operations will be affected by the temperature variations. Additionally, as K parameter is imposed by the CMOS process, the technological error will affect the functions implemented by the circuit.

The I_O currents that bias the V_O current-controlled voltage sources are generated, usually, by a current mirror. As the drain-source voltages of the MOS transistors that compose this current mirror are different, the channel-length modulation effect will introduce an undesired dependence of the circuit operation on the supply voltage. In order to improve the power supply rejection ration of the structure, cascode current sources for biasing the V_O voltage sources have to be used. The disadvantage of this replacement will be an increasing of the minimal required value for the supply voltage, the low-voltage of the circuit being affected.

Appendix 2

A multitude of continuous $f(x)$ mathematical functions can be third-order approximated using the following $g(x)$ function:

$$g(x) = \frac{a_1 x}{1 + a_2 x} + a_3 x + a_4. \tag{A2.1}$$

The a_1, a_2, a_3 and a_4 coefficients can be determined from the condition that the Taylor series of these functions to be identical in a third-order approximation. For thirteen usual mathematical functions, the $g(x)$ functions and the fourth-order approximation errors are presented in the following lines.

The exp (x) function

$$g(x) = \frac{3}{2} \frac{x}{1 - \frac{x}{3}} - \frac{x}{2} + 1 \tag{A2.2}$$

$$\varepsilon_{f(x)}^{g(x)}(x) \cong \frac{1}{72} \frac{x^4}{\exp(x)} \tag{A2.3}$$

The $\sqrt{1+x}$ function

$$g(x) = \frac{1}{4} \frac{x}{1 + \frac{x}{2}} + \frac{x}{4} + 1 \tag{A2.4}$$

$$\varepsilon_{f(x)}^{g(x)}(x) \cong \frac{x^4}{128\sqrt{1 + x}} \tag{A2.5}$$

The $\sqrt{1-x}$ function

$$g(x) = -\frac{1}{4}\frac{x}{1-\frac{x}{2}} - \frac{x}{4} + 1 \tag{A2.6}$$

$$\varepsilon_{f(x)}^{g(x)}(x) \cong \frac{x^4}{128\sqrt{1-x}} \tag{A2.7}$$

The $\sqrt[3]{1+x}$ function

$$g(x) = \frac{1}{5}\frac{x}{1+\frac{5}{9}x} + \frac{2x}{15} + 1 \tag{A2.8}$$

$$\varepsilon_{f(x)}^{g(x)}(x) \cong \frac{5x^4}{729\,(1+x)^{1/3}} \tag{A2.9}$$

The $\sqrt[3]{1-x}$ function

$$g(x) = -\frac{1}{5}\frac{x}{1-\frac{5}{9}x} - \frac{2x}{15} + 1 \tag{A2.10}$$

$$\varepsilon_{f(x)}^{g(x)}(x) \cong \frac{5x^4}{729\,(1-x)^{1/3}} \tag{A2.11}$$

The $\sqrt[4]{1+x}$ function

$$g(x) = \frac{9}{56}\frac{x}{1+\frac{7}{12}x} + \frac{5x}{56} + 1 \tag{A2.12}$$

$$\varepsilon_{f(x)}^{g(x)}(x) \cong \frac{35}{6144}\frac{x^4}{\sqrt{41+x}} \tag{A2.13}$$

The $\sqrt[4]{1-x}$ function

$$g(x) = -\frac{9}{56}\frac{x}{1-\frac{7}{12}x} - \frac{5x}{56} + 1 \tag{A2.14}$$

$$\varepsilon_{f(x)}^{g(x)}(x) \cong \frac{35}{6144}\frac{x^4}{\sqrt[4]{1-x}} \tag{A2.15}$$

The $1/(1+x)^2$ function

$$g(x) = -\frac{9}{4}\frac{x}{1+\frac{4x}{3}} + \frac{x}{4} + 1 \tag{A2.16}$$

$$\varepsilon_{f(x)}^{g(x)}(x) \cong \frac{x^4}{3}(1+x)^2 \tag{A2.17}$$

The $1/(1-x)^2$ function

$$g(x) = \frac{9}{4}\frac{x}{1-\frac{4x}{3}} - \frac{x}{4} + 1 \tag{A2.18}$$

$$\varepsilon_{f(x)}^{g(x)}(x) \cong \frac{x^4}{3}(1-x)^2 \tag{A2.19}$$

The $(1+x)^a$ function

$$g(x) = \frac{3a}{2}\frac{a-1}{a-2}\frac{x}{1-\frac{a-2}{3}x} - \frac{a}{2}\frac{a+1}{a-2}x + 1 \tag{A2.20}$$

$$\varepsilon_{f(x)}^{g(x)}(x) \cong \frac{a(a-2)}{72}(a^2-1)\frac{x^4}{(1+x)^a} \tag{A2.21}$$

The $(1-x)^a$ function

$$g(x) = -\frac{3a}{2}\frac{a-1}{a-2}\frac{x}{1+\frac{a-2}{3}x} + \frac{a}{2}\frac{a+1}{a-2}x + 1 \tag{A2.22}$$

$$\varepsilon_{f(x)}^{g(x)}(x) \cong \frac{a(a-2)}{72}(a^2-1)\frac{x^4}{(1-x)^a} \tag{A2.23}$$

The $\ln(1+x)$ function

$$g(x) = \frac{3}{4}\frac{x}{1+\frac{2x}{3}} + \frac{x}{4} \tag{A2.24}$$

$$\varepsilon_{f(x)}^{g(x)}(x) \cong \frac{x^4}{36\ln(1+x)} \tag{A2.25}$$

The $\ln(1-x)$ function

$$g(x) = -\frac{3}{4}\frac{x}{1-\frac{2x}{3}} - \frac{x}{4} \tag{A2.26}$$

$$\varepsilon_{f(x)}^{g(x)}(x) \cong \frac{x^4}{36\ln(1-x)} \tag{A2.27}$$

Index